NONLINEAR OPTICS

T0264959

NONLINEAR OPTICS

Jerome Moloney

Alan Newell

Advanced Book Program

CRC Press
Taylor & Francis Group
Boca Raton London New York

CRC Press is an imprint of the
Taylor & Francis Group, an **informa** business

First published 2004 by Westview Press

Published 2018 by CRC Press
Taylor & Francis Group
6000 Broken Sound Parkway NW, Suite 300
Boca Raton, FL 33487-2742

CRC Press is an imprint of the Taylor & Francis Group, an informa business

Visit the Taylor & Francis Web site at
http://www.taylorandfrancis.com

and the CRC Press Web site at
http://www.crcpress.com

A Cataloging-in-Publication data record for this book is available from the Library of Congress.

Typeface used in this text: Times

ISBN 13: 978-0-8133-4118-7 (pbk)

CONTENTS

PREFACE

This book is about Nonlinear Optics, the study of how high-intensity light propagates through and interacts with matter. It is a subject so scientifically rich and technologically promising that it is destined to become one of the most important areas of scientific research over the next quarter-century. The book is written for graduate students and the newcomer to nonlinear optics, or anyone who wants to get a unified picture of the whole subject. It takes the reader from the starting point of Maxwell's equations to some of the frontiers of modern research in the subject.

The modeling—how one pictures light and matter—and the mathematical methods are explained clearly and in great detail. In particular, the book starts from the point of view that light and matter each can be viewed as systems of oscillators, and the coupling between these oscillators is relatively weak. Essentially, the oscillators of light are plane electromagnetic waves and those of matter are electronic transitions, molecular vibrations and rotations, and acoustic waves. In semiconductors, they are Bloch wavefunctions. Because of weak coupling, we show how all the fast time (10^{-15} second) and space (10^{-6} meter) scales can be removed from the description of the interactions and how one is then left with well-known universal and canonical equations such as the coupled mode equations (useful for switches and in situations where linear and nonlinear birefringent effects are important), the nonlinear Schrödinger equation (useful for fiber optics and nonlinear waveguides), the three- and four-wave interaction equations (useful for understanding Raman and Brillouin scattering and phase conjugation), the Maxwell-Bloch equations (useful for lasers and understanding optical bistability), the Maxwell-Debye equations (for describing delayed response in transparent materials), the sine-Gordon equation (useful for describing the propagation of light pulses through active media), and the complex Ginzburg-Landau and complex Swift-Hohenberg equations (useful near the onset of lasing action).

The material is best taught in a year-long course. Students should have some familiarity with mathematical methods, Fourier series and Fourier integrals, elementary complex variables, elementary ordinary and partial differential equations, and vector calculus and should have had a course (undergraduate level is sufficient) in electromagnetic theory. Chapter 6, "Mathematical and Computational Methods," should be read concurrently with many of the sections in the other chapters. The correspondence between this chapter and the other chapters is indicated by the number contained in square brackets after the listing of each section in the table of contents. Important cross-reference sections are also indicated after the section headings in the text proper. A one semester course might cover Chapter 2, Chapter 3, Sections a, c-g, Chapter 4, Sections a-d, Chapter 5, Section a, b together with the relevant material from Chapter 6 including the uses of numerical simulations.

We begin in Chapter 2 with the propagation of light beams in weakly nonlinear (Kerr) media and, using elementary mathematics and heuristic arguments, introduce the reader to the notion of linear and nonlinear refractive index, intensity-dependent phase modulation, wavepackets, dispersion, diffraction and the nonlinear Schrödinger (NLS) equation. This is followed by a discussion of linear and nonlinear birefringence and three- and four-wave mixing. By the end of the first chapter, the reader will have met some of the most important equations describing propagation in passive nonlinear media and encountered some of their most important properties. Chapter 3 discusses communication in optical fibers using nonlinear pulses as information bits and the reflection and transmission characteristics (Snell's or Descartes' laws) of light beams at the interfaces of different nonlinear dielectrics. To this point, matter has played a more or less passive role. The effects of delay and field intensity are included through the use of a frequency and intensity-dependent refractive index. In Chapter 4, the oscillations of matter become active field variables, and we derive the Maxwell-Bloch and Maxwell-Debye equations from first principles. The basic mechanisms of Raman and Brillouin scattering are introduced. Chapter 5 is about applications, lasers, optically bistable cavities, co- and counterpropagating beam interactions, coherent pulse propagation in inhomogeneously broadened media, and finally Raman and Brillouin scattering.

We thank the Air Force Office of Scientific Research for their continuing support of the Nonlinear Optics Program at the Arizona Center for Mathematical Sciences. We also want to thank our colleagues Rob Indik, Alejandro Aceves, Jocelyn Lega, P. Varatharajah, John Geddes, Simon Wenden, and Claudia Röhner for discussions and help in proofreading and Linda Wilder for doing a superb typing job.

Wayne Hacker was kind enough to proofread parts of the second edition. We are also particularly indebted to Ildar Gabitor for bringing us up to date on all the developments in fiber optic communications. The book is dedicated to Els and Tish, good companions.

Chapter 1

INTRODUCTION

General Discussion

MAXWELL'S THEORY OF ELECTRIC AND magnetic fields and his idea that light is an electromagnetic wave were two of the great milestones of scientific thought, and unified understanding of a large and diverse set of phenomena. Indeed, by the late nineteenth century, the success of the classical electromagnetic theory of light led some to believe that there were few new fundamental discoveries to be made. This smug complacency was soon shattered by the inability of the wave theory to explain several observations: radiation spectra, the photoelectric effect, x-rays, radioactivity. These effects could only be understood by reviving the idea of the corpuscular nature of light, not in the original form conceived by Newton, but in a way that was compatible with the considerable success enjoyed by the classical wave theory. Out of this effort, modern quantum theory was born, and optical science settled once again into the complacency of a solved science. Rapid progress was made. The accuracy of the geometrical optics approximation, together with the linearity of the equations, which meant that complicated solution fields could be built by the linear superposition of much simpler solutions (e.g., plane waves), played important roles in this development. The amplitude of the electromagnetic field seemed to matter little.

There were, of course, rumblings, suspicions. Double refraction in isotropic media (which we will call nonlinear birefringence) and the Raman effect, in which a scattered wave whose frequency was the sum or difference of the frequencies of an applied field and a natural medium vibration, could not be explained on a linear basis and were indicators that the field intensity was perhaps important after all, but, for the most part, optics seemed to be a linear science. Where were the "far from linear" behaviors so richly manifested in other fields, the magnificent ferocity of a twenty-foot breaking water wave, the shock-wave boom of a supersonic jet, the majestic and thunderous cumulus cloud, the sudden implosion of a compressed shell, the surge of current in vacuum tubes at a critical applied voltage? They were there, all right, but were hidden because of the relatively low intensities that occurred naturally (e.g., sunlight, 600 volts per meter) or could be attained in the laboratory. Each was so much less than the binding fields of the hydrogen atom (10^{11} volts per meter) that it looked like the nonlinear tiger in optics would be forever contained.

The discovery of the laser in 1960 (an acronym for light amplification by stimulated emission of radiation), which was the naturel lightwave analog of the maser (substitute "microwave" for "light") developed in the early 1950s, changed all that. Now available was a source of highly coherent radiation that could be concentrated and focused to give extremely high local intensities (the latest laser pulses have peak intensities of up to 10^{18} watts per square centimeter!). The nonlinear tiger was released from its cage; a rich stream of fundamental new phenomena, plus several new manifestations of phenomena familiar from other fields, soon followed, and that stream continues to flow, becoming richer by the day. The relatively young subject of nonlinear optics, the study of how high-intensity light interacts with and propagates through matter, is so scientifically fertile and technologically promising that it is destined to be one of the most important areas of science for the next quarter century.

The field of nonlinear optics is partially driven by anticipation of enormous technological dividends. Already the use of lasers in modern technology is commonplace, ranging in application from high-density data storage on optical disks to greatly improved surgical techniques in ophthalmology, neurosurgery, and dermatology. However, for future uses, more sophisticated understanding will be required. Lasers are not the stable output intensity devices suggested by the simple models. In reality, they are highly complicated dynamical systems, which can display the full range of dynamical behavior, from the staid to the exotic, from fixed-point attractors to chaotic attractors, and the nature of this complicated behavior needs to be categorized and understood. The coupling of lasers is also important, in particular in semiconductor devices where hundreds of miniature lasers can be fabricated on a single substrate, providing in principle the capability of coherent high power output from a compact solid-state laser. But how will these arrays work in practice? Can they indeed be coupled in such a way as to give a phase-locked and coherent output? Nobody really knows. Preliminary models suggest that in many parameter ranges, the array will not behave as a single coherent unit but will display a rich mosaic of spatiotemporal patterns.

Nonlinear optics is also likely to revolutionize future telecommunications and computer technologies. The relatively long interaction lengths and small cross-sections available in waveguide and fiber materials means that low-energy optical pulses can achieve sufficiently high peak intensities to compensate for the intrinsically weak nonlinearities in many transparent optical materials. Today, using linear propagation methods, information races across continents and oceans on optical fibers as thin as a human hair at rates of gigabits (10^9 bits) per second. There is every indication that within the decade, the linear technology will be replaced by a nonlinear one in which trains of light pulses are transmitted as solitary waves. The enormous bandwidth for data transmission afforded by optical fibers allows much scope for original ideas on all-optical controlled multiplexing (integrating bits of information carried on different channels or wavelengths) of data bits represented as envelope pulses containing light at different wavelengths in a single fiber. Directional couplers, which exploit the overlap of evanescent tails of transversely guided field profiles, can switch soliton pulses from one fiber to its neighbor precisely because of the soliton's coherence. The semiconductor laser arrays mentioned earlier may well serve as the pump sources for the pulses used in optical fibers. They are also possible candidates for the power sources in all-optical signal processing and computing devices. The ultrafast, massively parallel (i.e., laser beams do not interact in free space), and global connectivity features make optical architecture an attractive alternative for computation. This architecture will likely exploit bistable behavior in optical feedback systems, such as ring and Fabry–Perot cavities, as the basic logic element. Lasers, communications, computing, image storage, beam cleanup ... the list of opportunities for useful applications of nonlinearity in optics goes on and on.

It is an ideal subject for the theoretician interested in nonlinear behavior and model building who is particularly well positioned to make major contributions to the development of the subject. First, it is incredibly diverse in that it displays the full spectrum of behavior associated with nonlinear equations, three- and four-wave resonant interactions, self-focusing, the development of singularities and weak solutions, solitons, pattern formation, phase locking, strange attractors, homoclinic tangles, the full range of bifurcation scenarios, turbulence—all familiar to the theoretician in a variety of contexts. Second, modeling or the art of the judicious approximation, the skill to recognize and then transfer ideas that run parallel in other fields, and the ability to see how the parts fit into a unified whole are key ingredients for success. Third, several new concepts of nonlinear science, including the *soliton* and the *strange attractor*, representing ideals at the opposite ends of the spectrum of dynamical behavior, are often encountered and require some depth of mathematical knowledge to understand. The soliton was discovered, with the aid of the computer experiment, by the mathematicians Kruskal and Zabusky, and we shall see in this book how this robust object is likely to play an ever-increasing role in propagation in fibers, surface waves in nonlinear dielectrics, switching in waveguides, optical bistability, and propagation through inhomogeneously broadened resonant media. The strange attractor was discovered by Lorenz, an atmospheric scientist with a distinctly mathematical bent, and developed as a fundamental concept in the theory of dynamics of dissipative systems by Ruelle and Takens. This idea, together with a revolution in our understanding about the nature of finite and low-dimensional dynamical systems, has had and will continue to have a broad impact, particularly in optical feedback devices such as lasers. Last but not least in the arsenal of tools the theoretician brings to bear are the techniques of modern computer simulation. Ideas and theories can be tested throughout whole parameter ranges, and quantitative support can be given to complement the qualitative understanding obtained through the use of general arguments and simple models. These attributes and tools, when combined with physical intuition and the keen realist's eye of the experimentalist, are and will continue to be the engines that drive the subject's development.

What do we mean when we use the word nonlinear or talk about nonlinear science (optics, mechanics, physics, waves, etc.) as a subject? The literal meaning "pertaining to things not linear" is not really satisfactory because it defines a subject by what it is not; and on the surface it has about the same degree of vagueness as a description of all American animals as "nonelephant." It would include all relations between quantities the graphs of which are not straight lines, and this interpretation covers many natural phenomena. Nevertheless, although it is difficult to come up with a precise definition, one can clearly indicate what the word and subject connote. In the section that follows, we discuss briefly the ingredients that make nonlinear systems so different. Generally one finds that the solutions or output data of nonlinear systems display behaviors that depend very sensitively on input data and parameters, and it is therefore very difficult not only to obtain expressions for the former in terms of the latter but even to gain any understanding using only analytical methods.

How is it, then, that nonlinear optics is so accessible to theoretical analysis, especially when compared to other branches of nonlinear physics? The key reasons are that (1) at currently available light intensities, the nonlinear coupling coefficients are small, and (2) the power spectrum of the electromagnetic field is concentrated in the neighborhoods of discrete frequencies. These properties allow one to remove all fast space (10^{-6} m) and time scales (10^{-15} s) from the equations using standard perturbation techniques, and this leads to considerable simplification. To a first approximation, light and matter can be considered as a system of uncoupled oscillators; light consists of wavetrains $\exp i(\vec{k}\cdot\vec{x}-\omega t)$ while the oscillators of matter are electronic transitions, molecular vibrations and rotations, and acoustic waves. Therefore, to a first approximation, the variables of light and matter obey linear equations and the nonlinear terms are an order of magnitude smaller. This does not mean that they have negligible long-time and long-distance effects. Weak coupling does mean, however, that only certain identifiable subsets of all possible linear and nonlinear interactions between the oscillators are important, namely the small number of sets that satisfy special resonance conditions. Because of this, the fields can be accurately approximated over long times and distances by a *finite* combination of oscillator modes, namely those modes that take part in resonant interactions. The loss of energy to the many other degrees of freedom can be accounted for by attenuation terms (called homogeneous broadening) that are linear in the matter variables. Further, since the spectra of the electromagnetic and matter fields are localized in wavenumber-frequency space, the fields can be represented as finite sums of discrete wavepackets, $A(x, y, z, t) \exp i(lx+my+kz-\omega t)+(*)$, where $*$ is the complex conjugate, the wavevector components (l, m, k) are related to the frequency ω through the dispersion relation, and the amplitude $A(x, y, z, t)$ is a slowly varying function of space and time, that is, $\partial^2 A/\partial t^2 \ll \omega \partial A/\partial t \ll \omega^2 A, \partial^2 A/\partial z^2 \ll k\partial A/\partial z \ll k^2 A$. This means that the envelope A of the wavepacket obeys an equation containing only low powers of the derivatives $\partial/\partial t, \partial/\partial x, \partial/\partial y, \partial/\partial z$. Typically, the basic time and distance units for light waves in the visible range, $2\pi\omega^{-1}$ and $2\pi k^{-1}$, are of the order of a femtosecond (fs, or 10^{-15} s) and a micrometer (μm, or 10^{-6} m) respectively. The times over which the amplitudes vary lie in the range between nanoseconds (ns, or 10^{-9} s) and picoseconds (ps, or 10^{-12} s), and the ratio $\frac{1}{\omega A}\frac{\partial A}{\partial t}$ is comparable to the width of the wavepacket $\frac{1}{kA}\frac{\partial A}{\partial z}$ and the magnitude of the coupling coefficient.

A principal goal of theory is to write down envelope equations for these amplitudes. These equations are nonlinear, but they often fall into categories of nonlinear equations about which much is known. Whereas in Chapters 2, 3, and 6 we introduce the standard perturbation procedures for deriving them from the governing Maxwell's equations, one can readily deduce what the inviscid or frictionless form of the envelope equations must be by applying simple symmetry arguments. For example, suppose one is following the evolution of a single wavepacket

$$\vec{E}(\vec{x}, t) = \hat{e}(A(\vec{x}, t) \exp i(\vec{k} \cdot \vec{x} - \omega t) + *), \quad \vec{x} = (x, y, z), \quad \vec{k} = (l, m, k), \tag{1.1}$$

in a nonlinear dielectric with a centrosymmetric crystal structure (the crystal has reflection symmetry about the origin) so that if $\vec{E}(\vec{x}, t)$ is a solution so is $-\vec{E}(\vec{x}, t)$. Therefore, if A satisfies the envelope equation, so does $-A$. The equation for $\vec{E}(\vec{x}, t)$, derived directly from Maxwell's equations, has the additional properties of space and time reversibility and translation, meaning that if $\vec{E}(\vec{x}, t)$ is a solution, so are $\vec{E}(-\vec{x}, -t)$ and $\vec{E}(\vec{x}+\vec{x}_0, t+t_0)$ for arbitrary $\vec{x}_0, t_0$. The last two properties mean that if $A(\vec{x}, t)$ is a solution of the equation which the envelope satisfies, so are $A^*(-\vec{x}, -t)$ and $A(\vec{x}, t) \exp i\phi_0$ for arbitrary constant ϕ_0. In addition, the fact that we are dealing with *weak* nonlinear coupling of *long* wavepackets means that the equation for the amplitude A is a multinomial expansion in powers of A and the gradient $\nabla(\partial/\partial x, \partial/\partial y, \partial/\partial z)$ and time derivative $\partial/\partial t$ operators. To leading order, the only nontrivial candidate satisfying all these properties must be a real, linear combination of $\partial A/\partial t, \nabla A, i$ times all possible second derivatives, iA, and $i|A|^2 A$, precisely the terms that compose the universal and ubiquitous nonlinear Schrödinger (NLS) equation. If A is constrained to depend on one direction only, as in a light fiber, its equation will be

$$\frac{\partial A}{\partial z} + k'\frac{\partial A}{\partial t} + \frac{i}{2}k''\frac{\partial^2 A}{\partial t^2} - i\frac{\delta n(|A|^2)}{n}kA = 0, \tag{1.2}$$

with the refractive index correction $\delta n/n$ given by the first two terms $a + b|A|^2$ in a Taylor expansion in the field intensity and where the coefficients k, k', and k'' all have a very natural interpretation. In Chapter 2, we show that the operator that gives rise to the multinomial is simply the *dispersion relation*, namely the equation that relates the wavevector $\vec{k}$, frequency ω, and intensity $|A|^2$ of a nonlinear wavetrain. Equation (1.2) is the one-dimensional NLS equation (the dimension of the NLS equation is defined by the number of variables appearing as second derivatives), which has very special mathematical properties reflected in a wonderful class of solutions called *solitons*. The soliton is to nonlinear science what the Fourier mode is to linear science, namely a fundamental "normal" mode of propagation of a nonlinear system, and we have much to say about it in this book.

Is it a fluke that many of the equations of mathematical physics like the NLS equation or the Korteweg–de Vries equation or the sine–Gordon equation, which are derived by standard perturbation analyses as the universal asymptotic description of a wide variety of physical systems, are integrable or close to being integrable? We don't know the answer to this, but the following comment may be relevant. A key observation is that if one starts with an exactly integrable system, then the asymptotic analysis leading to the equation that describes the long-time behavior of the envelopes of special types of solutions does not destroy this integrable character; rather, the integrability is preserved. Therefore if among the set of all equations that reduce to the same asymptotic description there is one equation that is integrable, then the asymptotic equation is integrable. Therefore whereas integrability is rare in general, the process of reduction to universal, asymptotic equations for wave envelopes increases the probability that the resulting equation has special properties. The reduction process introduces new symmetries and new constraints (conservation laws) and does not destroy existing ones.

For the most part, therefore, the theorist can gain access to nonlinear optics by decomposing the relevant field variables into a finite basis of weakly interacting wavepackets, and the evolution of the field is obtained by following the *fully nonlinear* evolution of the wavepacket envelopes, which equations, because of symmetries and their universal nature, tend to have rather special properties. Being able to get off the ground with a finite linear basis is a luxury denied to most other areas of continuum physics, like fluid mechanics for example, in which nonlinearity is strong and for which in most circumstances it is impossible to neglect any higher-order interactions and approximate the fields uniformly in time and space by a finite number of modes. In a sense we can think of the NLS equation as being the optical analog of the Euler equations, the high-Reynolds-number, inviscid limit of the Navier–Stokes equations, with the one (two) dimensional NLS equation corresponding to the two (three) dimensional Euler equations. Much more information is known about the NLS equation in both cases. Of course, when light intensities are so high they raise the temperature of matter to a point where it behaves like a plasma, then the weak coupling theory is no longer valid. Up to that point, however, there remains much to be explored and discovered in contexts where the weak coupling approximation is applicable.

For systems like lasers, which are confined by finite geometries, the spectrum is discrete, spatial shapes are determined, and the spatial gradient terms disappear. Again, however, because of the constraints of resonances, the dynamics is described by a relatively low-order set of ordinary differential equations of dissipative type for the oscillator amplitudes. The dissipation arises from two sources. The electric field loses energy mainly because of imperfect mirrors. Matter loses energy because of collisions and the slow irreversible decay of the number of excited atoms to levels that are not included in the approximation. These effects, called homogeneous broadening, are modeled by the inclusion of linear damping terms. As we have mentioned, the behavior of the solutions of these coupled sets of ordinary differential equations can be very complicated. While there is no general theory for determining the nature of the asymptotic states (whether fixed points, limit cycles, or strange attractors) for such sets of equations, there has nevertheless been built up much qualitative experience about their behavior. This has been greatly facilitated by the new generation of computer workstations with their high performance processing and graphical capabilities, which provide a very powerful analytical research tool. In Chapter 6, we describe some of the sophisticated software available on most modern computer workstations. In addition to the standard commercial scientific subroutine packages such as IMSL or NAG, and symbolic programming languages like MACSYMA or REDUCE for algebraic manipulation of complex expressions, one can gain access to bifurcation packages (AUTO, PITCON, etc.) and a host of very powerful UNIX-based public domain software. The student can, with little computational effort, realize the marvelous and deep mathematical insights of Poincaré regarding the geometry of phase space in real time on a high-resolution graphics screen. For example, numerical integrators for the systems of ordinary differential equations (o.d.e.'s) modeling standard laser systems can be found in the standard IMSL or NAG packages available on most workstations.

We begin in the next section by briefly describing behavior unique to nonlinear systems. We then discuss dielectrics and emphasize that the principal medium variable, the susceptibility or refractive index, which tells us how the polarization field depends on the applied electric field, is almost impossible to calculate in general. In a very real sense, therefore, we start from a constitutive relation that is at best an approximation. Consequently, it is clear that a top priority of the field is to develop a better understanding of how to calculate the susceptibility and how to design materials with susceptibilities having advantageous properties. We then briefly discuss two areas of great importance that are not treated in this edition, namely mode locking and pulse compression and the interaction of light with semiconductors, each of which has enormous practical potential and intellectual challenges. We felt that to omit any mention of them would signal that they were not important. To disabuse the reader of that conclusion, we decided to mention them up front.

The Nature of Nonlinearity

We look at the question of defining a nonlinear system two ways, first by the mathematical character of the laws (equations) that describe its behavior and second by its actual function as an input-output device. From the former point of view, the definition of linear and nonlinear is, at least on the surface, quite simple. Are the equations (we include in this set the equations of motion, initial and boundary conditions, constitutive relations between variables) themselves linear; namely, is each term in the equation either given or proportional to the first power of one of the dependent variables or its derivatives? For examples,

Maxwell's equations,

$$\nabla \times \vec{E} = -\frac{\partial \vec{B}}{\partial t}, \quad \nabla \times \vec{H} = \frac{\partial \vec{D}}{\partial t} + \vec{j}, \quad \nabla \cdot \vec{D} = \rho, \quad \nabla \cdot \vec{B} = 0 \quad (1.3)$$

with $\vec{B} = \mu \vec{H}$ and $\vec{D} = \epsilon \vec{E}$, μ, ϵ constant; Malthus' equation for population growth,

$$\frac{dP(t)}{dt} = \kappa P(t), \quad (1.4)$$

the equations for a forced, damped oscillator,

$$\frac{d^2 x}{dt^2} + 2\beta \frac{dx}{dt} + \omega^2 x = F \cos \Omega t$$

$$x(0) = x_0, \frac{dx}{dt}(0) = v_0, \quad (1.5)$$

and the equation for wave propagation,

$$\frac{\partial^2 u}{\partial t^2} - c^2 \frac{\partial^2 u}{\partial x^2} + \omega_0^2 u = 0, u(x, 0) = f(x), \frac{\partial u}{\partial t}(x, 0) = g(x), \quad (1.6)$$

are linear according to this definition while the same equations—(1.3) with the susceptibility ϵ depending on the electric field $\vec{E}$, (1.4) with the addition of the saturation term $-P^3(t)$, (1.5) with the restoring force $-\omega^2 x$ replaced by $-\omega^2 x + \beta x^3$, and (1.6) with $\omega_0^2 u$ replaced by either $\omega_0^2 u - \omega_0^2 u^3$ or $\omega_0^2 \sin u$—would be nonlinear. Another set of equations of much interest both in general and to us in this book are the Lorenz equations,

$$\frac{dA}{dt} = -\sigma A + \sigma P$$

$$\frac{dP}{dt} = \tau A - P - An$$

$$\frac{dn}{dt} = -bn + AP$$

$$A(0) = A_0, P(0) = P_0, n(0) = n_0. \quad (1.7)$$

The linearized system (drop the quadratic products $-An$ and AP) has general solutions

$$\begin{pmatrix} A \\ P \\ n \end{pmatrix} = b_1 \begin{pmatrix} \sigma \\ \sigma + \lambda_+ \\ 0 \end{pmatrix} e^{\lambda_+ t} + b_2 \begin{pmatrix} \sigma \\ \sigma + \lambda_- \\ 0 \end{pmatrix} e^{\lambda_- t} + b_3 \begin{pmatrix} 0 \\ 0 \\ 1 \end{pmatrix} e^{\lambda_3 t} \quad (1.8)$$

where $\lambda_\pm = -(\sigma + 1) \pm \sqrt{(\sigma + 1)^2 + 4\sigma(r - 1)}$, $\lambda_3 = -b$, which, depending on whether $r < 1$ or $r > 1$, approach either $\vec{0}$ or $\vec{\infty}$. We shall see that the solutions of the nonlinear system are much richer and cannot be expressed as a linear combination of simple solutions.

Now what is the great mathematical advantage of a linear system? It is that complicated or general solutions can be constructed by the linear superposition of simple or elementary ones. We have already seen this in the case of the linearized version of (1.7). In (1.6), we can decompose the general solution $u(x, t)$ into a sum (integral) of elementary wavetrains

$$A_s(k)e^{ikx - isw(k)t}, \quad \omega(k) = \sqrt{\omega_0^2 + c^2 k^2}, \quad s = \pm 1, \tag{1.9}$$

then determine from $f(x)$ and $g(x)$ how much of each wavetrain is present initially (i.e., find $A_+(k)$, $A_-(k)$), and compute the solution as the sum (integral) of the weighted wavetrains (1.9). We cannot do this with the nonlinear version, at least, not in any obvious way. Likewise, the solution to (1.5) can be written as a linear combination of decaying oscillations and a nondecaying response term

$$G\cos(\Omega t + \phi) \tag{1.10}$$

with the gain G given by

$$\frac{F}{((\omega^2 - \Omega^2)^2 + 4\beta^2\Omega^2)^{1/2}} \tag{1.11}$$

Observe that to each set of input values F, Ω, β there is one output gain G.

The fact therefore that general solutions can be written as a linear superposition of simple ones with the amount of each simple solution determined by initial or boundary conditions or external forcing means that in terms of an output-input curve the general solution is not very complicated at all. The graph of the solution itself—$x(t)$, $P(t)$, or $u(x, t)$—is not linear in t, but the dependence of a later state on an earlier one is simple and linear. This simple qualitative behavior of the output-input curve can change dramatically when the system's laws or equations are nonlinear. Consider what happens when we include the nonlinear restoring force $-\omega^2 x + \gamma x^3$ in (1.5). In that case the gain G as a function of the forcing frequency Ω looks like the curve shown in Figure 1.1, and then only for these values of the parameter for which the output is indeed a periodic function of time with period $2\pi/\Omega$. For certain ranges of Ω, there are three outputs for one input, two stable (*AB, CD*) and one unstable (*BC*). We also see the possibility of hysteresis. Imagine that we

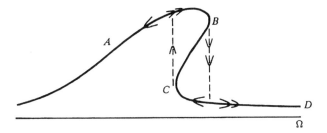

FIGURE 1.1   Input-output curve for a nonlinear system.

decrease Ω from D toward C. The gain lies along the stable branch DC. At C, the gain jumps to the second stable branch AB. Now if we again increase Ω, the gain increases along the stable branch B, at which point it returns to the branch CD. And again we stress that the $2\pi/\Omega$ periodic behavior only occurs for a certain range of parameters. As γ is increased even the stable branches can destabilize, to periodic solutions with period $4\pi/\Omega$, $8\pi/\Omega$, and eventually to chaotic states. All of this rich behavior of the output-input curve is a result of adding a little bit of nonlinearity.

Likewise for (1.7). For $r > 1$ and certain ranges of the parameters σ, r, b the solution tends to one or the other of the fixed points $A = \pm\sqrt{(r-1)} = P$, $n = r - 1$. This, we shall see, corresponds to laser action in a two-level laser model. On the other hand, for r much larger and $\sigma > b + 1$ (called in laser parlance the bad cavity limit), the fixed-point lasing solution is unstable and the output as a function of time is quite irregular and chaotic.

We have now seen two examples for which the output-input curve is quite complicated. Small changes in input can lead to drastic changes in the output. We emphasize, however, that while the nonlinearity in a system of equations is a necessary ingredient for a complicated input-output curve, it is not in itself sufficient. We see in Chapters 2, 3, and 6 that equations such as the nonlinear Schrödinger (NLS) equation

$$q_t - iq_{xx} - 2iq^2q^* = 0 \tag{1.12}$$

and the sine–Gordon equation

$$u_{tt} - c^2u_{xx} + \omega_0^2 \sin u = 0 \tag{1.13}$$

have very simple solutions because, through a (nonlinear) change of variables, they can be converted to uncoupled linear equations. Whereas the general solution in the new coordinates can be constructed as a linear combination of normal modes, such is not the case for the nonlinear fields $q(x, t)$ and $u(x, t)$. However, their general solution can still be constructed as a nonlinear superposition of elementary solutions, and these nonlinear superpositions can be built through series of transformations called Bäcklund transformations.

An analogous situation occurs if we try to solve Burgers' equation,

$$u_t - 2uu_x = u_{xx}. \tag{1.14}$$

As an exercise show that if

$$v_x = u(x, t)v \tag{1.15}$$

and

$$v_t = v_{xx} \tag{1.16}$$

then $u(x, t)$ satisfies (1.14). Therefore we can construct general solutions of Burgers' equation by first constructing solutions of the linear heat equation for $v(x, t)$ and then using the nonlinear transformations (1.15) to construct $u(x, t)$.

So there are lots of nonlinear equations that behave essentially as linear ones, although it is not always obvious whether or not a given nonlinear equation belongs to this integrable class. However, they are relatively rare when considered as members of the set of all nonlinear equations that one can possibly write down, and the solutions of most equations in the noninte-grable class display a behavior in which the output data are a complicated and sensitive function of the input data and parameters. Moreover, this dynamics is erratic, slipping capriciously between very irregular output to intermittent bursts of regular behavior and back again, so that the system defies standard treatment by the methods of equilibrium statistical mechanics. Most of the complicated behavior seen in turbulent fluid flows falls into this category, and so does some of the behavior seen in nonlinear optics. However, as we have said, because one can get off the ground in optics with a balance of linear forces, the nonlinear response of optical systems is somewhat more manageable but none the less violent than in the case in high-Reynolds-number fluid flows.

Dielectrics

The Maxwell equations (1.3) connect the six basic electromagnetic quantities $\vec{E}, \vec{H}, \vec{B}, \vec{D}, \rho$, and $\vec{j}$. To allow a unique determination of the field vectors from a given distribution of currents and charges, these equations must be supplemented by relations that describe the behavior of materials under the influence of the field. Such relations provide material equations (or constitutive relations) that generally are very complicated. If the field is time-harmonic and if the bodies are at rest or in very slow motion relative to each other, and if the material is isotropic, the constitutive relations usually take a relatively simple form

$$\vec{j} = \sigma \vec{E}, \quad \vec{D} = \epsilon \vec{E}, \quad \vec{B} = \mu \vec{H},$$

where the scalar coefficient σ is called the specific conductivity, ϵ is known as the dielectric constant (or permittivity), and μ is called the magnetic permeability. For anisotropic materials, these relations are replaced by tensor equations.

Substances for which $\sigma \neq 0$ (or more precisely is not negligibly small) are called conductors. Metals are very good conductors, as are ionic solutions in liquids and in solids. In metals the outermost electrons in the orbits of the constituent atoms making up a regular array are essentially free to move and give rise to an electric current. Other classes of materials that play an increasingly important role in modern technology are semiconductors, with silicon and gallium arsenide being the best known. Semiconductors are distinguished from metals in the sense that at very low temperature their conductivity is zero. When analyzed as a periodic crystalline array of atoms, we find that the lowest energy states occupied by electrons, the so-called valence bands, are filled. The energy gap between the filled valence band and the next highest unoccupied energy state (the conduction band) is small enough that increased temperature or incident light photons can easily promote some electrons from the filled valence band to the unoccupied conduction band, giving rise to a finite conductivity.

Insulators or dielectrics are materials for which σ is negligibly small; their electric and magnetic properties are then completely determined by ϵ and μ. For most substances and for the range of frequencies we consider, the magnetic permeability μ is essentially constant, and such nonmagnetic substances are the focus of much of the discussion in the book.

An intense quasi-monochromatic light pulse can, either directly or indirectly, excite material oscillation modes covering an enormous frequency bandwidth ranging from 10^{15} s^{-1} (electronic oscillations) to 1 s^{-1} (thermal oscillations). A proper theoretical treatment of these many diverse linear and nonlinear material oscillations must include a proper description of the polarization wave vector $\vec{P}(\vec{r}, t)$ induced by the optical wave vector $\vec{E}(\vec{r}, t)$. These quantities are linked through Maxwell's equations via the constitutive relation where $\vec{D} = \epsilon_0 \vec{E}(\vec{r}, t) + \vec{P}(\vec{r}, t)$, the first term being the vacuum and the second the material contribution. Fortunately, dielectric materials have certain characteristics that allow lasers in many instances to isolate and probe specific types of interactions of light with matter. Under other circumstances the interactions may be extremely complex. A case in point is the collapse of an intense laser pulse into multiple self-focused filaments as a consequence of the modulational instability discussed in Section 2h. Whereas the initial phase of the instability is well understood, in the latter stages of the collapse as peak intensities grow rapidly it is known that the light field drives acoustic waves via stimulated Brillouin scattering (as evidenced by the loud bang heard during beam collapse in glasses) and so-called optical lattice phonon modes via stimulated Raman scattering, generates an electron plasma via ionization, and furthermore induces large thermal gradients near the collapsing foci. In other cases, the physical description of the light-matter coupling may be much simpler, as is the case for an optical fiber, where transverse waveguiding constraints prevent catastrophic collapse and the longer available interaction lengths significantly reduce the light wave intensities needed for observations of nonlinear phenomena. Care must be exercised here too as the basic material oscillations excited during collapse in bulk dielectrics (acoustic, optical lattice phonons) can still be easily excited in a fiber. Thus, for example, in soliton communications systems (discussed in Chapter 3) there is a relatively narrow pulse duration window, of the order of tens of picoseconds, where clean solitary wave pulse transmission becomes feasible. If the pulse is too long (of the order of tens of nanoseconds), then acoustic modes can be easily excited, and the consequent stimulated Brillouin scattering can lead to a backward-propagating intense frequency-downshifted optical pulse that literally consumes the original forward-propagating soliton pulse. On the other hand, if the pulse is too short (fractions of a picosecond), then optical lattice modes can be excited via stimulated Raman scattering leading to a degradation of the soliton pulse. Optical fibers allow a quantitative study of the complex interplay between these various material oscillation modes, a luxury not available for bulk interactions. In the telecommunication window the material oscillation dynamic response is so fast ($\sim 10^{-14}$ s) that it may be considered instantaneous. In this case the coupling to the dielectric material is lumped into a single parameter, the so-called "n_2" or Kerr coefficient, and the NLS type equation provides a complete description of the nonlinear dynamics.

Light can couple in a nonresonant or resonant fashion to material waves by inducing dipoles in the atoms or molecules making up the material. A simple picture of a nonresonant coupling is provided by the reorientation of anisotropic polarizable molecules in a light field, as discussed in Section 4h. The idea is simple, as the sketch in Figure 1.2 shows. A polarizable anisotropic molecule such as carbon disulfide (CS_2) is electrically neutral in the absence of an applied field, and its anisotropy arises from the fact that the electron cloud binding the atoms together in the molecule can be more easily displaced along the molecule's axis than at right angles to it. We can introduce a polarizability tensor $\overline{\overline{\alpha}}$ with two

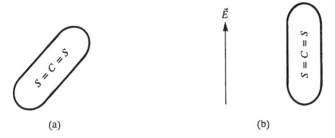

FIGURE 1.2 Polarization of neutral carbon disulfide molecules: (a) no field; (b) with static field.

unequal components $\alpha_\parallel$ and $\alpha_\perp$ reflecting the symmetry of the molecule. An applied static field displaces the electron cloud and as a consequence introduces a finite charge displacement. As the displacement is easiest along the molecule's axis, the largest dipole can be induced in this direction, and the molecule tends to align along the field direction. Of course there are $\sim 10^{18}$ molecules occupying a cubic centimeter in liquid CS_2, and in thermal equilibrium these long molecules are oriented at random. One can easily imagine that as the static field strength is increased, more and more molecules will align along the field direction until eventually all are aligned. The complete alignment corresponds to saturation of the induced polarization wave. In many practical situations field strengths are well below their saturation values, and the random (Boltzmann) orientational distribution function can be approximated by a Taylor series expansion of its argument.

Consider now an alternating (AC) optical field. The $\vec{E}$ vector switches orientation approximately every 10^{-15} second, and the molecules cannot possibly orient at this rate. There is therefore no first order effect in $\vec{E}$. However, the molecular reorientation can respond to the average of $\vec{E}^2$ as this contains a static (DC) component, which can induce a finite orientation effect. Mathematically, the nonresonant reorientation effect is captured dynamically by the Debye equation derived in Section 4h,

$$\tau \frac{\partial}{\partial t} \delta n + \delta n = \alpha \langle E^2 \rangle,$$

where the angle bracket implies that the refractive index perturbation appearing in the polarization wave is driven by the average of the intensity. This interaction is termed nonresonant as no internal material oscillation is being driven; turn off the field and the aligned molecules return to their random orientation in an exponential decaying manner.

In a resonant coupling mechanism such as encountered in classic three-wave interactions processes (such as stimulated Raman scattering, discussed in Chapters 4 and 5), a natural oscillation within the dielectric material with a frequency many orders of magnitude less than the optical wave frequency can only respond to the total field intensity $E^2(t)$ and not to the field itself. The mechanism therefore is identical to that for the reorientational effect, but now we think of the total field as a linear superposition of a main optical wave and a weak frequency-shifted sideband; the frequency shift corresponds to the material oscillation frequency. This resonant interaction is captured mathematically as a driven damped harmonic oscillator (1.3), rewritten here for convenience

$$\frac{d^2 Q}{dt^2} + 2\beta \frac{dQ}{dt} + \Omega_0^2 Q = \text{const. } E(t)^2,$$

where Q is the appropriate normal mode amplitude, 2β is a damping constant, and Ω_0 is the natural oscillation frequency of the material (i.e., acoustic or optical phonon). The resonant driving term comes from

$$E^2(t) = (A_i e^{i(k_i z - \omega_i t)} + A_j e^{i(k_j z - \omega_j t)} + (*))^2$$

$$= (|A_i|^2 + |A_j|^2 + 2A_i A_j^* e^{i[(k_i - k_j)z - (\omega_i - \omega_j)t]} + (*)),$$

where A_i, A_j are the optical and sideband field envelopes. The third term on the second line drives the material oscillation at Ω_0 when $\omega_i - \omega_j \simeq \Omega_0$.

What We Have Left out

This textbook aims to provide the reader with a glimpse of the extremely rich and varied nature of the challenges associated with nonlinear optical processes. This exciting field is so broad in scope, and new developments are occurring so rapidly, that it would be impossible to do justice to the entire field in a monograph of this nature. Inevitably some areas have been covered rather superficially and others not at all. In the former case we guide readers to specific articles in the research literature that will enable them to carry through the ideas briefly enunciated in the text with (we hope) the least possible pain. Of the topics that we have not dealt with explicitly, two deserve special mention, namely the interaction of light with semiconductor media and what can be loosely grouped under the heading "mode-locking/coupled cavity dynamics." Both of these areas are of great significance, as can be appreciated by glancing through journals such as _Optics Letters_ or the _Journal of the Optical Society of America B_. Basically, the tools developed in this monograph should enable the reader to acquire a familiarity with both topics in a relatively short time.

Existing semiconductor laser theories are based on a simple phenomenological model that accounts qualitatively for many of the gross dynamical features of a single laser. Such theories have proved essential to describing transient relaxation oscillations to a steady-state output; current modulation, whereby the injection current that achieves inversion through carrier (electron) generation in the conduction band of the semiconductor is periodically varied at high frequencies close to the relaxation oscillation frequency in order to produce pulsed laser outputs; injection-locking by an external driving optical field; and noise characteristics close to threshold that lead to multilongitudinal mode oscillation. A semiconductor laser differs fundamentally in a number of ways from the discrete energy level (two-and-three-level atom) lasers that we discuss. The periodic array of atoms in the crystalline lattice of a semiconductor significantly alters characteristics of the energy states accessed for laser operation. Both the lower state, fully occupied with electrons (valence band), and the empty upper state (conduction band) are quasi-continuous energy bands showing a parabolic dependence ($E \sim k^2$) on the particle (electron) wavenumber k. Mathematically, these states are essentially the Floquet solutions to a second-order o.d.e. with periodic coefficients; they are called Bloch states in a semiconductor. An external voltage is applied in order to generate electrons (carriers) in the upper laser state (conduction band), and as the electron density in the band increases there is a strong Coulomb repulsion between these identically charged species, leading to a strong perturbation of the bands themselves. These important aspects of particle-particle interaction are ignored in phenomenological theories. A detailed treatment of the phenomenological semiconductor laser model can be found in the textbook by Yariv [3].

This theory has many shortcomings that have become apparent from recent experimental observations on both single laser diodes and arrays of such lasers coupled together in differing geometries. One of the most controversial "ad hoc" parameters introduced in the phenomenological theory is the so-called linewidth enhancement factor. Numerical values for this quantity, which measures the carrier density dependence of the ratio of the refractive and gain contributions to the optical susceptibility (see Chapter 4 for a definition of this quantity), range in magnitude between 2 and 6 typically, based on experimental observations. Unfortunately, the dynamical behavior of semiconductor laser arrays is a very sensitive function of the magnitude of this parameter. A proper theoretical treatment of the interaction of light with semiconductor media requires the introduction of the many-body aspects of the fundamental particle interactions (carriers and holes). This latter theory removes the need to introduce ad hoc parameters such as the linewidth enhancement factor at the added expense of a more complicated many-body approach. One important result from a steady-state many-body laser theory is that the "linewidth enhancement factor" cannot be viewed as a parameter at all but instead is a sensitive function of carrier density and frequency. In fact, this function typically oscillates in magnitude in the range between 2 and 6 as a function of carrier density, which probably accounts for the great disparity between experimentally measured quantities. A many-body dynamical theory of semiconductor lasers is yet to be developed, and the reader who wishes to understand the various issues glossed over above is strongly urged to read the recent textbook by Haug and Koch [19]. One of the most fascinating breakthroughs in semiconductor physics over the past few years has been the fabrication of "d-dimensional" structures where the dimension d can range between zero (quantum dots) and three (bulk semiconductors). The light absorption and gain characteristics of the semiconductor change dramatically with dimension.

The basic laser operating principles are introduced in Section 5a, where we discuss and elaborate on the simplest nonlinear dynamical features of the laser. A laser in general is a complex and highly flexible nonlinear dynamical system that can be controlled to operate in a variety of ways. Of great importance is "mode-locking" whereby many hundreds or perhaps thousands of longitudinal cavity modes (see the discussion in Chapter 5), separated in frequency by $\Delta\omega = \pi c/L$ (L is length of the cavity, c is velocity of light), conspire to oscillate with a fixed relative phase under a very broad gain curve (see Chapter 4 for a definition of gain). The result is a periodic train of optical pulses whose peak amplitude is N times the amplitude of a single mode; this can be simply seen by summing up N in-phase frequency-shifted Fourier modes. Moreover, the width of each mode-locked pulse is approximately the inverse of the gain bandwidth. Consequently, short intense pulse trains can be produced by the laser. We do not wish to elaborate on the means by which mode-locking is achieved except to say that the laser is operating in a multi-longitudinal mode regime in contrast with the single or multi-transverse mode case that we focus on in Chapter 5. Combining mode-locked lasers with optical fibers in coupled cavity arrangements has led to enhanced stabilization of mode-locked pulse trains, the development of the soliton laser, and compression of optical pulses. The most dramatic compression of an optical pulse down to an 8 fs duration (3 optical cycles!) has been achieved with such a coupled cavity arrangement with additional linear optical elements cleverly arranged to compensate for higher order group velocity dispersion. The reader is strongly urged to read the monograph *The Supercontinuum Laser Source* [20] to gain an appreciation of the possible applications of ultrashort laser pulses as probes in physics, chemistry, and biology.

Physical Units

We adopt the MKS (meters (m), kilogram (M), second (s)) system. This system derives directly from Newton's second law

$$\vec{F} = M\vec{a} = M\frac{d\vec{v}}{dt}$$

relating the force acting on a body to its acceleration. The proportionality constant is the mass of the body, which we measure in kilograms. Given that the velocity is the rate of change of position, we see that the definition of force in terms of the basic quantities mass, length, and time is

$$\text{unit of force} = \frac{1M \times 1m}{1s^2}$$

which we call the newton. When dealing with electromagnetic quantities it is convenient to introduce one additional basic quantity, the unit of charge e called the coulomb (C) (the charge on an electron $e = 1.602 \times 10^{-19}C$). Using these four fundamental quantities we can establish the physical units of all other physical quantities of interest by using the basic established laws of physics. Thus, for example, the field intensities $\vec{E}$ and $\vec{H}$ are commonly defined in terms of the force experienced by a small, unit test charge or a free, unit magnetic pole if it could be placed at a given point without significantly disturbing the field. As magnetic test entities exist only in the form of dipoles associated with motion, circulation, or rotation of electric charges, it is convenient to define $\vec{E}$ and $\vec{H}$ using the Lorentz force law, which describes the force acting on a unit test charge undergoing some arbitrary motion:

$$\vec{F} = e(\vec{E} + \vec{v} \times \vec{B})$$

in which $\vec{v}$ is the velocity of the test charge and $\vec{B}$ is the magnetic induction; the magnetic induction contributes an additional, velocity-dependent force on moving charges. In vacuum $\vec{B} = \mu_0 \vec{H}$. Now the unit of $\vec{E}$ is obviously the newton per coulomb or equivalently, the joule meter^{-1} coulomb^{-1}, where the joule is the unit of energy in the MKS system (energy is work done $= \int \vec{F} \cdot d\vec{r}$ and so the unit of energy, the joule, is newton · meter). To save on awkward strings of symbols we abbreviate this definition as volt per meter (obviously the volt (V) is joule coulomb^{-1}). The unit of $\vec{B}$ follows from the Lorentz force law and is newton coulomb^{-1} meter^{-1} second, usually abbreviated as volt second meter^{-2}. The constitutive relations link the four fundamental vector quantities $\vec{E}, \vec{H}, \vec{D}, \vec{B}$ appearing in Maxwell's equations. In vacuum,

$\vec{D} = \epsilon_0 \vec{E}$ and $\vec{B} = \mu_0 \vec{H}$. The coefficient of proportionality ϵ_0 in the first relation has the dimensions of coulomb volt^{-1} meter^{-1} or farad meter^{-1} where one farad (F) is a coulomb volt^{-1}. The value of ϵ_0 is 8.85×10^{-12} farads meter^{-1}. It is straightforward to see from Coulomb's law (which determines the dimension of E) and Ampere's law that the product $(\mu_0\epsilon_0)^{-1}$ has the dimension of velocity squared, which in vacuum corresponds to c^2 (c is the velocity of light in vacuum). This basically establishes a value for the magnetic permeability μ_0 which in MKS units is $4\pi \times 10^{-7}$V s A^{-1} m^{-1} where the dimension of ampere (A) measuring current is coulomb second^{-1}. Two other physical quantities of great importance are the energy density per unit volume

$$\varepsilon = \tfrac{1}{2}\text{E.D} + \tfrac{1}{2}\text{B.H}$$

and the energy flux, measured by the Poynting vector,

$$\vec{S} = \vec{E} \times \vec{H}.$$

The latter measures the rate at which energy is transported across a unit surface area whose normal is parallel to $\vec{S}$. The units of ε are coulomb volts (joules) per cubic meter. The units of S are coulomb volts per second per square meter or watts per square meter. A watt is a coulomb volt (joule) per second.

If one considers the propagation of plane waves ($\vec{E} = \vec{E}_0 e^{i(\vec{k}_0 \cdot \vec{r} - \omega_0 t)} +$ (∗), $\vec{H} = \vec{H}_0 e^{i(\vec{k}_0 \cdot \vec{r} - \omega_0 t)} +$ (∗)), then the Poynting vector (in vacuum) is

$$\vec{S} = \vec{k}_0 \left(\frac{4E_0^2}{\mu_0 \omega_0} \right) \cos^2(\vec{k}_0 \cdot \vec{r} - \omega_0 t),$$

where we have used the fact that the $\vec{E}$ and $\vec{H}$ vectors are mutually perpendicular and are in turn perpendicular to the direction of propagation defined by $\vec{k}_0$. The energy density ε in a vacuum, averaged over a wave period, is $2\epsilon_0 E_0^2$. Because $\vec{S} = \vec{E} \times \vec{H}$ has the direction $\vec{k}_0$, energy always flows in the direction of propagation of the wave, and its time average over a wave period is

$$S_{\text{av}} = 2\epsilon_0 c E_0^2 \text{ watts per square meter.}$$

which is the energy density ε multiplied by the speed of light. When the electromagnetic wave propagates in a transparent medium of refractive index n, then the average energy is modified to

$$S_{av} = 2\epsilon_0 n c E_0^2 \text{ watts per square meter.}$$

This quantity measures the average energy transported by a laser beam, for example, and is often referred to as the laser intensity.

Summary Table

quantity	symbol	units
mass	M	kilogram (kg.)
length	m	meter
time	s	second
electric charge	C	coulombs
velocity	ms^{-1}	meters per second
electric current	$A = Cs^{-1}$	ampere = coulomb per second
force	$N = Mms^{-2}$	newton = kg. meter per second squared
energy	$Nm = J$	newton meter = joule
electric field $\vec{E}$	$Vm^{-1} = JC^{-1}\,m^{-1}$	volts (joules per coulomb) per meter
electric induction $\vec{D}$	Cm^{-2}	coulombs per square meter
magnetic field $\vec{H}$	$Cm^{-1}s^{-1}$	coulombs per meter per second
magnetic induction $\vec{B}$	$Vs\,m^{-2}$	volt seconds per square meter
energy density ε	Jm^{-3}	joules per cubic meter
energy flux $\vec{S}$	$Js^{-1}\,m^{-2} = Wm^{-2}$	joules per second (watts) per square meter

Table of Values

quantity	symbol	units
electron charge	e	-1.610^{-19} C
Planck's constant	h	1.0510^{-34} Js
photon energy for $\omega = 10^{+15}\,s^{-1}$	$h\omega$	1.0510^{-19} J = .66eV
dielectric constant in vacuum	ϵ_0	8.8510^{-12} $CV^{-1}m^{-1}(CV^{-1}$ = F, farad)
magnetic constant	μ_0	$4\pi 10^{-7}$ Vs $A^{-1}\,m^{-1}$ = $4\pi 10^{-7}$ $Vs^2C^{-1}\,m^{-1}$
speed of light in vacuum	$c = \frac{1}{\sqrt{\mu_0\epsilon_0}}$	310^8ms^{-1}
typical refraction index of glass	n	1.5
nonlinear refraction index	n_2	$1.210^{-22}m^2V^{-2}$

EXERCISE 1.1

Imagine a pulse of 1 mW (one milliwatt or 10^{-3} Js^{-1}) travelling in an optical fiber. From the table, estimate the number of photons per second which would produce such a flux (Answer: 10^{16} per second). This enormous number gives one confidence that one can deal collectively with light photons as continuous fields.   □

THEORY OF LIGHT PROPAGATION IN LINEAR AND WEAKLY NONLINEAR (KERR MEDIUM) DIELECTRICS

Overview

IN THIS CHAPTER, WE PRESENT the theory of light wave propagation in linear and weakly nonlinear (Kerr) dielectrics. A principal goal is to become familiar with the nonlinear Schrödinger (NLS) equation, which governs the evolution of the envelope A of a wavepacket in a medium for which the refractive index depends on the field intensity. The outline of the chapter is as follows:

2a. We present Maxwell's equations, show how they give rise to the wave equation, and introduce the *refractive index* function.

2b. Here we introduce the effects of delay, the noninstantaneous response of the induced polarization to the applied electric field, and show how it gives rise to a *frequency-dependent refractive index*. The refractive index is necessarily complex and includes the effects of absorption and amplification. The real and imaginary parts are related through the well-known Kramers–Krönig relations.

2c. We discuss the elementary plane wave or wavetrain solutions of Maxwell's equations, the building blocks of more general solutions, and introduce the notions of phase, wavevector, frequency, and *dispersion relation*, which connects the latter two.

2d. We discuss the nature of the relations between the components of the susceptibility tensors when the polarization is expressed as a power series in the electric field.

2e. We show that an intensity-dependent refractive index of a wavetrain gives rise to a nonlinear dispersion relation involving the field intensity in addition to the wavevector and frequency. The net effect is that the *wavenumber and propagation characteristics of the nonlinear wavetrain depend on field intensity*.

2f. We review the main ideas of geometric optics, in which a plane wave travels in a medium of slowly varying refractive index. We learn that light tends to concentrate in regions of elevated refractive index.

2g. We meet the idea of a *wavepacket*, a linear superposition of wavetrains whose wavevectors and frequencies lie in a narrow band, and show how it can be conveniently described by a wavetrain whose amplitude (envelope) varies slowly in space and time. The notions of *group velocity*, *dispersion*, and *diffraction* are introduced, and we show how these quantities are simply coefficients in the partial differential equation that governs the evolution of the slowly varying wavepacket envelope. As a special case, we meet what is often called the paraxial approximation. At this stage, we prove a very important result on phase conjugated waves.

2h. We derive the general equation, the nonlinear Schrödinger (NLS) equation, governing the evolution of wavepackets through a weakly nonlinear medium. This equation is simply a juxtaposition of those derived in 2e and 2g. In particular, we discuss the idea of the *self-focusing* of a light beam and show how in these cases the nonlinear wavetrain solution is unstable. *Light beams, in a medium where the refractive index increases with increasing intensity, have a natural tendency to focus.*

2i. We next introduce *linear* and *nonlinear birefringence*. The former is an effect that occurs in anisotropic dielectrics because the phase velocities of different directions of polarization of the electric field are different. The latter does not necessarily require anisotropy in the medium but is the result of nonlinear coupling between two different (often transverse) modes of the electric field.

2j. In this section, we review three- and four-wave mixing processes important for the transfer of energy between different wavetrains. They occur in a wide variety of contexts, in Raman and Brillouin scattering, in atmospheric, plasma, and ocean wave interactions, and are the mechanism used to construct a phase conjugated wave.

2k. Finally, in contrast to the heuristic derivation of 2h, we give a formal derivation of NLS taking account of the vector nature of Maxwell's equations. We also discuss in what sense the equation is universal and point out circumstances whereby a new D.C. field can be generated by quadratic nonlinearities. The ramification of these effects in the context of optics is explored and shown to be potentially important in creating a large effective nonlinear refractive index.

2a.   Maxwell's Equations, the Wave Equation, and Refractive Index

A rational theory for the propagation of light, combining for the first time a unified treatment of electric and magnetic fields, was developed by Maxwell in the 1860s. His equations for the electric and magnetic intensity fields $\vec{E}$ and $\vec{H}$ and the electric and magnetic induction fields $\vec{D}$ and $\vec{B}$ were the differential forms of Gauss', Faraday's, and Ampere's laws and in MKS units are

$$\nabla \cdot \vec{D} = \rho, \tag{2.1}$$

$$\nabla \cdot \vec{B} = 0, \tag{2.2}$$

$$\nabla \times \vec{E} = -\frac{\partial \vec{B}}{\partial t}, \tag{2.3}$$

$$\nabla \times \vec{H} = \frac{\partial \vec{D}}{\partial t} + \vec{j}, \tag{2.4}$$

where ρ and $\vec{j}$ are the electric charge and current densities. Connected with these equations is the law of conservation of charge. Adding the time derivative of (2.1) and the divergence of (2.4), we find

$$\frac{\partial \rho}{\partial t} + \nabla \cdot \vec{j} = 0, \tag{2.5}$$

which expresses the fact that an increase or decrease of charge in a given volume V with enclosing surface S

$$\frac{\partial}{\partial t} \int_V \rho \, dV = -\int_V \nabla \cdot \vec{j} dV = -\int_S \vec{j} \cdot \hat{n} \, dS \tag{2.6}$$

is effected by the transport of charge, namely flow of current, through the boundaries of V.

In this book, for the most part, we deal with dielectrics in which there are no free charges and in which no current flows. Therefore ρ and $\vec{j}$ will be zero. The fundamental electromagnetic "atoms" of a dielectric medium are (effective) dipoles, and one of our principal goals is to understand how these dipoles affect the electric induction field $\vec{D}$ in response to an applied electric field $\vec{E}$. Equations (2.1) through (2.4) together with constitutive relations connecting $\vec{D}$ and $\vec{E}$ and $\vec{B}$ and $\vec{H}$ are the field equations. At boundaries, it is simple to show from these equations that the normal components of $\vec{D}$ and $\vec{B}$ and the tangential components of $\vec{E}$ and $\vec{H}$ are continuous at interfaces that separate distinct media.

EXERCISE 2.1

By integrating (2.2) over a circular cylinder astride the boundary whose axis is in the direction of the boundary normal, show that $\vec{B} \cdot \hat{n}$ is continuous across the boundary. Using a similar approach, derive the other conditions.

(*Hint*: you will need to use the divergence theorem and Stokes' theorem.)
(See refs [1, 2].)   □

The fields are related through constitutive relations

$$\vec{B} = \mu \vec{H}, \quad \vec{D} = \epsilon_0 \vec{E} + \vec{P} = \epsilon \vec{E}, \tag{2.7}$$

where in the optical frequency range μ can be taken constant and equal to $(\epsilon_0 c^{-2})^{-1}$, where c is the speed of light in a vacuum. $\vec{P}$ is called the polarization and ϵ the dielectric constant. In a vacuum, $\vec{P}$ is zero. When light propagates in a dielectric medium, however, the electric field causes distortion in the atomic structure, creating local dipole moments, and thereby induces a polarization field $\vec{P}$ which depends on $\vec{E}$. It is the dependence of $\vec{P}$ and ϵ on time history, spatial inhomogeneities, medium density fluctuations, and field intensity that leads to interesting and nontrivial behavior in the propagation of light. For small to moderate values of the electric field amplitude, and when there are no resonances between the electric field and the medium, $\vec{P}$ is linear in $\vec{E}$. In the simplest case when the medium is isotropic and responds instantaneously to the electric field, $\vec{P}$ is simply a scalar times $\vec{E}$, $\vec{P} = \epsilon_0 \chi \vec{E}$, and χ is called the electric susceptibility. In this case, a second conservation law, the conservation of energy, obtains,

$$\frac{\partial}{\partial t} \left(\tfrac{1}{2} \vec{D} \cdot \vec{E} + \tfrac{1}{2} \vec{H} \cdot \vec{B} \right) + \nabla \cdot (\vec{E} \times \vec{H}) = 0 \tag{2.8}$$

which says that the change of energy per unit volume ε (same as symbol used in Physical Units Section) $\tfrac{1}{2} \vec{D} \cdot \vec{E} + \tfrac{1}{2} \vec{H} \cdot \vec{B}$ is controlled by the flux vector $\vec{S} = \vec{E} \times \vec{H}$, called the Poynting vector. Observe that we have used $\vec{D} = \epsilon \vec{E}$ and $\vec{B} = \mu \vec{H}$, ϵ, and μ constant, in deriving (2.8).

We can obtain a single equation for the propagation of the electric field $\vec{E}$, valid for both linear and nonlinear propagation, by taking the curl of (2.3) and using (2.1) and (2.4), to obtain

$$\nabla^2 \vec{E} - \frac{1}{c^2} \frac{\partial^2 \vec{E}}{\partial t^2} - \nabla (\nabla \cdot \vec{E}) = \frac{1}{\epsilon_0 c^2} \frac{\partial^2 \vec{P}}{\partial t^2}. \tag{2.9}$$

Throughout this text, this will be the central equation for describing the propagation of the electric field.* If $\vec{P} = \epsilon_0 \chi \vec{E}$ and χ is constant, then (2.9)

---

*If we wish to write the dependent variable $\vec{E}$ in curvilinear coordinates (e.g., cylindrical (E_τ, E_θ, E_z) or spherical $(E_\tau, E_\theta, E_\phi)$), the operation $\nabla^2 \vec{E}$ is not defined and $\nabla^2 \vec{E} - \nabla(\nabla \cdot \vec{E})$ must be replaced by $-\nabla \times \nabla \times \vec{E}$. All terms in (2.9) are defined in Cartesian (E_x, E_y, E_z) coordinates even though we may choose to express each of the components in the cylindrical or spherical coordinates—for example, $E_x(\tau, \theta, z)$ or $E_x(\tau, \theta, \phi)$—of the independent variables x, y, z.

becomes

$$\nabla^2 \vec{E} - \frac{n^2}{c^2} \frac{\partial^2 \vec{E}}{\partial t^2} = 0 \qquad (2.10)$$

where $n = \sqrt{1 + \chi}$, a dimensionless number greater than unity, is called the *refractive index* of the medium. The electric field propagates according to the wave equation except that its effective velocity is c/n.

2b. Frequency Dependence of the Refractive Index

Matter does not respond instantaneously to stimulation by light. In an isotropic medium the delay is captured by writing

$$\vec{P}(t) = \epsilon_0 \int_{-\infty}^{\infty} \chi(t - \tau) \vec{E}(\tau) d\tau \qquad (2.11)$$

where, because of causality,

$$\chi(t - \tau) = 0, \quad t - \tau < 0. \qquad (2.12)$$

Defining the Fourier (time) transform and its inverse of a field $F(t)$ to be

$$\hat{F}(\omega) = \int_{-\infty}^{\infty} F(t) e^{i\omega t} dt, \quad F(t) = \frac{1}{2\pi} \int_{-\infty}^{\infty} \hat{F}(\omega) e^{-i\omega t} d\omega, \qquad (2.13)$$

we see from (2.11) that

$$\hat{P}(\omega) = \epsilon_0 \hat{\chi}(\omega) \hat{E}(\omega). \qquad (2.14)$$

EXERCISE 2.2

Substitute (2.13) in (2.11), use the fact that $\int_{-\infty}^{\infty} e^{-i(\omega_2 - \omega_1)\tau} d\tau = 2\pi \delta(\omega_2 - \omega_1)$, where δ is the Dirac delta function, and show (2.14) holds. □

The equation for the Fourier time transform of the electric field is

$$\nabla^2 \hat{E} + \frac{\omega^2}{c^2} (1 + \hat{\chi}(\omega)) \hat{E} = 0, \qquad (2.15)$$

and the refractive index $n = \sqrt{1 + \hat{\chi}(\omega)}$ is frequency dependent and, as we shall see, necessarily complex. The reason for this is that the causality condition (2.12) on $\chi(t)$ implies a relationship between the real and imaginary parts of its Fourier transform:

$$\hat{\chi}(\omega) = \int_{-\infty}^{\infty} \chi(t) e^{iwt} dt = \int_{0}^{\infty} \chi(t) e^{iwt} dt. \qquad (2.16)$$

EXERCISE 2.3

Show that, if $\chi(t)$ is real, $\hat{\chi}(-\omega) = \hat{\chi}^*(\omega^*)$.   □

Since the dependence of $\vec{P}$ via (2.11) on the time history of $\vec{E}$ gets smaller as $t - \tau \to \infty$, we expect $\chi(t) \to 0$ as $t \to \infty$ at a sufficiently fast rate such that $\hat{\chi}(\omega)$ exists for all real ω. All we require is that $\int_{-\infty}^{\infty} |\chi(t)| dt < \infty$, but in most cases $\chi(t)$ also decays exponentially. However, we also observe that the convergence of the integral is enhanced when ω is not real but complex lying in the upper half complex plane, that is, $\omega = \omega_r + i\omega_i$ with $\omega_i > 0$, because then there is the additional factor $e^{-\omega_i t}$ in the integrand of (2.16). So $\hat{\chi}(\omega)$ not only exists for ω in the upper half complex ω plane but is also differentiable and therefore analytic there. It also decays to zero as $\omega \to \infty$, $\text{Im}\,\omega > 0$. These ingredients allow us use of Cauchy's theorem to write

$$\int_{C_A} \frac{\hat{\chi}(\omega')}{\omega' - \omega} d\omega' = 0, \tag{2.17}$$

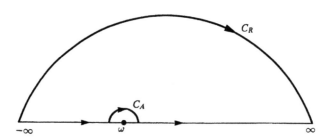

FIGURE 2.1   The contours C_A and C_R.

where C_A is the contour shown in Figure 2.1. The reason that the integral is zero is that the integral along C_A can be deformed to an integral along C_R, $\omega' = Re^{i\theta}, 0 < \theta < \pi, R \gg 1$, the semicircle at infinity, without changing its value because the integrand is analytic in the region between C_A and C_R, and the integral along C_R is zero as $R \to \infty$. We can now write (2.17) as

$$\int_{C_A} \frac{\hat{\chi}(\omega')d\omega'}{\omega' - \omega} = P \int_{-\infty}^{\infty} \frac{\hat{\chi}(\omega')d\omega'}{\omega' - \omega} - \pi i \hat{\chi}(\omega) = 0$$

where the Cauchy Principal value is

$$P \int_{-\infty}^{\infty} = \lim_{\epsilon \to 0} \left(\int_{-\infty}^{\omega - \epsilon} + \int_{\omega + \epsilon}^{\infty} \right)$$

and the second term comes from evaluating the integral on the semicircle $\omega' - \omega = \epsilon\, e^{i\theta}$, $0 < \theta < \pi$, indented over ω. If

$$\hat{\chi}(\omega) = \hat{\chi}'(\omega) + i\,\hat{\chi}''(\omega)$$

we obtain

$$\hat{\chi}'(\omega) = \frac{1}{\pi} P \int_{-\infty}^{\infty} \frac{\hat{\chi}''(\omega')d\omega'}{\omega' - \omega}$$

$$\hat{\chi}''(\omega) = -\frac{1}{\pi} P \int_{-\infty}^{\infty} \frac{\hat{\chi}'(\omega')d\omega'}{\omega' - \omega}. \tag{2.18}$$

The real and imaginary parts of the susceptibility are related through what are called the *Kramers–Krönig* relations. In Chapter 5 we calculate $\hat{\chi}(\omega)$ in the case of a medium of two-level atoms and find that, for ω near ω_{12},

$$\hat{\chi}(\omega) = \hat{\chi}'(\omega) + i\,\hat{\chi}''(\omega) \propto \frac{\omega_{12} - \omega + i\gamma_{12}}{\gamma_{12}^2 + (\omega_{12} - \omega)^2} = \frac{1}{\omega_{12} - \omega - i\gamma_{12}}. \tag{2.19}$$

In (2.19), ω_{12} is $1/\hbar(E_1 - E_2)$, the frequency difference corresponding to the two energy levels E_1 and E_2, and $\gamma_{12} > 0$ is an attenuation term associated with irreversible losses to other states in the medium. Note that $\hat{\chi}(\omega) = \frac{-1}{\omega - \omega_{12} + i\gamma_{12}}$ is analytic in the upper half complex ω plane, Im$\omega > 0$ because its only singularity is a pole at $\omega = \omega_{12} - i\gamma_{12}$.

EXERCISE 2.4

Show that (2.18) holds for $\hat{\chi}(\omega)$ given by the right-hand side of (2.19). ☐

What we have learned is that, due to the finite response time of the medium, the susceptibility is in general complex and frequency dependent. The real part $\hat{\chi}'(\omega)$ modifies the propagation rate, and the imaginary part $\hat{\chi}''(\omega)$ is associated with absorption or amplification of the electric field energy by the medium. The fact that $\hat{\chi}(\omega)$ is frequency dependent is very important because it means that different frequencies travel at different speeds. For most frequencies, the dependence is weak, but at frequencies at which the frequency of light matches a natural oscillation frequency of matter (an atomic transition, molecular vibration or rotation) the response is strong and, locally, $\hat{\chi}(\omega)$ looks like (2.19). In Figure 2.2, we show the typical graphs of $\hat{\chi}'(\omega)$ and $\hat{\chi}''(\omega)$.

The dependence of refractive index on frequency should not be surprising. Most of us have seen how ordinary light divides into a rainbow of colors after passing through a glass prism or a moist atmosphere.

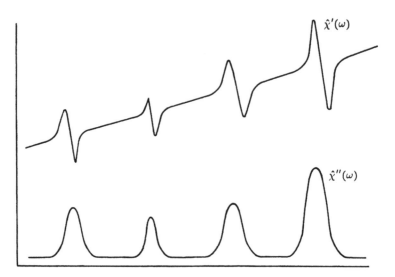

$\hat{\chi}'(\omega)$

$\hat{\chi}''(\omega)$

FIGURE 2.2   Graphs of the frequency dependence of the susceptibility ($\hat{\chi}'(\omega)$) and absorption ($\hat{\chi}''(\omega)$) for typical materials.

2c.   Linear Plane Waves or Wavetrains and the Dispersion Relation

The most elementary and familiar solutions, the building blocks that form the basis of most of our discussions, are singly polarized plane waves, sometimes called wavetrains,

$$\vec{E} = \hat{e}(A \exp i\theta + (*)) \tag{2.20}$$

where $\hat{e}$ is a unit vector in the direction of polarization of $\vec{E}$, A is a complex amplitude equal to $ae^{i\varphi}$, $\theta = \vec{k} \cdot \vec{x} - \omega t$, $\vec{k} = (l, m, k)$ is the phase, and $*$ is the complex conjugate. They are called monochromatic because they contain only one wavevector and one frequency. The field repeats itself at lattice points whose coordinates are integer multiples of $2\pi/l$, $2\pi/m$, $2\pi/k$ and after a time $2\pi/\omega$. The vector $\vec{k}$ is called the wavevector and ω the frequency. The plane of constant phase moves in the direction of $\vec{k}$ and the distance between planes on which $\vec{E}$ is equal is the wavelength $\lambda = 2\pi/\sqrt{l^2 + m^2 + k^2}$. The condition $\nabla \cdot \vec{E} = 0$ means that $\vec{k} \cdot \hat{e} = 0$, so that the electric field lies in the plane of constant phase. The magnetic field $\vec{H}$ is also in this plane. The energy density $\varepsilon = \frac{1}{2}(\vec{D} \cdot \vec{E} + \vec{H} \cdot \vec{B})$ in a vacuum is $4a^2\varepsilon_o \cos^2(\theta + \varphi)$ and the energy flux, measured by the Poynting Vector $\vec{S} = \vec{E} \times \vec{H}$ is $\frac{\vec{k}}{|\vec{k}|} 4a^2\varepsilon_o c \cos^2(\theta + \varphi)$. The average energy per unit wavelength λ is $2a^2\varepsilon_o$ and the average flux in the direction $\vec{k}$ is $2a^2\varepsilon_o c$, namely the average energy per wavelength transported with speed c. For a medium with dielectric constant $\varepsilon = \varepsilon_o(1 + \hat{x}(\omega))$, the flux is $2a^2 \varepsilon c/n$ or $2a^2 \varepsilon_o nc$. The direction of the wavevector $\vec{k}$, the normal to the plane of constant phase, is called the direction of the light ray. In order that (2.20) satisfies (2.9), the following equation

$$D(l, m, k, \omega)A = \left(l^2 + m^2 + k^2 - \frac{n^2 \omega^2}{c^2} \right) A = 0 \qquad (2.21)$$

must hold. For the time being, we will assume that we are well away from a material resonance, that the absorption (the imaginary part of $\hat{\chi}(\omega)$) is small and ignorable so that the refractive index $n(\omega)$ is a real function of the frequency ω. In general, however, we must also take the wavevector $\vec{k}(l, m, k)$ complex, and this leads to wave attenuation in the direction of propagation. For example, if $l = m = 0$, then

$$k = \frac{\omega}{c} \sqrt{1 + \hat{\chi}'(\omega) + i\hat{\chi}''(\omega)} = \frac{\omega}{\sqrt{2}c} \sqrt{((1 + \hat{\chi}')^2 + \hat{\chi}''^2)^{1/2} + (1 + \hat{\chi}')}$$

$$+ \frac{i\omega}{\sqrt{2}c} \sqrt{((1 + \hat{\chi}')^2 + \hat{\chi}''^2)^{1/2} - (1 + \hat{\chi}')}.$$

Typically for glass and wavelengths of about 1.5 μm, $\chi' \simeq 1$ (the refractive index of glass is 1.5). For ordinary glass, χ'' is of order 10^{-6} corresponding to attenuation distances of about one meter or less. In optical fibers, χ'' is much less, of the order of 10^{-11}, corresponding to attenuation distances of tens of kilometers.

For nonlinear media, the function D also depends on the field amplitude. Although one can solve the dispersion relation $D = 0$ for any one of the variables in terms of the other three, in optics it is customary to solve for wavenumber in the propagation direction (which we shall usually take as the z direction) in terms of frequency because the refractive index $n(\omega)$ also depends on frequency. For linear and nonlinear wavetrains, the dispersion relation is algebraic and given by $D = 0$. For wavepackets in a linear medium, (2.21) is a linear partial differential equation in A that describes how the wavepacket envelope A translates with the group velocity and is modified by dispersion and diffraction. For wavepackets in a nonlinear medium, (2.21) becomes a nonlinear partial differential equation for the complex wave envelope A; in the simplest nontrivial case it is the Schrödinger (NLS) equation, which is central to nonlinear optics. We see, therefore, that the *dispersion relation* is a very basic concept in the theory of wave propagation.

2d. The General Relation between P and E

In general, the jth component of the polarization field is given by

$$\frac{1}{\epsilon_0}P_j = \int_{-\infty}^{\infty} \chi_{jk}^{(1)}(t-\tau)E_k(\tau)d\tau$$

$$+ \int_{-\infty}^{\infty} \chi_{jkl}^{(2)}(t-\tau_1, t-\tau_2)E_k(\tau_1)E_l(\tau_2)d\tau_1 d\tau_2$$

$$+ \int_{-\infty}^{\infty} \chi_{jklm}^{(3)}(t-\tau_1, t-\tau_2, t-\tau_3)E_k(\tau_1)E_l(\tau_2)E_m(\tau_3)d\tau_1 d\tau_2 d\tau_3.$$

$$(2.22)$$

In (2.22) summation convention over repeated indices is implied. Observe that because $\chi_{jkl}^{(2)}$ multiplies $E_k(\tau_1)E_l(\tau_2)$ and we integrate over τ_1, τ_2, we can always relabel τ_1 as τ_2 and vice versa. Hence without loss of generality, we can arrange for $\chi_{jkl}^{(2)}(t_1, t_2)$ and likewise $\chi_{jklm}^{(3)}$ to be symmetric in their arguments, that is, $\chi_{jkl}^{(2)}(t_1, t_2) = \chi_{jkl}^{(2)}(t_2, t_1)$. Similarly, by relabeling the summation indices we can also arrange that $\chi_{jkl}^{(2)} = \chi_{jlk}^{(2)}$ and that $\chi_{jklm}^{(3)}$ is symmetric under any rearrangement of klm. Next define the Fourier transform of the susceptibilities:

$$\hat{\chi}_{jk}^{(1)}(\omega) = \int_{-\infty}^{\infty} \chi_{jk}^{(1)}(t)e^{i\omega t}dt, \quad \chi_{jk}^{(1)}(t) = \frac{1}{2\pi}\int_{-\infty}^{\infty} \hat{\chi}_{jk}(\omega)e^{-i\omega t}d\omega,$$

$$\hat{\chi}_{jkl}^{(2)}(\omega_1, \omega_2) = \int_{-\infty}^{\infty} \chi_{jkl}^{(2)}(t_1, t_2)e^{i\omega_1 t_1 + i\omega_2 t_2}dt_1 dt_2$$

$$\hat{\chi}_{jklm}^{(3)}(\omega_1, \omega_2, \omega_3) = \int_{-\infty}^{\infty} \chi_{jklm}^{(3)}(t_1, t_2, t_3)e^{i\omega_1 t_1 + i\omega_2 t_2 + i\omega_3 t_3}dt_1 dt_2 dt_3, \quad (2.23)$$

and then the corresponding relation between $\hat{P}$ and $\hat{E}$ is

$$\frac{1}{\epsilon_0}\hat{P}_j = \hat{\chi}_{jk}^{(1)}(\omega)\hat{E}_k(\omega)$$

$$+ \frac{1}{2\pi}\int_{-\infty}^{\infty} d\omega_1 d\omega_2 \hat{\chi}_{jkl}^{(2)}(\omega_1, \omega_2)\hat{E}_k(\omega_1)\hat{E}_l(\omega_2)\delta(\omega_1 + \omega_2 - \omega)$$

$$+ \frac{1}{(2\pi)^2}\int_{-\infty}^{\infty} d\omega_1 d\omega_2 d\omega_3 \hat{\chi}_{jklm}^{(3)}(\omega_1, \omega_2, \omega_3)$$

$$\hat{E}_k(\omega_1)\hat{E}_l(\omega_2)\hat{E}_m(\omega_3)\delta(\omega_1 + \omega_2 + \omega_3 - \omega)$$

$$+ \cdots.$$

$$(2.24)$$

Again $\hat{\chi}_{jkl}^{(2)}(\omega_1, \omega_2)$ and $\hat{\chi}_{jklm}^{(3)}(\omega_1, \omega_2, \omega_3)$ are symmetric in all their arguments and over the last two (kl) and three (klm) indices respectively. The reality of $\chi^{(1)}$, $\chi^{(2)}$, and $\chi^{(3)}$ implies that for real values of the frequencies ω,

$$\hat{\chi}_{jk}^{(1)}(-\omega) = \hat{\chi}_{jk}^{(1)*}(\omega)$$

$$\hat{\chi}_{jkl}^{(2)}(-\omega_1, -\omega_2) = \hat{\chi}_{jkl}^{(2)*}(\omega_1, \omega_2)$$

$$\hat{\chi}_{jklm}^{(3)}(-\omega_1, -\omega_2, -\omega_3) = \hat{\chi}_{jklm}^{(3)*}(\omega_1, \omega_2, \omega_3). \tag{2.25}$$

At the moment we are dealing with many, many components of the susceptibility tensors: $\chi_{jk}^{(1)}$ has 3^2, $\chi_{jkl}^{(2)}$ has 3^3, and $\chi_{jklm}^{(2)}$ has 3^4. However, depending on the crystal symmetry of the dielectric of interest, many of these components are related. The main idea in finding these relations is to realize that the relations (2.22) and (2.24) should remain invariant under the various rotations and reflections associated with the group of symmetries enjoyed by the crystal. Let $R = (R_{\mu\nu})$ be a rotation matrix so that $R^T R = I$, $R^{-1} = R^T$. Its determinant can be ± 1. The transformation between the coordinates of the polarization and electric fields (and their Fourier transforms) from one set of coordinate axes (x_1, x_2, x_3) to another (x_1', x_2', x_3') is given by

$$\hat{E}_\mu' = R_{\mu\nu}\hat{E}_\nu$$

and the transformations between the susceptibilities are

$$\hat{\chi}_{\mu\nu}^{(1)'} = R_{\mu r} R_{\nu s} \hat{\chi}_{rs}^{(1)}$$

$$\hat{\chi}_{\mu\nu\tau}^{(2)'} = R_{\mu r} R_{\nu s} R_{\tau p} \hat{\chi}_{rsp}^{(2)}$$

$$\hat{\chi}_{\mu\nu\tau\rho}^{(3)'} = R_{\mu r} R_{\nu s} R_{\tau p} R_{\rho q} \hat{\chi}_{rspq}^{(3)} \tag{2.26}$$

When the rotation defined by R belongs to the symmetry group of the crystal, the susceptibilities must be identical in the two coordinate systems so that

$$\hat{\chi}_{\mu\nu}^{(1)'} = \hat{\chi}_{\mu\nu}^{(1)}, \quad \hat{\chi}_{\mu\nu\tau}^{(2)'} = \hat{\chi}_{\mu\nu\tau}^{(2)}, \quad \hat{\chi}_{\mu\nu\tau\rho}^{(3)'} = \hat{\chi}_{\mu\nu\tau\rho}^{(3)} \tag{2.27}$$

and leads to relations between the various tensor components. By considering the groups associated with the 32 crystal classes, these relationships can be established [21].

However, that is way beyond the scope of this book. We will, for the most part, deal with the simplest case where the crystal is centrosymmetric and isotropic. The first condition means that if $\vec{E} \to -\vec{E}$, then $\vec{P} \to -\vec{P}$ so that we can take $R = -I$. It is clear that in this case $\hat{\chi}^{(2)}$ is identically zero. Isotropy means that of the 21 nonzero elements in $\hat{\chi}^{(3)}_{jklm}$ (only those elements for which all indices are equal or the indices are equal in pairs, e.g., $j = k, l = m$ are nonzero), only three are independent.

$$\hat{\chi}^{(3)}_{1111} = \hat{\chi}^{(3)}_{2222} = \hat{\chi}^{(3)}_{3333}$$

$$\hat{\chi}^{(3)}_{2233} = \hat{\chi}^{(3)}_{3322} = \hat{\chi}^{(3)}_{1133} = \hat{\chi}^{(3)}_{3311} = \chi^{(3)}_{1122} = \chi^{(3)}_{2211}$$

$$\hat{\chi}^{(3)}_{2332} = \hat{\chi}^{(3)}_{3223} = \hat{\chi}^{(3)}_{3113} = \hat{\chi}^{(3)}_{1313} = \chi^{(3)}_{1212} = \chi^{(3)}_{2121}$$

$$\hat{\chi}^{(3)}_{1111} = \hat{\chi}^{(3)}_{1122} + \chi^{(3)}_{1212} + \chi^{(3)}_{1221} \tag{2.28}$$

We are going to leave it as an exercise to the reader to show that in this case,

$$\frac{1}{\epsilon_0}\vec{P}(t) = \int_{-\infty}^{\infty} \chi^{(1)}(t - \tau)\vec{E}(\tau)d\tau$$

$$+ \int_{-\infty}^{\infty} \chi^{(3)}(t - \tau_1, t - \tau_2, t - \tau_3)(\vec{E}(\tau_1) \cdot \vec{E}(\tau_2))\vec{E}(\tau_3)d\tau_1 d\tau_2 d\tau_3 \tag{2.29}$$

and

$$\frac{1}{\epsilon_0}\hat{P}(\omega) = \hat{\chi}^{(1)}(\omega)\hat{E}(\omega)$$

$$+ \int_{-\infty}^{\infty} d\omega_1 d\omega_2 d\omega_3 \hat{\chi}^{(3)}(\omega_1, \omega_2, \omega_3)(\hat{E}(\omega_1)$$

$$\cdot \hat{E}(\omega_2))\hat{E}(\omega_3)\delta(\omega_1 + \omega_2 + \omega_3 - \omega), \tag{2.30}$$

where $\vec{E}(\tau_1) \cdot \vec{E}(\tau_2)$ and $\hat{E}(\omega_1) \cdot \hat{E}(\omega_2)$ are the inner products $\vec{E}_j(\tau_1)\vec{E}_j(\tau_2)$ and $\hat{E}_j(\omega_1)\hat{E}_j(\omega_2)$, and

$$\chi^{(3)}(t_1, t_2, t_3) = \chi^{(3)}_{1111}(t_1, t_2, t_3), \tag{2.31}$$

$$\hat{\chi}^{(3)}(\omega_1, \omega_2, \omega_3) = \frac{1}{(2\pi)^2}\hat{\chi}^{(3)}_{1111}(\omega_1, \omega_2, \omega_3). \tag{2.32}$$

2e.   Nonlinear Wavetrains in a Kerr Medium. [6a, 6b, 6g]

Let $\vec{E}$ and $\vec{P}$ be singly polarized, $\vec{E} = \hat{e}E$, $\vec{P} = \hat{e}P$ and be independent of the transverse coordinates x, y. The unit vector $\hat{e}$ is in the x, y plane. Then

(2.9) becomes

$$\frac{\partial^2 E(z,t)}{\partial z^2} - \frac{1}{c^2}\frac{\partial^2 E(z,t)}{\partial t^2} - \frac{1}{c^2}\frac{\partial^2}{\partial t^2}\int_{-\infty}^{\infty}\chi^{(1)}(t-\tau)E(z,\tau)d\tau$$

$$= \frac{1}{\epsilon_0 c^2}\frac{\partial^2}{\partial t^2}P_{NL}(E) \tag{2.33}$$

where $1/\epsilon_0 P_{NL}$ is the nonlinear term on the right-hand side of (2.29). We assume that the nonlinear terms are small, that is, $|\chi^{(3)}||E|^3 \ll \chi^{(1)}|E|$, and solve (2.33) iteratively

$$E = E_0 + E_1 + \cdots \tag{2.34}$$

where E_0 is chosen to be the wavetrain solution

$$E_0 = F(z)e^{-i\omega t} + F^*(z)e^{i\omega t}. \tag{2.35}$$

For a right-going wave,

$$F(z) = Ae^{ikz}, k = \frac{\sqrt{1 + \hat{\chi}^{(1)}(\omega)}}{c}\omega, \tag{2.36}$$

with A constant.

EXERCISE 2.5

Show that $\int_{-\infty}^{\infty}\chi^{(1)}(t-\tau)e^{-i\omega\tau}d\tau = \hat{\chi}^{(1)}(\omega)e^{-i\omega t}$. $\square$

In general $\hat{\chi}^{(1)}(\omega) = \hat{\chi}'^{(1)}(\omega) + i\hat{\chi}''^{(1)}(\omega)$, but we are assuming that $\hat{\chi}''(\omega)$ is small and positive so that

$$k \simeq \sqrt{1 + \hat{\chi}'^{(1)}(\omega)}\left(1 + \frac{i\hat{\chi}^{(1)''}(\omega)}{2(1 + \hat{\chi}'^{(1)}(\omega))}\right)\frac{\omega}{c}. \tag{2.37}$$

The imaginary term in k leads to a slow attenuation of the signal when $\hat{\chi}''(\omega)$ is positive. For the moment we will ignore this and take k to be given only by the real part.

The equation for E_1 reads

$$\frac{\partial^2 E_1}{\partial z^2} - \frac{1}{c^2}\frac{\partial^2 E_1}{\partial t^2} - \frac{1}{c^2}\frac{\partial^2}{\partial t^2}\int_{-\infty}^{\infty}\chi^{(1)}(t-\tau)E_1(\tau)d\tau = \frac{1}{\epsilon_0 c^2}\frac{\partial^2}{\partial t^2}P_{NL}(E_0).$$

$$\tag{2.38}$$

The right-hand side of (2.38) is

$$-\frac{9\omega^2}{c^2}\hat{\chi}^{(3)}(\omega,\omega,\omega)A^3e^{3ikz-3i\omega t} - \frac{3\omega^2}{c^2}\hat{\chi}^{(3)}(-\omega,\omega,\omega)A^2A^*e^{ikz-i\omega t}$$

$$-\frac{3\omega^2}{c^2}\hat{\chi}^{(3)}(\omega,-\omega,-\omega)AA^{*2}e^{-ikz+i\omega t}$$

$$-\frac{9\omega^2}{c^2}\hat{\chi}^{(3)}(-\omega,-\omega,-\omega)A^{*3}e^{-3ikz+3i\omega t}$$

We look for solutions to (2.38) in the form

$$E_1 = F_1(z)e^{-i\omega t} + F_1^*(z)e^{i\omega t} + F_3(z)e^{-3i\omega t} + F_3^*(z)e^{3i\omega t}$$

and find

$$\frac{d^2F_3}{dz^2} + \frac{1+\hat{\chi}^{(1)}(3\omega)}{c^2}9\omega^2 F_3 = -\frac{9\omega^2}{c^2}\hat{\chi}^{(3)}(\omega,\omega,\omega)A^3e^{3ik(\omega)z} \qquad (2.39)$$

$$\frac{d^2F_1}{dz^2} + \frac{1+\hat{\chi}^{(1)}(\omega)}{c^2}\omega^2 F_1 = -\frac{3\omega^2}{c^2}\hat{\chi}^{(3)}(-\omega,\omega,\omega)A^2A^*e^{ik(\omega)z}. \qquad (2.40)$$

Since $k(3\omega) = (3\omega\sqrt{1+\hat{\chi}^{(1)}(3\omega)}/c \neq 3\omega\sqrt{1+\hat{\chi}^{(1)}(\omega)}/c$, F_3 is bounded. But, because the exponential term $e^{ik(\omega)z}$ on the right-hand side of (2.40) is a solution of the homogeneous equation, F_1 is proportional to $\hat{\chi}^{(3)}A^2A^*ze^{ik(\omega)z}$ and then the asymptotic expansion (2.34) is nonuniform because the second term is as big as the first after a distance z of the order of $(\chi^{(3)}|A|^2)^{-1}$. (See the discussion on asymptotic expansions and multiple time scale methods in Section 6g.). We remove this bad behavior by allowing A, up to now a constant, to vary slowly with z so that the additional term

$$-2ik\frac{dA}{dz}e^{ikz-i\omega t} + 2ik\frac{dA^*}{dz}e^{-ikz+i\omega t}, \qquad (2.41)$$

arising from $\partial^2 E_0/\partial z^2$, appears on the right-hand side of (2.38). We can choose the dependence of A on z to suppress the term on the right-hand side of (2.40) that gives rise to the resonance. Choose (write $\hat{\chi}^{(3)}(-\omega,\omega,\omega)$ as $\hat{\chi}^{(3)}(\omega)$)

$$\frac{dA}{dz} = \frac{3i}{2k}\frac{\omega^2}{c^2}\hat{\chi}^{(3)}A^2A^* = \frac{3i\omega}{2c}\frac{\hat{\chi}^{(3)}(\omega)}{\sqrt{1+\hat{\chi}^{(1)}(\omega)}}A^2A^* \qquad (2.42)$$

or, integrating,

$$A = A_0 \exp\left(\frac{3i\omega}{2c} \frac{\hat{\chi}^{(3)}(\omega)}{\sqrt{1 + \hat{\chi}^{(1)}(\omega)}} |A_0|^2 z\right) \tag{2.43}$$

so that

$$E_0 = A_0 e^{-i\omega t + ik(\omega, |A_0|^2)z} + (*). \tag{2.44}$$

The "new" wavenumber $k(\omega, |A_0|^2)$ is given by

$$\frac{c}{\omega} k(\omega, |A_0|^2) = \sqrt{1 + \chi^{(1)}(\omega)} + \frac{3}{2} \frac{\hat{\chi}^{(3)}(\omega)}{\sqrt{1 + \hat{\chi}^{(1)}(\omega)}} |A_0|^2.$$

Therefore the nonlinear refractive index is

$$n(\omega, |A_0|^2) = n_0(\omega) + n_2(\omega)|A_0|^2 \tag{2.45}$$

where $2n_0 n_2 = 3\hat{\chi}^{(3)}(\omega)$.

We generally treat the attenuation $\hat{\chi}''^{(1)}(\omega)$ term (and therefore the imaginary part of $n_0(\omega)$) as being about the same order of magnitude as the small nonlinear correction $n_2(\omega)|A_0|^2$. We ignore altogether the imaginary part of $n_2(\omega)$. Its frequency dependence is only important for very narrow pulses.

What we have learned then is that finite-amplitude wavetrains have a nonlinear dispersion relation

$$k = \frac{n(\omega, |A_0|^2)\omega}{c} \tag{2.46}$$

where n is given by (2.45). If we had included the transverse directions x and y, then (2.46) would read

$$D(l, m, k, \omega) = l^2 + m^2 + k^2 - \frac{n^2(\omega, |A_0|^2)\omega^2}{c^2} = 0. \tag{2.47}$$

We want to stress the importance of the frequency dependence of $\hat{\chi}^{(1)}(\omega)$. If the susceptibility were not frequency dependent, then $k(3\omega) = 3k(\omega)$ and the right-hand side of (2.39) would also be resonant, and a much more complicated evolution of the wavetrain would occur.

2f. Geometric Optics. [6g]

The dispersion relation (2.21) also holds when the refractive index of the medium changes slowly; that is, $\nabla_n = O(n/L) \ll n/\lambda$, the distance L over which n changes, is many wavelengths of the lightwave. One sees this as follows. Let X, the electric susceptibility and therefore n the refractive index, be a slowly varying function of the coordinate x, and let $\vec{E} = \hat{y}E$ be polarized in the y direction so that, without any change, (2.10) holds for the scalar field E. For this exercise, we will ignore delay effects and the frequency dependence of n. Look for a solution,

$$E(x, z, t) = A \exp i\theta + (*) + \epsilon E_1 + \cdots \tag{2.48}$$

where $\vec{k} = (l, o, k) = \nabla\theta$ and $\omega = -\theta_t$ are the wavevector and frequency of the electromagnetic wave and ϵ is $\lambda/L \ll 1$.

Because n changes slowly with x, that is, $n = n(X = \epsilon x)$ where $\epsilon = \lambda/L \ll 1$, so, in principle, can $\vec{k}$ and ω. However, ω is chosen constant and, because $l = \partial\theta/\partial x$ and $k = \partial\theta/\partial z$ implies $\partial k/\partial x = \partial l/\partial z$, we must have k also constant since l is independent of z. Therefore, only l, the x wavenumber, can change in the direction of changing refractive index. Substitute (2.48) into (2.10) and treat n, A, and l as slowly varying functions of x, that is, as functions of $X = \epsilon x$. To leading order we obtain the dispersion relation (2.21),

$$D(l, 0, k, \omega)A = \left(l^2 + k^2 - \frac{n^2(X)\omega^2}{c^2}\right) A = 0. \tag{2.49}$$

At order ϵ, we find

$$\nabla^2 E_1 - \frac{n^2}{c^2}\frac{\partial^2 E_1}{\partial t^2} = \left(-2il\frac{dA}{dX} - i\frac{dl}{dX}A\right) e^{i\theta} + (*). \tag{2.50}$$

In order to suppress solutions E_1 that grow as $ze^{\pm i\theta}$, we must choose

$$2l\frac{dA}{dX} + \frac{dl}{dX}A = 0 \tag{2.51}$$

or

$$lA^2 = \text{constant} = l_0 A_0^2. \tag{2.52}$$

Hence $l(x)$ is given by

$$l(x) = \sqrt{n^2(X)\frac{\omega^2}{c^2} - k^2} \tag{2.53}$$

and $A(x)$ is given by

$$A(x) = \left(\frac{l_0}{l}\right)^{1/2} A_0 \tag{2.54}$$

as long as l is nonzero. This method of analysis was first introduced by Wentzel, Kramers, Brillouin, and Jeffries and is known as the WKBJ method. Equation (2.53) is known as the eikonal equation and (2.52) as conservation of wave action. Now suppose $n(X)$ is as shown in Figure 2.3. The path of the light ray in the (x, z) plane is shown. Initially $n\omega/c > k$ (or the wave would not propagate) and l is positive. However, if no $\omega/c < k$, n_0 the background refractive index (see Fig. 2.3), there is a point x_1 (called a caustic) at which $n(x_1)\omega/c = k$ and $l = 0$. At this point the light rays turn around and the electromagnetic wave propagates back into the medium with $l = -\sqrt{(n^2\omega^2/c^2)} - k^2$. The exact behavior near $x = x_1$ cannot be described by the WKBJ analysis because as $l \to 0$, the amplitude A increases as $l^{-1/2}$ and the WKBJ analysis breaks down. Near $x = x_1$, the point where the WKBJ analysis breaks down, we approximate n^2 locally by $n^2(x_1) - \alpha(x - x_1)$ and look for a solution

$$E = B(x)e^{ikz - i\omega t} + (*), \quad k = n(x_1)\frac{\omega}{c}. \tag{2.55}$$

The amplitude $B(x)$ satisfies the Airy equation

$$\frac{d^2 B}{dx^2} - \frac{\omega^2}{c^2}\alpha(x - x_1)B = 0. \tag{2.56}$$

For $x > x_1$, we choose the solution of (2.56), which decays exponentially. For $x < x_1$, this solution has two parts, one corresponding to an incoming wave with wavevector $(\sqrt{(n^2\omega^2/c^2)} - k^2, 0, k)$ and amplitude given by (2.54) and the other to the reflected wave with wavevector $(-\sqrt{(n^2\omega^2/c^2)} - k^2, 0, k)$.

The key thing to emphasize is that *light rays will turn away from regions of smaller and toward regions of greater refractive index.*

Another way to see this is to imagine the crest of the electric wave given by $\theta = 0$ and shown by the slightly curved crest AB in Figure 2.3. Each portion travels with speed c/n in a direction locally normal to the crest. But c/n at A is less than c/n at B, and therefore the portion of the crest lying in the region of lesser refractive index travels faster. Hence the light ray turns back toward regions of higher refractive index.

The fact that light rays turn toward regions of higher refractive index and away from regions of lower refractive index is one of the principles used to guide one's intuition on the propagation of light in dielectric materials. First,

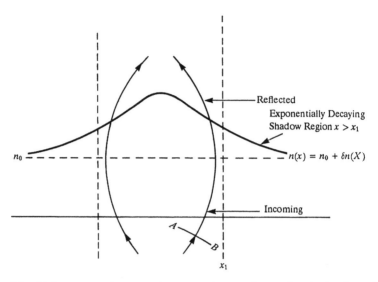

FIGURE 2.3   Light rays turning around at a caustic $x = \pm x_1$ and gathering in regions of higher refractive index.

it is the fundamental idea used in the design of configurations that confine light to waveguides. Second, it has enormous ramifications if the correction δn to the refractive index is intensity dependent and increases with increasing intensity; that is, $n = n_0 + n_2|A|^2, n_2 > 0$. Suppose the amplitude of the wavetrain is perturbed $A_0 \rightarrow A_0 + a(x, y, 0)$ at $z = 0$ so that at some (x, y) locations the intensity of the light is greater and at others it is smaller. (See Figure 2.4.) If $n_2 > 0$, the light propagates toward the regions A and C of higher refractive index and away from the region B where the refractive index is lower. This leads to a positive feedback effect, because now there is more light at A and C and therefore its intensity and refractive index

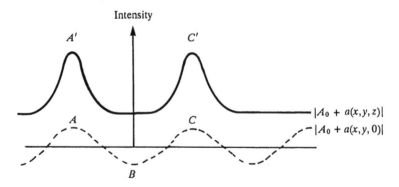

FIGURE 2.4   Graph of initial perturbation.

$n_0 + n_2|A|^2$ become even greater. As the light wave propagates forward in the z direction, the light continues to focus in regions where the intensity is greatest. This instability, called the self-focusing instability in optics, is widespread in nature. We shall discuss it again in Section 2g. What it means is that in certain circumstances, a wavetrain, whose amplitude is uniform in space and time, is unstable and the light intensity concentrates in collapsing filaments. If there is only one transverse dimension, x say, then as z increases, there comes a point at which diffraction effects, which we describe in the next section, balance the compression due to nonlinearity. This balance leads to the formation of a soliton, a very special type of nonlinear wavepacket. In two transverse dimensions, the collapse is not arrested, and the light filament continues to intensify. More about this later.

2g. Wavepackets, Group Velocity, Diffraction and Dispersion. [6c, 6h]

It is rare that lightwaves are exactly monochromatic and more usual that they consist of a linear combination of plane waves whose wavevectors and frequencies lie in a narrow band, say of order $\mu, 0 < \mu \ll 1$ about some central values $\vec{k}_0, \omega_0$. Such objects are called *wavepackets* and are conveniently represented in a very simple mathematical form

$$\vec{E} = \hat{e}(A(x, y, z, t) \exp(i\vec{k}_0 \cdot \vec{x} - i\omega_0 t) + (*))$$

$$+ O(\mu), \quad D(l_0, m_0, k_0, \omega_0) = 0 \qquad (2.57)$$

where the envelope $A(x, y, z, t)$ is considered as a slowly varying function of space and time; that is, $\partial A/\partial x$ is of order $\mu|\vec{k}_0|A$ and therefore much smaller than $|\vec{k}_0|A$. How can we see this? Suppose we did represent the field as a linear combination of modes with wavevectors and frequencies $\vec{k}_0 + \mu \vec{K}_j, \omega_0 + \mu \Omega_j$ with $\vec{K}_j = (L_j, M_j, K_j)$, each set satisfying the dispersion relation (2.21),

$$\vec{E} = \sum_j \hat{e}_j A_j \exp(i(\vec{k}_0 + \mu \vec{K}_j) \cdot \vec{x} - i(\omega_0 + \mu \Omega_j)t)$$

$$+ (*), \quad (\vec{k}_0 + \mu \vec{K}_j) \cdot \hat{e}_j = 0 \qquad (2.58)$$

However, note that this can be rewritten as (2.57) where $A(x, y, z, t) = \sum_j A_j \exp(i\mu \vec{K}_j \cdot \vec{x} - i\Omega_j \mu t)$, namely where A is a function of $\mu \vec{x}$ and μt and therefore slowly varying in $\vec{x}$ and t and $\hat{e}$ is a unit vector perpendicular to $\vec{k}_0$. The small correction term in (2.57) must be added in order to ensure that the electric field is divergence free, that is, $\nabla \cdot \vec{E} = 0$. In general, this correction will contain contributions to $\vec{E}$ not parallel to $\hat{e}$. We calculate the correction in Section 2k. The reason for this, of course, is that the light rays of the plane wavetrains from which the wavepacket is built are not quite parallel. To avoid the technical difficulties associated with the correction, we will first derive the envelope equation by taking $\vec{E}$ to be in the y direction only and ask that all the wavevectors $\vec{k}_0 + \mu \vec{K}_j$ lie in the (x, z) plane so that A is a slowly varying function of x, z, and t only.

Before we do this, let us point out that one can define the notion of local wavevector $\vec{k}$ and frequency ω. If the complex envelope A is written in polar form as $ae^{i\phi}$, a, ϕ real, then the total phase of the leading order contribution to the electric field is

$$\theta = \vec{k}_0 \cdot \vec{x} - \omega_0 t + \phi(x, y, z, t). \tag{2.59}$$

It is natural to define the local wavevector as the gradient and the local frequency as the negative of the time derivative of the total phase $\theta(x, y, z, t)$,

$$\vec{k} = \nabla\theta = \vec{k}_0 + \nabla\phi, \qquad \omega = -\theta_t = \omega_0 - \phi_t. \tag{2.60}$$

Observe that

$$\vec{k}_t + \nabla\omega = 0, \tag{2.61}$$

which has the form of a conservation law and is known as the conservation of wavenumber.

Let us first look at the evolution of the wavepacket (2.57) with $\vec{k}_0 = (0, 0, k_0)$ and $\hat{e} = (0, 1, 0)$ of an electric field $\vec{E}$ that satisfies (2.10):

$$\nabla^2 \vec{E} - \frac{n^2}{c^2}\frac{\partial^2 \vec{E}}{\partial t^2} = 0.$$

For simplicity, we have assumed in this first calculation that the electric susceptibility, and therefore the refractive index n, is constant. Substitute (2.57) into (2.10) and obtain

$$2\,ik_0\frac{\partial A}{\partial z} + \frac{\partial^2 A}{\partial z^2} + \nabla_\perp^2 A + 2\,i\omega_0\frac{n^2}{c^2}\frac{\partial A}{\partial t} - \frac{n^2}{c^2}\frac{\partial^2 A}{\partial t^2} = 0, \tag{2.62}$$

where $k_0 = n\omega_0/c$ and $\nabla_\perp^2$ is the one-dimensional horizontal Laplacian $\partial^2/\partial x^2$. Now recall that each derivative is of order μ. Therefore, the dominant terms in

(2.62) are the first and fourth and so to leading order,

$$\frac{\partial A}{\partial z} + \frac{n}{c}\frac{\partial A}{\partial t} = 0. \tag{2.63}$$

To a first approximation, A is a function of $t - (n/c)z$ and is constant when $t - (n/c)z$ is constant, that is, in a frame of reference moving with the group velocity

$$\vec{v}_g = \nabla_{\vec{k}}\omega = \left(0, 0, \frac{c}{n}\right). \tag{2.64}$$

Using (2.63), we see that $\partial^2 A/\partial z^2 = (n^2/c^2)(\partial^2/\partial t^2 A)$, and so to order μ^2, equation (2.62) is simply

$$\frac{\partial A}{\partial z} + \frac{n}{c}\frac{\partial A}{\partial t} - \frac{i}{2k_0}\nabla_\perp^2 A = 0. \tag{2.65}$$

This equation is often called the *paraxial* approximation because the rays, the normals to the planes of constant phase, of all the plane waves in the wavepacket are almost parallel. Under the change of variables to a frame of reference moving with the group velocity,

$$z = z, \quad \tau = t - \frac{n}{c}z,$$

(2.65) becomes

$$\frac{\partial A}{\partial z} - \frac{i}{2k_0}\nabla_\perp^2 A = 0. \tag{2.66}$$

Equation (2.66) describes the dispersion or spreading out of a wavepacket in the transverse direction, a process called wave *diffraction*. We will now ask the reader to believe (we will derive it later) that (2.65) and (2.66) also hold when the envelope $A(x, y, z, t)$ depends on both transverse directions x and y. In this case, however, the order μ correction to the electric field vector $\vec{E}$ contains components in directions other than $\hat{e}$ in order to make it divergence-free. Only the leading order contribution is transverse (see Section 2k).

EXERCISE 2.6

Verify, by direct calculation, that

$$A(x, z) = \frac{1}{\left(1 + \frac{2iz}{k_0 w_0^2}\right)^{1/2}} \exp\left[-\left(\frac{x^2}{w_0^2 + \frac{2iz}{k_0}}\right)\right] \tag{2.67}$$

satisfies (2.66). Here w_0 is the minimum radius of the beam, which value occurs at $z = 0$.   □

EXERCISE 2.7

Verify, by direct calculation, that

$$A(x, y, z) = \frac{1}{1 + \frac{2iz}{k_0 w_0^2}} \exp\left[-\left(\frac{x^2 + y^2}{w_0^2 + \frac{2iz}{k_0}}\right)\right] \tag{2.68}$$

satisfies (2.66) if we take $\nabla_1^2 = \frac{\partial^2}{\partial x^2} + \frac{\partial^2}{\partial y^2}$.   □

EXERCISE 2.8

Solve the initial value problem (2.66) with initial condition given by the Gaussian beam of width w_0

$$A(x, y, 0) = \exp\left[-\left(\frac{x^2 + y^2}{w_0^2}\right)\right] \tag{2.69}$$

and obtain the answer cited in the previous exercise.   □

What we learn from these exercises is that a Gaussian beam of width w_0 will double its width in a distance equal to

$$z_{\text{diffraction}} = \frac{\sqrt{3}}{2} k_0 w_0^2 = \sqrt{3}\pi \frac{w_0^2}{\lambda_0} \tag{2.70}$$

where $\lambda_0 = 2\pi/k_0$. The waist at a point z is $w_0\left(1 + \frac{4z^2}{k_0^2 w_0^4}\right)^{1/2}$ (see (2.71)) which doubles from w_0 to $2w_0$ when z is given by (2.70). For example, a light beam with a wavelength of about 1μm and a width of a millimeter will double its width in about 5 meters. We also learn that because the Gaussian beam contains power over a range of different transverse wavenumbers l, m and because each of the wavevectors (l, m, k_0) propagates in different directions with different speeds, the original Gaussian pulse spreads out with diminishing amplitude. We can rewrite the solution (2.68) as

$$A(x, y, z) = \frac{1}{\sqrt{1 + \frac{4z^2}{k_0^2 w_0^4}}}$$

$$\times \exp\left\{-\frac{x^2 + y^2}{w_0^2\left(1 + \frac{4z^2}{k_0^2 w_0^4}\right)} + \frac{ik_0(x^2 + y^2)}{2\left(z + \frac{1}{4}\frac{w_0^4 k_0^2}{z}\right)} - i\tan^{-1}\frac{2z}{k_0 w_0^2}\right\} \tag{2.71}$$

which shows that the amplitude of the pulse has decreased and the width increased by the same factor. (See Section 6f.)

EXERCISE 2.9

Show that the power $P = \int AA^* dx\, dy$ contained in the beam is independent of z by (1) direct calculation and (2) by using equation (2.66) to show $dP/dz = 0$ (assume $A \to 0$ as $x^2 + y^2 \to \infty$). $\square$

EXERCISE 2.10

Derive the analog to the paraxial equation when finite delay effects are included.

$$\nabla^2 \vec{E} - \frac{1}{c^2}\frac{\partial^2 \vec{E}}{\partial t^2} - \frac{1}{c^2}\frac{\partial^2}{\partial t^2}\int_{-\infty}^{\infty}\chi^{(1)}(t-\tau)\vec{E}(\tau)d\tau = 0. \tag{2.72}$$

For those who need help, we include some details. The key step is in evaluating the convolution integral $\int_{-\infty}^{\infty}\chi^{(1)}(t-\tau)\,\vec{E}(x,y,z,\tau)\,d\tau$ when $\vec{E}$ is no longer a wavetrain but a wavepacket given by (2.57). The answer is

$$\int \chi^{(1)}(t-\tau)\vec{E}(\tau)d\tau$$

$$= \hat{e}\left(e^{i\vec{k}_0\cdot\vec{x}-i\omega_0 t}\,\hat{\chi}^{(1)}\left(\omega_0+i\frac{\partial}{\partial t}\right)A(x,y,z,t)\right.$$

$$\left. +e^{-i\vec{k}_0\cdot\vec{x}+i\omega_0 t}\,\hat{\chi}^{(1)*}\left(\omega_0-i\frac{\partial}{\partial t}\right)A^*(x,y,z,t)\right). \tag{2.73}$$

Also,

$$\frac{\partial^2}{\partial t^2}\int \chi^{(1)}(t-\tau)\vec{E}(\tau)d\tau$$

$$= -\hat{e}(e^{i\vec{k}_0\cdot\vec{x}-i\omega_0 t}(\omega_0+i\frac{\partial}{\partial t})^2\hat{\chi}^{(1)}(\omega_0+i\frac{\partial}{\partial t})A(x,y,z,t)$$

$$+ e^{-i\vec{k}_0\cdot\vec{x}+i\omega_0 t}(\omega_0-i\frac{\partial}{\partial t})^2\hat{\chi}^{(1)*}(\omega_0-i\frac{\partial}{\partial t})A^*(x,y,z,t)). \tag{2.74}$$

Expand (2.74) in a Taylor series and keep terms up to second order in the derivatives. Defining

$$k(\omega) = \frac{n(\omega)\omega}{c}, \quad n(\omega) = \sqrt{1+\hat{\chi}^{(1)}(\omega)}, \tag{2.75}$$

one then obtains from (2.72) that

$$0 = -k_0^2 A + 2ik_0 \frac{\partial A}{\partial z} + \frac{\partial^2 A}{\partial z^2} + \nabla_\perp^2 A + k^2 \left(\omega_0 + i \frac{\partial}{\partial t} \right) A$$

$$= 2ik_0 \frac{\partial A}{\partial z} + \frac{\partial^2 A}{\partial z^2} + \nabla_\perp^2 A + (k^2(\omega_0))' i \frac{\partial A}{\partial t} - \frac{1}{2}(k^2(\omega_0))'' \frac{\partial^2 A}{\partial t^2}. \qquad (2.76)$$

To leading order (we again ignore attenuation and assume $\hat{\chi}^{(1)}$ is real),

$$\frac{\partial A}{\partial z} + \frac{\partial k}{\partial \omega}(\omega_0) \frac{\partial A}{\partial t} = 0. \qquad (2.77)$$

Using (2.77) to write $\partial^2 A/\partial z^2$ in terms of $\partial^2 A/\partial t^2$, (2.76) becomes

$$\frac{\partial A}{\partial z} + \frac{\partial k}{\partial \omega}(\omega_0) \frac{\partial A}{\partial t} - \frac{i}{2k_0} \nabla_\perp^2 A + \frac{i}{2} \frac{\partial^2 k}{\partial \omega^2}(\omega_0) \frac{\partial^2 A}{\partial t^2} = 0. \qquad (2.78)$$

$\square$

EXERCISE 2.11

The reader who would like to understand (2.73) should look for details in Section 3c. Here we give the idea. Take the Fourier transform

$$\text{F.T.} \int \chi^{(1)}(t - \tau)\vec{E}(\tau)dr = \hat{\chi}^{(1)}(\omega)\hat{E}(\omega). \qquad (2.79)$$

Since $\hat{E}(\omega)$ is the Fourier transform of a wavepacket, it is localized about $\omega = \omega_0$ and $\omega = -\omega_0$. Since $\hat{E}^*(\omega) = \hat{E}(-\omega)$ for ω real, we write

$$\hat{E}(\omega) = \hat{A}(\omega - \omega_0)e^{i\vec{k}_0 \cdot \vec{x}} + \hat{A}^*(-\omega - \omega_0)e^{-i\vec{k}_0 \cdot \vec{x}}. \qquad (2.80)$$

(For example, if $E(\tau)$ is a wavetrain, then A in (2.57) is constant and $\hat{E}(\omega) = A\delta(\omega - \omega_0) + A^*\delta(-\omega - \omega_0)$.) Substitute (2.80) into (2.79) and expand $\hat{\chi}^{(1)}(\omega)$ in a Taylor series about ω_0 to obtain

$$\left(\hat{\chi}^{(1)}(\omega_0) + (\omega - \omega_0)\hat{\chi}^{(1)'}(\omega_0) + \frac{(\omega - \omega_0)^2}{2}\hat{\chi}^{(1)''}(\omega_0) + \cdots \right)$$

$$\times \hat{A}(\omega - \omega_0)e^{i\vec{k}_0 \cdot \vec{x}} + \left(\hat{\chi}^{(1)}(-\omega_0) + (\omega + \omega_0)\hat{\chi}^{(1)'}(-\omega_0) \right.$$

$$\left. + \frac{(\omega + \omega_0)^2}{2}\hat{\chi}^{(1)''}(-\omega_0) + \cdots \right) \hat{A}^*(-\omega - \omega_0)e^{-i\vec{k}_0 \cdot \vec{x}} \qquad (2.81)$$

Next, take the inverse Fourier transform of (2.81) knowing that

$$A(t) = \frac{1}{2\pi} \int \hat{A}(\Omega) e^{-i\Omega t} d\Omega \qquad (2.82)$$

and obtain

$$e^{i\vec{k}_0 \cdot \vec{x} - i\omega_0 t} \hat{\chi}^{(1)} \left(\omega_0 + i\frac{\partial}{\partial t} \right) A(t) + e^{-i\vec{k}_0 \vec{x} + i\omega_0 t} \hat{\chi}^{(1)*} \left(\omega_0 - i\frac{\partial}{\partial t} \right) A^*(t). \qquad (2.83)$$

$\square$

We will now rederive (2.78) from another, and more general, point of view. In doing so we introduce a general algorithm for deriving envelope equations that is useful in a broad variety of contexts. Recall that the dispersion relation (2.21) with constant refractive index was originally obtained by substituting $A \exp i(\vec{k} \cdot \vec{x} - \omega t)$ into (2.10). If instead of inserting a plane wave for $\vec{E}$ we substitute the wavepacket (2.57) with $\vec{k}_0 = (0, 0, k_0)$, we obtain a modified dispersion relation

$$D\left(-i\frac{\partial}{\partial x}, -\frac{\partial}{\partial y}, k_0 - i\frac{\partial}{\partial z}, \omega_0 + i\frac{\partial}{\partial t} \right) A(x, y, z, t) = 0, \qquad (2.84)$$

which is a partial differential equation for the slowly varying envelope $A(x, y, z, t)$. The reason for this is that if we differentiate $A(x, y, z, t) \exp i(k_0 z - \omega_0 t)$ with respect to z, we obtain $\exp i(k_0 z - \omega_0 t)(ik_0 + \partial/\partial z)A$. For a plane wave with A constant, the action of $\partial/\partial z$ on $A \exp i(k_0 z - \omega_0 t)$ is just to multiply A by the algebraic factor ik_0. For a wavepacket, the effect of applying the operator $\partial/\partial z$ to the electric field is to multiply the envelope A by the differential operator $i(k_0 - i(\partial/\partial z))$. Similarly,

$$\frac{\partial}{\partial x} A \exp i(k_0 z - \omega_0 t) = \exp i(k_0 z - \omega_0 t) \cdot i\left(0 - i\frac{\partial}{\partial x} \right) A$$

$$\frac{\partial}{\partial y} A \exp i(k_0 z - \omega_0 t) = A \exp i(k_0 z - \omega_0 t) \cdot i\left(0 - i\frac{\partial}{\partial y} \right) A$$

$$\frac{\partial}{\partial t} A \exp i(k_0 z - \omega_0 t) = A \exp i(k_0 z - \omega_0 t) \cdot -i\left(\omega_0 + i\frac{\partial}{\partial t} \right) A.$$

Therefore the action of $\nabla^2 - (n^2/c^2)(\partial^2/\partial t^2)$ on the electric field is equivalent to the action of $D(l, m, k, \omega)$ on A with $l = -i(\partial/\partial x)$, $m = -i(\partial/\partial y)$, $k = k_0 - i(\partial/\partial z)$, $\omega = \omega_0 + i(\partial/\partial t)$, namely (2.84).

We will now also take account of the frequency dependence of the refractive index. With $D = l^2 + m^2 + k^2 - (n(\omega)\omega/c)^2$, equation (2.84) is

$$
\left\{ \left(-i\frac{\partial}{\partial x} \right)^2 + \left(-i\frac{\partial}{\partial y} \right)^2 + \left(k_0 - i\frac{\partial}{\partial z} \right)^2 \right.
$$

$$
\left. - \left(\frac{n\left(\omega_0 + i\frac{\partial}{\partial t} \right)\left(\omega_0 + i\frac{\partial}{\partial t} \right)}{c} \right)^2 \right\} A(x, y, z, t) = 0. \qquad (2.85)
$$

At order one, we regain $k_0^2 = n^2(\omega_0)\omega_0^2/c^2$, the algebraic dispersion relation for the plane wave. At order μ (recall $\partial/\partial z = 0(\mu)$, $\partial^2/\partial z^2 = 0(\mu^2)$), we obtain

$$
\frac{\partial A}{\partial z} + \left(\frac{\partial}{\partial \omega} \cdot \left(\frac{n(\omega)\omega}{c} \right) \right)_{\omega = \omega_0} \frac{\partial A}{\partial t} = 0, \qquad (2.86)
$$

consistent with the idea that, to leading order, the envelope A of a wavepacket moves with the group velocity v_g with $v_g^{-1} = \partial k/\partial \omega|_0$. Expanding out (2.84) to order μ^2, that is, including everything up to second derivatives, we obtain

$$
2ik_0\frac{\partial A}{\partial z} + \nabla_1^2 A + 2i\frac{n(\omega_0)\omega_0}{c^2} \cdot \left(\frac{\partial}{\partial \omega}(n\omega) \right)_0 \frac{\partial A}{\partial t}
$$

$$
+ \frac{\partial^2 A}{\partial z^2} - \frac{1}{c^2}\left\{ \left(\frac{\partial}{\partial \omega}(n\omega) \right)_0^2 + n_0\omega_0 \left(\frac{\partial^2}{\partial \omega^2}(n\omega) \right)_0 \right\} \frac{\partial^2 A}{\partial t^2} = 0, \qquad (2.87)
$$

which can be rewritten, using (2.86), as

$$
\frac{\partial A}{\partial z} + \left(\frac{\partial k}{\partial \omega} \right) \frac{\partial A}{\partial t} + \frac{i}{2}\left(\frac{\partial^2 k}{\partial l^2} \right)_0 \frac{\partial^2 A}{\partial x^2}
$$

$$
+ \frac{i}{2}\left(\frac{\partial^2 A}{\partial m^2} \right)_0 \frac{\partial^2 A}{\partial y^2} + \frac{i}{2}\left(\frac{\partial^2 k}{\partial \omega^2} \right)_0 \frac{\partial^2 A}{\partial t^2} = 0 \qquad (2.88)
$$

EXERCISE 2.12

Given $k = \sqrt{(n^2(\omega)\omega^2/c^2) - l^2 - m^2}$. Show that

$$
\frac{\partial k}{\partial l} = \frac{\partial k}{\partial m} = 0, \qquad \frac{\partial^2 k}{\partial l^2} = \frac{\partial^2 k}{\partial m^2} = -\frac{1}{k_0},
$$

$$
\frac{\partial k}{\partial \omega} = \left(\frac{\partial}{\partial \omega}\left(\frac{n\omega}{c} \right) \right)_0, \qquad \frac{\partial^2 k}{\partial \omega^2} = \left(\frac{\partial^2}{\partial \omega^2}\left(\frac{n\omega}{c} \right) \right)_0,
$$

where the derivatives are estimated at $l = m = 0$, $k = k_0 = n(\omega_0)\omega_0/c$. $\square$

The effects of a small variation δn of order μ^2 in the refractive index can readily be incorporated into (2.88). Simply replace n in (2.85) by $n + \delta n$. This will add an extra term

$$\frac{2n\delta n\omega_0^2}{c^2} A$$

to (2.87). When we divide across by $2ik_0$, we get the additional term $-i\delta n(\omega_0/c)A = -i(\delta n/n)k_0 A$ in equation (2.88), which may also be written as $-i(\partial k/\partial n)_0\delta n A$. Therefore

$$\frac{\partial A}{\partial z} + \left(\frac{\partial k}{\partial \omega}\right)_0 \frac{\partial A}{\partial t} + \frac{i}{2}\left(\frac{\partial^2 k}{\partial l^2}\right)_0 \frac{\partial^2 A}{\partial x^2}$$

$$+ \frac{i}{2}\left(\frac{\partial^2 k}{\partial m^2}\right)_0 \frac{\partial^2 A}{\partial y^2} + \frac{i}{2}\left(\frac{\partial^2 k}{\partial \omega^2}\right) \frac{\partial^2 A}{\partial t^2} - i\left(\frac{\partial k}{\partial n}\right)_0 \delta n A = 0, \qquad (2.89)$$

when the subscript 0 means we evaluate the derivatives of k with respect to l, m, k, w and n at $l = m = 0$, $k = k_0 = \frac{n_0\omega}{c}$, $\omega = \omega_0$, $n = n_0$.

Equation (2.89) has an interesting and extremely useful property. Imagine that we are propagating a wavepacket through a medium with random variation $\delta n(x, y, z)$ in the refractive index. The situation is shown in Figure 2.5 and is described by (2.89) with $\partial^2 k/\partial l^2 = \partial^2 k/\partial m^2 = -1/k$ and $\partial^2 k/\partial \omega^2 = 0$.

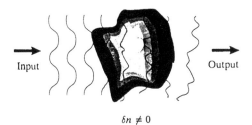

$$\delta n \neq 0$$

FIGURE 2.5   Beam distortion by a random medium.

A smooth input $A(x, y, z)$ at $z = z_1$ is highly distorted by the time it reaches z_2. However, note that the envelope B of a wave of wavenumber k traveling in the $-z$ direction obeys the same equation (2.89) as A except that the sign of k is reversed. Therefore, its complex conjugate $B^*(x, y, z)$ obeys the same equation as A does. Suppose now we can arrange that the right-traveling wave is converted with its phase intact to a left-traveling one so that at some $z = z_2$, $B^*(x, y, z_2) = A(x, y, z_2)$. From the uniqueness of solutions of (2.89), $B^*(x, y, z)$ must be equal to $A(x, y, z)$ everywhere, and therefore when the left-traveling wave re-emerges at $z = z_1$ from the distorting region, all the distortions are removed and the information carried on $A(x, y, z)$ for $z < z_1$ is also carried by $B^*(x, y, z)$. The cleaning up of light beams using this idea of *phase conjugation* is of great practical importance, and we include here in Figure 2.6(a) a picture of a beam containing a message, (b) the distortion as seen at $z = z_2$, (c) the cleaned up beam.

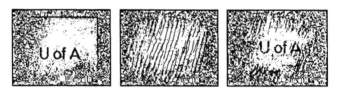

FIGURE 2.6   Beam clean-up by phase conjugation.

The mechanism by which one creates a phase conjugated wave is a non-linear one (a four-wave interaction process) and is discussed in Section 2j.

2h.   Nonlinear Wavepackets and the Nonlinear Schrödinger (NLS) Equation. [3b, 6j]

We now combine the results of Sections 2e and 2g to derive the equation for the evolution of a wavepacket envelope $A(x, y, z, t)$ in a medium with a nonlinear (intensity dependent) refractive index. For a nonlinear wavetrain, the dispersion relation is

$$D(l, m, k, \omega, |A|^2) \, A = \left(l^2 + m^2 + k^2 - \left(\frac{n\omega}{c} \right)^2 \right) A = 0 \qquad (2.90)$$

with $n = n_0 + n_2|A|^2$. For amplitudes A of order μ (i.e., we scale the amplitudes like derivatives, e.g., $\partial^2 A/\partial z^2 = 0(\mu^3)$, $|A|^2 A = 0(\mu^3)$), so that the variation δn in refractive index is of order μ^2, the resulting equation for the envelope A is exactly (2.89) with δn given by $n_2|A|^2$. Combining the effects of a variation $\delta n(x, y)$ in refractive index due to small spatially dependent variations and amplitude dependent variations; that is, we write δn in (2.89) as $\delta n(x, y) + n_2|A|^2$ and obtain

$$\frac{\partial A}{\partial z} + \frac{\partial k}{\partial \omega} \frac{\partial A}{\partial t} + \frac{i}{2} \frac{\partial^2 k}{\partial l^2} \frac{\partial^2 A}{\partial x^2} + \frac{i}{2} \frac{\partial^2 k}{\partial m^2} \frac{\partial^2 A}{\partial y^2}$$

$$+ \frac{i}{2} \frac{\partial^2 k}{\partial \omega^2} \frac{\partial^2 A}{\partial t^2} - ik \left(\frac{\delta n}{n} + \frac{n_2}{n}|A|^2 \right) A = 0. \qquad (2.91)$$

In (2.91), we have omitted the subscripts zero on k and the coefficients. We remind the reader that if the underlying carrier wave is $\exp(i\vec{k} \cdot \vec{x} - i\omega t)$ with $\vec{k} = (0, 0, k = n(\omega)\omega/c)$, then $\partial k/\partial \omega = 1/c(\partial/\partial\omega)(n\omega)$, $\partial^2 k/\partial l^2 = \partial^2 k/\partial m^2 = -1/k$. We mention here and prove later that in a guided wave with a fixed transverse structure, there is no diffraction and the frequency dependence of the effective refractive index $\beta(\omega)$, called the mode propagation constant, is determined by the eigenvalue problem which determines the mode shape as well as by the material properties of the local refractive index. In that case, there are no diffraction terms in (2.91), and $k = \beta(\omega)\omega/c$.

Equation (2.91) is central to much of the analysis in this book. It has many important properties that we discuss in detail in the next chapter and again in Section 6k. Here we simply emphasize a few. First and foremost, we emphasize the self-focusing property, in which, for certain conditions on the coefficients, wave envelopes A cannot remain independent of x, y, and t even if they start out that way. The light insists on becoming more concentrated in some regions and less in others. It cannot remain uniform and absorb the effect of nonlinearity with an amplitude dependent wavenumber. This last state is given by the solution obtained by taking A independent of x, y, and t (assume here that $\delta n = 0$) and solving (2.91)

$$A(z) = A_0 \exp ik\frac{n_2}{n}| A_0|^2 z, \qquad (2.92)$$

which makes the wavenumber of the electric field $\vec{E} = \hat{E}(A(z) \exp i(kz - \omega t) + (*))$ equal to $k(1 + \frac{n_2}{n}| A|^2)$ or $\frac{\omega}{c}(n + n_2| A|^2)$. We shall see that if the product of kn_2 with any one of the three dispersion coefficients $\partial^2 k/\partial l^2$, $\partial^2 k/\partial m^2$, or $\partial^2 k/\partial \omega^2$ is negative (the diffraction coefficients $\partial^2 k/\partial l^2 = \partial^2 k/\partial m^2 = -1/k$, and so the sign of the product of kn_2 and $-1/k$ is the negative of the sign of n_2), then the amplitude modulated wavenumber solution (2.92) is unstable and the light concentrates in local patches in x, y and t space.

EXERCISE 2.13

Consider

$$A_z - i\gamma \nabla_1^2 A - i\beta A^2 A^* = 0.$$

Show that the solution $A = A_0 \exp i\beta| A_0|^2 z$ is unstable by substituting

$$A(x, y, z) = (A_0 + ae^{i\vec{K} \cdot \vec{x} + \sigma z}) \exp i\beta| A_0|^2 z$$

and proving that $\sigma^2 = 2\beta\gamma K^2| A_0|^2 - \gamma^2 K^4$. Draw the range of K^2 for which $\sigma^2 > 0$ and show that the maximum growth occurs for $K =$

$\sqrt{\beta/\gamma}|\,A_0|$. The reader should also consult Exercise 6.11, where the instability is interpreted as an energy exchange due to a four-wave mixing process.   □

In order to provide an intuitive proof of this theorem, let us consider the envelope $A(x, y, t)$ to be independent of t and write (2.91) as

$$\frac{\partial A}{\partial z} - \frac{i}{2k}\nabla_1^2 A - ik\left(\frac{\delta n}{n} + \frac{n_2}{n}|\,A|^2\right)A = 0. \tag{2.93}$$

We have already argued in Section 2f that if δn is positive locally (see Figure 2.3), then light concentrates in this region of higher refractive index. Now suppose $\delta n = 0$ but the nonlinear term is present with positive n_2. Suppose one begins with a field envelope $A = A_0$ that is independent of x and y. A small perturbation δA localized at some point x, y causes the refractive index $n_2|\,A|^2$ to increase there. The light rays from neighboring (x, y) sites converge toward this region, enhancing the intensity there and depleting it at neighboring sites. The refractive index perturbation responds by becoming bigger. More light focuses there, the refractive index becomes even higher, and even more light converges.

This instability, namely the instability of a monochromatic wavetrain in a dispersive, weakly nonlinear medium, was discovered almost simultaneously in several contexts in the mid-1960s. In deep-water gravity waves, it is called the Benjamin–Feir instability [22, 23, 24] after Brooke Benjamin and Jim Feir, who found they were experimentally unable to sustain a nonlinear, constant-amplitude wavetrain of gravity waves in deep water. In plasma physics, it was called the modulational instability and leads to the collapse of Langmuir waves [25, 26]. In optics, where the instability is most easily deduced by appealing, as we have done, to one's intuitive notion of the effect of an increased refractive index, it is known as the focusing instability.

It is not constrained to transverse modulations. In the next chapter we derive the nonlinear Schrödinger equation for a confined mode propagating in a waveguide and, for these structures, there is no transverse dispersion but only dispersion in the direction of propagation of the wave. If we take $A(z, t)$ independent of x and y, then for $\delta n = 0$,

$$\frac{\partial A}{\partial z} + \frac{\partial k}{\partial \omega}\frac{\partial A}{\partial t} + \frac{i}{2}\frac{\partial^2 k}{\partial \omega^2}\frac{\partial^2 A}{\partial t^2} - ik\frac{n_2}{n}|\,A|^2 A = 0. \tag{2.94}$$

The condition for the focusing instability is that the product of the nonlinear coefficient $-k(n_2/n)$ with the coefficient $\frac{1}{2}\partial^2 k/\partial \omega^2$ is positive. For the diffraction case, the product is n_2/n whose sign depends on that of n_2. In the case of wave propagation in a fiber, we will see that if $n_2 > 0$ this product is positive only if $\partial^2 k/\partial \omega^2$ is negative. We will see, in fact, that the dispersion $\partial^2 k/\partial \omega^2$ can be both positive and negative in the vicinity of resonances where the real part of the susceptibility has a nontrivial dependence on frequency (see Figure 2.1). These cases are called normal dispersion and anomalous dispersion respectively.

EXERCISE 2.14

Show that if $n_2 = 0$, a pulse with initial Gaussian shape

$$A(0, t) = \exp\left(-\frac{t^2}{t_0^2}\right)$$

will evolve to

$$A(z, t) = \frac{1}{\left(1 - \frac{2\ ik''z}{t_0^2}\right)^{1/2}} \exp - \left(\frac{(t - k'z)^2}{t_0^2 - 2\ ik''z}\right).$$

□

Without nonlinearity, a pulse doubles its width in a distance

$$z_{\text{dispersion}} = \frac{\sqrt{3}}{2} \frac{t_0^2}{|k''|}. \tag{2.95}$$

We now examine the effects of nonlinearity. Let us imagine that we have a very *weakly* modulated pulse as shown in Figure 2.7 so that initially the size of the dispersion term $i(k''/2)\partial^2 A/\partial t^2$ is very small. Its local nonlinear wavenumber k is $\omega/c(n_0 + n_2 I)$, $I = |A|^2$ because, from (2.94), if we neglect dispersion, $A = A_0 \exp \frac{ikn_2}{n}|A_0^2|z$.

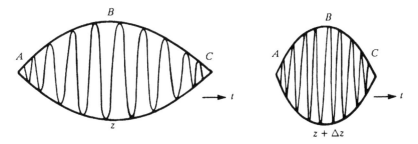

FIGURE 2.7   Wavepacket compression.

For $n_2 > 0$, the pulse intensity increases in AB, the leading edge of the wavepacket as seen by an observer at a fixed position z. Hence k increases from A to B, namely $\partial k / \partial t > 0$, and therefore from the third component of (2.61), $\partial \omega / \partial z < 0$. Another way of seeing this is as follows. If $k_B > k_A$, then the number of waves per unit length at B is greater than the number at A, and, because of this, the frequency one would measure at a fixed point z is greater when the B part of the wavepacket passes than when the A part passes. But the waves at time B arrive at the point z_A later than those at time A because $t_B > t_A$ and therefore are at a point $z_B < z_A$ when the leading edge A is at z_A. Thus the frequency ω at z_B is greater than that at z_A; that is, $\omega(z_B) > \omega(z_A)$ for $z_A > z_B$. Hence $\partial \omega / \partial z < 0$ and ω decreases between z_B and z_A. But if ω decreases in z and if $\partial^2 k / \partial \omega^2 < 0$, then $\partial k / \partial \omega$ increases, and the group velocity $v_g = \partial \omega / \partial k$ in the front part AB of the wavepacket decreases. A similar argument gives that the group velocity in the trailing edge of the pulse increases. Hence as the wavepacket ABC moves forward in z, it is compressed until the second derivative term, the dispersion $(ik''/2)(\partial^2 A / \partial t^2)$, becomes important. At this point the effects of pulse compression and pulse dispersion balance and can lead to a robust, stable finite width pulse called a soliton (substitute into (2.94))

$$A(z, t) = 2\eta \, \text{sech} \, 2\eta \sqrt{\frac{2\pi n_2}{\lambda_0 |k_0''|}} (t - k_0' z - k_0'' \Omega z)$$

$$\exp\left(-i\Omega t + i\left(k_0' \Omega + \frac{k_0''}{2} \Omega^2\right) z + i \frac{z}{z_0}\right), \quad (2.96)$$

for arbitrary η and Ω and where $z_0 = \lambda_0 / 4\eta^2 \pi n_2$ and $\lambda_0 = 2\pi c / \omega_0$, the vacuum wavelength. Writing $\omega = \omega_0 + \Omega$, the corresponding electric field is (note $k(\omega) = k(\omega_0) + \Omega k'(\omega_0) + \Omega^2 / 2k''(\omega_0) + \cdots$, $k'(\omega) = k'(\omega_0) + k''(\omega_0)\Omega + \cdots$)

$$\vec{E} = \hat{e} 2\eta U(x, y, \omega) \, \text{sech}\left(\frac{t - k'(\omega)z}{t_0}\right) \exp\left(ik(\omega)z - i\omega t + i\frac{z}{z_0}\right) + (*)$$

$$(2.97)$$

where $U(x, y, \omega)$ is the (fixed) transverse shape of the field and

$$t_0 = 1/2\eta(\lambda_0 |k_0''| / 2\pi n_2)^2$$

is the pulse width. Two solitons with different frequencies have different group velocities and pass through each other without distortion, in the sense that each reemerges from an extremely nontrivial collision and regains its former shape exactly. In principle, therefore, information could be carried along an optical fiber by trains of solitons with different frequencies. When two solitons have the same frequency and therefore the same speed, they undergo a periodic oscillation. The behavior of interacting solitons is discussed in more detail in Chapters 3 and 6. Similar types of solutions obtain if the envelope $A(z, x)$ is independent of one transverse direction y and time t, whence (2.91) describes the change along the direction of propagation of a light beam in which diffraction is balanced by nonlinearity. In this case, a soliton has the form

$$A(z, x) = 2\eta \, \text{sech} \, \frac{4\pi \eta}{\lambda} \sqrt{n_0 n_2} (x - \bar{x}(z)) e^{i(v/2)x + i\sigma} \qquad (2.98)$$

where $d\bar{x}/dz = v/2k$, $d\sigma/dz = -v^2/8k + 2\eta^2 k(n_2/n)$, and the electric field is given by $\vec{E} = \hat{E}(A(z, x) \exp(i\frac{n\omega}{c} - i\omega t) + (*))$. These solitons will be the building blocks for the nonlinear surface waves that one finds trapped at interfaces between slabs of dielectrics with different refractive indices.

There is a dramatic change in the behavior of solutions to (2.91) when we include two transverse dimensions or two transverse dimensions and time. Somewhat counter to intuition, the extra dimension does not increase the diffraction so that it overcomes compression; rather, pulse compression wins, the wavepacket continues to focus and compress, and solutions blow up in finite time. In the collapse phase, each filament takes on a very special spatial shape of self-similar form and circular cross-section whose amplitude grows as its width decreases, the former without bound. Of course, once the light gets intense, the approximation that $n = n_0 + n_2|A|^2$ no longer holds. In fact, it is often the case that this refractive index law is the small-amplitude expansion of the saturable nonlinearity for which the real susceptibility $\hat{\chi}(\omega)$ has the form $a/(b + c|A|^2)$. In that case, a filament that begins to accelerate toward collapse reaches a point of saturation. We will meet both one- and two-dimensional solitary waves and filaments when we study the behavior of light in bistable cavities.

In the defocusing case when $n_2 < 0$ and $k'' < 0$ (or if $n_2 > 0$ and we are in a region with normal dispersion $k'' > 0$), $n_2 k'' > 0$ and uniform wavetrains are neutrally stable. Deformations in the uniform wavetrain in which the intensity is locally decreased neither grow nor decay but evolve into special shapes that propagate as solitons of a different nature, which, because they have intensities lower than the ambient wavetrain, are called dark solitons.

2i. Linear and Nonlinear Birefringence; the Coupling of Wavetrains and Wavepackets

So far we have assumed, except for the discussion in Section 2d, that the medium is isotropic and that the induced polarization is parallel to the electric field. This is not the case in dielectric crystals. The crystal consists of periodic arrays of atoms and molecules so that the induced polarization also depends on the direction of the electric field with respect to the lattice, that is,

$$P_i = \epsilon_0 \sum_{j=1}^{3} \chi_{ij} E_j, \quad i = 1, 2, 3, \tag{2.99}$$

the polarization in the ith direction (the ith component of the vector $\vec{P}$) depends on all components E_j, $j = 1, 2, 3$ of the electric field vector. The tensor χ_{ij}, $j = 1, 2, 3$ is called the electric susceptibility tensor. We will take its entries to be real and symmetric; that is, $\chi_{ij} = \chi_{ji}$, $i \neq j$. Because of this, one can choose the coordinates $(x_1, x_2, x_3) = (x, y, z)$ relative to that of the crystal structure so that the off-diagonal terms vanish and

$$P_i = \epsilon_0 \chi_{ii} E_i, \quad i = 1, 2, 3 \text{ (no summation convention).} \tag{2.100}$$

Because in general $\chi_{11} \neq \chi_{22} \neq \chi_{33}$, the speed of propagation of the phase of an electromagnetic wave in the crystal depends on its direction of polarization. To illustrate, imagine an electromagnetic wave propagating along one of the principal axes of the crystal, which we take to be the z direction. We write $\vec{E} = E_x \hat{i} + E_y \hat{j}(\hat{i} = (1, 0, 0), \hat{j} = (0, 1, 0))$, so that it lies in the x, y plane. Observe that $\nabla \cdot \vec{D} = 0$ and that E_x and E_y satisfy

$$\nabla^2 E_x - \frac{n_1^2}{c^2} \frac{\partial^2 E_x}{\partial t^2} = 0, \quad \nabla^2 E_y - \frac{n_2^2}{c^2} \frac{\partial^2 E_y}{\partial t^2} = 0 \tag{2.101}$$

respectively with

$$n_1^2 = 1 + \chi_{11}, \quad n_2^2 = 1 + \chi_{22}. \tag{2.102}$$

Because the refractive indices are different, the phase velocities c/n_1 and c/n_2 of the two waves are unequal. If E_x and E_y are plane waves of frequency ω, then

$$E_x = a \, \exp(ik_1 z - i\omega t) + (*), \quad k_1 = \frac{n_1 \omega}{c},$$

$$E_y = b \, \exp(ik_2 z - i\omega t) + (*), \quad k_2 = \frac{n_2 \omega}{c}. \tag{2.103}$$

Birefringence is the name we give to the phenomenon in which the phase velocity of an optical beam propagating in a crystal depends on the direction of polarization of the electric field. This leads to a change in the type of polarization of the electric field at different points in the medium. Suppose at

$z = 0$, we take a and b real and equal. Then on any plane $z = \text{constant}$, the electric field vector can be written:

$$2a \, \cos\left(\omega t - \frac{k_1 + k_2}{2} z - \frac{k_1 - k_2}{2} z\right),$$

$$2a \, \cos\left(\omega t - \frac{k_1 + k_2}{2} z + \frac{k_1 - k_2}{2} z\right).$$

At $z = 0$, the phases are the same and the vector $\vec{E}$ executes a linear motion in the E_x, E_y plane along $E_y = E_x$ as t goes through a full optical cycle $0 \le t \le 2\pi/\omega$. The light is said to be linearly polarized. At $z = \pi/2(k_1 - k_2)$, however, the electric field vector is

$$2a \, \sin\left(\omega t - \frac{k_1 + k_2}{4(k_1 - k_2)}\pi + \frac{\pi}{4}\right), \quad 2a \, \cos\left(\omega t - \frac{k_1 + k_2}{4(k_1 - k_2)}\pi + \frac{\pi}{4}\right),$$

which traces out a counterclockwise circular path as t goes through the full optical cycle. The light is said to be circularly polarized.

EXERCISE 2.15

Draw the motion of the top of the vector $\vec{E}$ on the planes $(k_1 - k_2)z = r\pi$, $r = 0, 1/4, 1/2, 3/4, 1$. For $r = 1/4, 3/4$ the light is elliptically polarized. What are the major and minor axes of the ellipse? □

Often the amount of birefringence is very small. For example, in birefringent fibers the index difference divided by the index sum, $(n_1 - n_2)/(n_1 + n_2)$, may be as low as 10^{-4} to 10^{-7}. The corresponding delay difference Δt between the times of arrival of two initially coincident fronts over a distance L is

$$\Delta t = \frac{k_1 - k_2}{\omega} L = \frac{n_1 - n_2}{c} L$$

and can lie between 1,000 and 1 picoseconds per kilometer. Therefore it is convenient to think of this difference as a perturbation and write

$$\vec{E} = (\hat{i} A(z) + \hat{j} B(z)) e^{i(kz - \omega t)} + (*), \tag{2.104}$$

with $k = (k_1 + k_2)/2$, $\Delta k = (k_1 - k_2)/2$, and

$$\frac{dA}{dz} = i \Delta k \, A, \quad \frac{dB}{dz} = -i \Delta k B. \tag{2.105}$$

Solving (2.105), one obtains

$$A = a \exp i \Delta kz, \quad B = b \exp -i \Delta kz$$

and then (2.104) is the exact solution (2.103).

We have just seen how anisotropy in a crystal can result in different phase velocities for different modes of polarization of the electric field. We will now look at the effects of isotropic nonlinearity in a linearly anisotropic dielectric crystal. We will take the electric field to lie in the (x, y) plane and the nonlinear polarization $\vec{P}$ to be given by the second term on the right-hand side of (2.29). For the same reason discussed in the last paragraph of Section 2e, it will be important that the basic plane-wave solutions are dispersive in character. Therefore the linear susceptibilities $\chi_{11}^{(1)}$ and $\chi_{22}^{(1)}$ each have finite delay (they are not proportional to the Dirac delta function $\delta(t - \tau)$), and their Fourier transforms are frequency dependent. We will assume that $\hat{\chi}^{(1)}(\omega) = 1/2(\hat{\chi}_{11}^{(1)} + \hat{\chi}_{22}^{(1)})$ is much bigger than the difference and that the difference $\hat{\chi}_{11}^{(1)} - \hat{\chi}_{22}^{(1)}$ is of the same order as $\chi^{(3)}|\vec{E}|^2$. For simplicity we will assume that there is no delay in the nonlinear term. The equation (2.9) is

$$\nabla^2 \vec{E} - \frac{1}{c^2} \frac{\partial^2 \vec{E}}{\partial t^2} - \frac{1}{c^2} \frac{\partial^2}{\partial t^2} \int_{-\infty}^{\infty} \chi^{(1)}(t - \tau) \vec{E}(\tau) dz$$

$$= \frac{1}{c^2} \frac{\partial^2}{\partial t^2} \int_{-\infty}^{\infty} \begin{Bmatrix} (\chi_{11}^{(1)} - \chi^{(1)})(t - \tau) E_1(\tau) \\ (\chi_{22}^{(1)} - \chi^{(1)})(t - \tau) E_2(\tau) \end{Bmatrix}$$

$$+ \frac{1}{c^2} \frac{\partial^2}{\partial t^2} \begin{Bmatrix} \chi^{(3)}(E_1^2 + E_2^2) E_1 \\ \chi^{(3)}(E_1^2 + E_2^2) E_2 \end{Bmatrix} \qquad (2.106)$$

where $\vec{E} = (E_1, E_2, 0)$. To leading order, $\vec{E}^{(0)}$ is given by

$$\vec{E}^{(0)} = (A(z)\hat{i} + B(z)\hat{j})e^{i\theta} + (*) \qquad (2.107)$$

where $A(z)$ and $B(z)$ are slowly varying functions of z, the direction of propagation and $\theta = kz - \omega t$ with $k = n(\omega)\omega/c, n = \sqrt{1 + \hat{\chi}^{(1)}(\omega)}$. The correction $\vec{E}^{(1)}$ to $\vec{E}$ satisfies (2.106) with E_1, E_2 on the right-hand side of (2.106) given by the $\hat{i}$ and $\hat{j}$ components in (2.107) and with the additional term

$$\left(-2ik \frac{dA}{dz} \hat{i} - 2ik \frac{dB}{dz} \hat{j} \right) e^{i\theta} + (*)$$

on the right-hand side coming from $d^2 \vec{E}^{(0)}/dz^2$. The dependence of $A(z)$ and $B(z)$ on z are chosen to suppress all terms $\exp(\pm i\theta)$ appearing on the right-hand side of the equation for $\vec{E}^{(1)}$ as these terms would lead to $\vec{E}^{(1)}$ being proportional to $ze^{\pm i\theta}$. Terms proportional to $e^{\pm 3i\theta}$ do not lead to a secular growth in $\vec{E}^{(1)}$ because $k(3\omega) \neq 3k(\omega)$. We find

$$\frac{d\vec{e}}{dz} = i\Delta k\sigma_3\vec{e} + \frac{i\omega}{2c}\frac{\chi^{(3)}}{n}(2(\vec{e}\cdot\vec{e}^*)\vec{e} + (\vec{e}\cdot\vec{e})\vec{e}^*) \qquad (2.108)$$

where

$$\Delta k = \frac{i\omega}{4nc}\left(\hat{\chi}^{(1)}_{11} - \hat{\chi}^{(1)}_{22}\right), \quad \vec{e} = A(z)\hat{i} + B(z)\hat{j}, \quad \sigma_3 = \begin{pmatrix} 1 & 0 \\ 0 & -1 \end{pmatrix} \quad (2.109)$$

Observe that for $\chi^{(3)} \equiv 0$, we recover the result for linear birefringence (2.105). Also observe that if we set $B \equiv \Delta k \equiv 0$, we obtain (2.42), the modulation of the envelope $A(z)$ of a singly polarized electric field.

What we see, then, is that nonlinearity can couple two modes of polarization of a wavetrain. These coupled mode equations (2.108) have extremely rich classes of solutions, even more so when combined with the effects of wavepacket dispersion, and the reader can look at suitable references (for example, [27, 28]) on their possible uses. To add wavepacket effects we simply rewrite

$$\frac{d}{dz} \rightarrow \frac{\partial}{\partial z} + k'\frac{\partial}{\partial t} - \frac{i}{2k}\nabla^2_\perp + \frac{i}{2}k''\frac{\partial^2}{\partial t^2}. \qquad (2.110)$$

EXERCISE 2.16

Work out how the two modes of polarization exchange energy as a function of z. Substitute $\vec{e} = A\hat{i} + B\hat{j}$ in (2.108) and write individual equations for A and B:

$$\frac{dA}{dz} = i\Delta kA + \frac{i\omega\chi^{(3)}}{2nc}(3A^2A^* + 2BB^*A + A^*B^2)$$

$$\frac{dB}{dz} = -i\Delta kB + \frac{i\omega\chi^{(3)}}{2nc}(3B^2B^* + 2AA^*B + B^*A^2). \qquad (2.111)$$

Define

$$AA^* + BB^* = E, \qquad AA^* - BB^* = N,$$

$$AB^* + A^*B = P_+, \qquad AB^* - A^*B = iP_- \qquad (2.112)$$

and show (write $\omega \chi^{(3)}/nc = \beta$)

$$\frac{dE}{dz} = 0, \qquad \frac{dN}{dz} = \beta P_+ P_-, \qquad \frac{dP_+}{dz} = -2\Delta k P_- - \beta P_- N,$$

$$\frac{dP_-}{dz} = 2\Delta k P_+ \tag{2.113}$$

Substitute for P_+ and integrate the second equation to obtain

$$N = N_0 + \frac{\beta}{4\Delta k}\left(P_-^2 - P_{-0}^2\right), \tag{2.114}$$

where N_0 and P_{-0} are values of N and P_- at some initial position $z = z_0$. Then the third equation of (2.113) becomes

$$\frac{d^2 P_-}{dz^2} + \left(4(\Delta k)^2 + 2\beta N_0 \Delta k - \frac{\beta}{2}P_{-0}^2\right)P_- + \frac{\beta^2}{2}P_-^3 = 0, \tag{2.115}$$

which can be written as the motion of a particle (with position P_-) in a potential well $V(P_-)$:

$$\frac{d^2 P_-}{dz^2} = -\frac{\partial}{\partial P_-}V(P_-)$$

$$V(P_-) = \frac{1}{2}\left(4(\Delta k)^2 + 2\beta N_0 \Delta k - \frac{\beta}{2}P_{-0}^2\right)P_-^2 + \frac{\beta^2}{8}P_-^4. \tag{2.116}$$

$\square$

From the results of Exercise 2.16, observe the following:

- For zero nonlinear susceptibility $\chi^{(3)} = \beta = 0$, the energy is shared back and forth with period $z = \pi/\Delta k$.

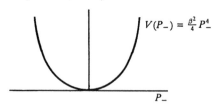

$$V(P_-) = \frac{\beta^2}{4}P_-^4$$

P_-

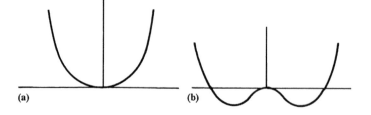

(a)   (b)

- For zero linear birefringence (i.e., $\Delta k = 0$), the graph of $V(P_-)$ is as shown below, and the solution oscillates symmetrically about the state $N = AA^* - BB^* = 0$. Again the energy is shared periodically in z.

- For linear and nonlinear coupling, the graph of $V(P_-)$ changes from that of figure a to figure b, depending on whether

$$\mu = 4(\Delta k)^2 + 2\beta \, N_0 \, \Delta k - \frac{\beta}{2} P_{-0}^2 \qquad (2.117)$$

is greater than or less than zero. In the former case, the energy is shared periodically. In the latter, it is possible to have a broken symmetry where the particle oscillates near the bottom of one of the two wells, in which case the energy remains almost entirely in one polarization field or the other.

- In summary, one does not need an anisotropic medium to have coupling between two directions of polarization. A nonlinear *isotropic* medium produces birefringence.

The interaction of two modes of orthogonal polarization, which we have just treated, is somewhat special because each shares the same phase, that is, wavenumber and frequency. In general, if there are two wavetrains with different phases $\theta_j = \vec{k}_j \cdot \vec{x} - \omega_j t$, $j = 1, 2$, and polarization $\hat{e}_j$, $j = 1, 2$, namely

$$\vec{E}_0 = \hat{e}_1 \left(A_1 e^{i\theta_1} + A_1^* e^{-i\theta_1} \right) + \hat{e}_2 \left(A_2 e^{i\theta_2} + A_2^* e^{-i\theta_2} \right), \qquad (2.118)$$

then the terms in the cubic product $(\vec{E}_0 \cdot \vec{E}_0)\vec{E}_0$ proportional to $e^{i\theta_1}$ are $(3A_1^2 A_1^* + 2A_1 A_2 A_2^*)e_1^* + 4A_1 A_2 A_2^*(\hat{e}_1 \cdot \hat{e}_2)\hat{e}_2$. There is, however, another term $\hat{e}_1 A_1^* A_2^2 e^{i(2\theta_2 - \theta_1)}$, which, only in the case $\theta_1 = \theta_2$, is added to the $\hat{e}_1$ component and gives rise to the $A^* B^2$ and $B^* A^2$ in terms in (2.111).

Note that the presence of a second mode not only affects the propagation of the envelope of the first but also produces a component of the first with the direction of polarization of the second. Accordingly, in general, it is necessary to take

$$\vec{E}_0 = \hat{e}_1 \left(A_1 e^{i\theta_1} + B_2 e^{i\theta_2} + (*) \right) + \hat{e}_2 \left(A_2 e^{i\theta_2} + B_1 e^{i\theta_1} + (*) \right),$$

and write a full set of coupled equations for A_1, A_2, B_1, B_2. If $\hat{e}_1 = \hat{e}_2 = \hat{e}$, things simplify because we only have to deal with a single direction of polarization for each wavetrain. In that case, we obtain (2.120), given below, with $\beta = 2$. In order to make a special point, we include in (2.120) effects of diffraction by taking A_1 and A_2 to be functions of z and the transverse coordinates x, y.

EXERCISE 2.17

A surprise! The coupled set of envelope equations

$$\frac{\partial A_1}{\partial z} - \frac{i}{2k}\nabla_1^2 A_1 - i\frac{k_1 n_2}{n(\omega_0)}\left(|A_1|^2 + \beta|A_2|^2\right) A_1 = 0$$

$$\frac{\partial A_2}{\partial z} - \frac{i}{2k}\nabla_1^2 A_2 - i\frac{k_2 n_2}{n(\omega_0)}\left(\beta|A_1|^2 + |A_2|^2\right) A_2 = 0 \qquad (2.119)$$

in the defocusing case, $n_2 < 0$ can still exhibit a modulational instability if the cross-coupling coefficient β is greater than unity. Show that the solutions

$$A_1 = A_{10}\exp i\frac{k_1 n_2}{n(\omega_0)}\left(|A_{10}|^2 + \beta|A_{20}|^2\right) z$$

$$A_2 = A_{20}\exp i\frac{k_2 n_2}{n(\omega_0)}\left(\beta|A_{10}|^2 + |A_{20}|^2\right) z \qquad (2.120)$$

are unstable to (x, y) dependent perturbations when $\beta > 1$. The reason is that if one has perturbations in the intensities I_{A_1} and I_{A_2} of A_1 and A_2 which look like the curves below,

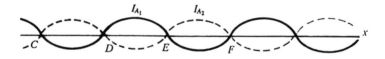

then the increase of intensity in the region DE where I_{A_1} is initially increased is brought about by the fact that the light of the A_2 mode is channeled away from regions CD and EF and, when $\beta > 1$, more light reaches DE then is channeled away from it. This simple observation has enormous consequences for both co- and counterpropagating light beams and, just as the (self) modulational instability is responsible for interesting spatial patterns in the single mode, focusing nonlinearity case, so this (mutual) modulational instability is responsible for rich spatial patterns in the coupled mode case. ☐

2j. Three- and Four-wave Mixing. [6g]

We now return to the general expression for the jth component of the polarization as a function of the electric field,

$$\frac{1}{\epsilon_0} P_j$$

$$= \int \chi_{jk}^{(1)}(t - \tau) E_k(\tau) d\tau + \int \chi_{jkl}^{(2)}(t - \tau_1, t - \tau_2) E_k(\tau_1) E_l(\tau_2) d\tau_1 d\tau_2$$

$$+ \int \chi_{jklm}^{(3)}(t - \tau_1, t - \tau_2, t - \tau_3) E_k(\tau_1) E_l(\tau_2) E_m(\tau_3) d\tau_1 d\tau_2 d\tau_3$$

$$+ \cdots \tag{2.121}$$

or, in Fourier transform variables,

$$= \frac{1}{2\pi} \int \hat{\chi}_{jk}^{(1)}(\omega) \hat{E}_k(\omega) e^{-i\omega t} d\omega$$

$$+ \frac{1}{(2\pi)^2} \int \hat{\chi}_{jkl}^{(2)} \hat{E}_k(\omega_1) \hat{E}_k(\omega_2) e^{-i(\omega_1 + \omega_2)t} d\omega_1 d\omega_2$$

$$+ \frac{1}{(2\pi)^3} \int \hat{\chi}_{jklm}^{(3)}(\omega_1, \omega_2, \omega_3)$$

$$\hat{E}_k(\omega_1) \hat{E}_l(\omega_2) \hat{E}_m(\omega_3) e^{-i(\omega_1 + \omega_2 + \omega_3)t} d\omega_1 d\omega_2 d\omega_3$$

$$+ \cdots . \tag{2.122}$$

As before, the convention is that repeated indices are summed over 1, 2, 3. Often, we write (2.121) schematically as

$$\frac{1}{\epsilon_0} \vec{P} = \chi_\bullet^{(1)} \vec{E} + \chi_\bullet^{(2)} \vec{E} \vec{E} + \chi_\bullet^{(3)} \vec{E} \vec{E} \vec{E} + \cdots \tag{2.123}$$

and the operation denoted by $\bullet$ is a convolution. The governing equation for the evolution of $\vec{E}$ is (2.9),

$$\nabla^2 E - \nabla(\nabla \cdot \vec{E}) - \frac{1}{c^2} \frac{\partial^2}{\partial t^2} \left(1 + \chi^{(1)}\right) \bullet \vec{E} = \frac{1}{c^2} \frac{\partial^2}{\partial t^2} \left(\chi_\bullet^{(2)} \vec{E} \vec{E} + \chi_\bullet^{(3)} \vec{E} \vec{E} \vec{E}\right). \tag{2.124}$$

To leading order, it admits plane wavetrain solutions

$$\vec{E} \simeq \vec{E}^{(0)} = \sum_{j=1}^{N} \hat{e}_j A_j(z) \exp(i\vec{k}_j \cdot \vec{x} - i\omega_j t) + (*), \qquad (2.125)$$

where $\hat{e}_j$ is the direction of polarization of the jth wavetrain and A_j is its amplitude, which, to account for the long-distance cumulative effects of the nonlinear terms, is allowed to be slowly varying in z. The wavevector and polarization directions are chosen to ensure that $\nabla \cdot \vec{D} = 0$ for each of the wavetrains. Each wavevector or frequency pair $(\vec{k}_j, \omega_j)$ satisfies a linear dispersion relation (no summation convention implied)

$$D(\vec{k}_j, \omega_j) = 0. \qquad (2.126)$$

Our goal is to write down evolution equations for the complex amplitudes A_j as functions of z. The structure of these equations, which describe how almost monochromatic, weakly nonlinear, dispersive wavetrains interact, is universal and so many of the properties of lightwaves can be inferred from parallel situations in other fields. The envelope equations are obtained by applying the same philosophy as was used in Section 2e. By solving (2.124) iteratively

$$\vec{E} = \vec{E}^{(0)} + \vec{E}^{(1)} + \vec{E}^{(2)} + \cdots \qquad (2.127)$$

with (2.125) as a first approximation, we observe that the quadratic nonlinearity gives rise to all sums and differences of the phases in (2.125), and the resulting $E^{(1)}$ contains factors like $\pm\omega_j \pm\omega_l \pm\omega_m$, where $\pm\vec{k}_j \pm\vec{k}_l \pm\vec{k}_m = 0$, in its denominator and therefore the strongest response occurs at and near values of the wavevectors when $\pm\omega_j \pm \omega_l \pm \omega_m = 0$.* These particular interactions are called resonant and dominate the exchange of energy between three modes. All the other terms left in $E^{(1)}$ have factors that are oscillatory exponentials with large phases, and these contributions may be ignored. In optics, this is called the rotating wave approximation. We remove the terms with small or zero denominators from $E^{(1)}$ by choosing the time and space dependence of the envelopes A_j. This method for deriving the envelope equation as an asymptotic solvability condition (choosing A_j as function of $\vec{x}$ and t so as to keep the iterates in (2.127) well ordered $\vec{E}^{(0)} \gg \vec{E}^{(1)} \gg \vec{E}^{(2)}$ for all $\vec{x}, t$) is used widely in mathematical physics and is often called the averaging or multiple time scales method. It is discussed in Chapter 6, Section 6g.

---

*From here on, we omit writing $\pm$ and $\vec{k}, \omega$ will stand for either their plus or minus values.

Significant quadratic nonlinear interactions between the wavepackets take place, therefore, when triads of wavevectors and frequencies $\vec{k}_j, \omega_j$ obey

$$\vec{k}_1 + \vec{k}_2 + \vec{k}_3 = 0$$

$$\omega_1 + \omega_2 + \omega_3 = 0 \qquad (2.128)$$

where each $\vec{k}_j, \omega_j$ satisfies its dispersion relation (2.21) or (2.126). While these conditions arise mathematically as small denominators in a perturbation expansion, they have the physical interpretation that both the total momentum $(\vec{k})$ and energy (ω) of the photons involved in the most important scattering process are conserved. The envelopes A_1, A_2, and A_3 of the three wavepackets thus obey the three-wave interaction equations of universal type,

$$\frac{dA_j}{dt} + \vec{c}_j \cdot \nabla A_j = \theta_j A_l^* A_m^* \qquad (2.129)$$

with j, l, m cycled over 1, 2, 3, where $\vec{c}_j$ is the linear group velocity $\nabla_{\vec{k}}\omega$ of the jth wavepacket and θ_j is a coupling coefficient proportional to $\chi^{(2)}$. For wavetrains in which the amplitudes are simply slowly varying functions of the direction z of propagation, the left-hand sides of (2.129) are replaced by

$$c_{jz} \frac{d A_j}{dz},$$

when c_{jz} is the z component of $\vec{c}_j$.

Equations very similar to (2.129) arise in the analysis of Raman and Brillouin scattering processes, except that in the present discussion all three waves are electromagnetic whereas in Raman scattering one of the modes is a molecular vibration of the medium, and in Brillouin scattering it is an acoustical mode. A special case of the three-wave interaction equations arises when (2.128) can be satisfied with $\vec{k}_3 = \vec{k}_1, \omega_3 = \omega_1, \vec{k}_2 = -2\vec{k}_1, \omega_2 = -2\omega_1$. In that case, (2.129) are the equations for the second harmonic interaction. It should be emphasized, however, that the satisfaction of (2.129) is not automatic, and depending on the dispersion relation (which here depends crucially on how the refractive index n varies with frequency), there may be no strong quadratic interactions at all. Furthermore, often the medium consists of crystals that are centrosymmetric, which means that $\vec{P}(-\vec{E})$ should be $-\vec{P}(\vec{E})$. This reflection symmetry demands that $\chi^{(2)}$ is identically zero. In either of these cases, any significant exchange of energy between wavepackets is governed by the cubic terms in (2.121), although now the exchange of energy takes place over much longer times.

Again, significant exchange of energy only takes place between resonant quartets $(\vec{k}_j, \omega_j)_{j=1}^4$, each obeying its dispersion relation (2.22) and satisfying

$$\vec{k}_1 + \vec{k}_2 + \vec{k}_3 + \vec{k}_4 = 0$$

$$\omega_1 + \omega_2 + \omega_3 + \omega_4 = 0. \qquad (2.130)$$

In the envelope equation for A_1, these resonances give rise to a nonlinear term $\theta_1 A_2^* A_3^* A_4^*$. Contrary to the triad resonance conditions, the conditions (2.130) can always be satisfied with several trivial choices.

1. $\vec{k}_4, \omega_4 = -\vec{k}_2, -\omega_2; \vec{k}_3, \omega_3 = -\vec{k}_1, -\omega_1$. This interaction, called a *mutual modal interaction*, is captured by the nonlinear term $A_1 A_2 A_2^*$ in the equation for A_1. When A_1 and A_2 represent the envelopes of wavepackets of different polarizations, such interactions give rise to nonlinear birefringence, as we discussed in Section 2i.

2. $-\vec{k}_4, -\omega_4 = -\vec{k}_3, -\omega_3 = \vec{k}_2, \omega_2 = \vec{k}_1, \omega_1$. This interaction, called a *self-modal interaction*, gives rise to the focusing nonlinearity $A_1^2 A_1^*$ familiar in the nonlinear Schrödinger equation.

Because quartet resonances take longer to exchange energy than triad ones and because of the Benjamin–Feir modulational or self-focusing instability discussed in Section 2h, in general the linear terms in the envelope equation for A_1 must also include dispersion and diffraction effects.

Four-wave mixing can be used to create a *phase conjugated* wave. Imagine the four wavetrains $A_j \exp i(\vec{k}_j \cdot \vec{x} - \omega_j t) + (*) j = 1, \ldots, 4$, of Figure 2.8 interacting in a nonlinear medium with $\chi^{(3)}$ nonlinearity. Wavetrains 1 and 2 are taken to be pump beams with amplitudes A_1 and A_2 sufficiently large that they can be taken constant over the interaction. We choose $\vec{k}_1 + \vec{k}_2 = 0$ and $\omega_1 = \omega_2 = \omega$. Wavetrain 3 is the signal and 4 is the phase conjugated wave $\vec{k}_4 = -\vec{k}_3$ and $\omega_4 = \omega_3 = \omega$. We can create the phase $\vec{k}_4 \cdot \vec{x} - \omega t$

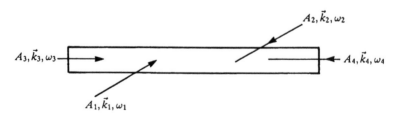

FIGURE 2.8   The making of a phase conjugated wave.

of the phase conjugated wave by the cubic interactions, (1) self-modal with coefficient proportional to $i\chi^{(3)}|A_4|^2 A_4$, (2) mutual modal with coefficients $i\chi^{(3)}(|A_1|^2, |A_2|^2, |A_3|^2)A_4$, and finally (3) four-wave resonance with coefficient proportional to $i\chi^{(3)} A_1 A_2 A_3^*$. The first two of these interactions serve merely to change the phase of A_4. The last gives a genuine energy exchange between the signal and the phase conjugated wave. We find the coupled mode equation

$$\frac{d\,A_4}{dz} = -i\alpha A_3^*, \quad \frac{d\,A_3}{dz} = -i\alpha^* A_4^*$$

with boundary conditions $A_4(L)$ and $A_3(0)$ given and $\alpha = \omega/2\epsilon_0 c\chi^{(3)}$ is treated as a constant. Note that the relative phase between the pump waves can make α complex and energy can be transferred from the pumps (although we have not treated the loss of energy from them) to the signal so that in certain circumstances the output of the phase conjugated wave $|A_4(0)|$ can exceed its input $|A_4(L)|$. For more details, the reader should consult [3].

We discuss three- and four-wave mixing processes again in Section 6g. The reader who is interested in their relevance in gravity waves and plasma physics should consult references [28, 29].

2k. Derivation and "Universality" of the NLS Equations. [3b, 6j]

In this section, we cover three points: (1) the formal as opposed to the heuristic derivation of the NLS equation for the envelope of a traveling wave in a weakly nonlinear and dispersive medium; (2) the direct expansion of $D(-i\partial/\partial x, -i\partial/\partial y, k-i\partial/\partial z, \omega+i\partial/\partial t, |A|^2)A = 0$ and the universal nature of the NLS equation; and (3) the effects of a nonzero $\chi^{(2)}$ and the obstruction to universality.

We write (2.9) as

$$L\vec{E} \equiv \nabla^2 \vec{E} - \nabla(\nabla \cdot \vec{E}) - \frac{1}{c^2}\frac{\partial^2 \vec{E}}{\partial t^2} - \frac{1}{c^2}\frac{\partial^2}{\partial t^2}\int \chi^{(1)}(t-\tau)\vec{E}(\tau)d\tau$$

$$= \frac{1}{\epsilon_0 c^2}\frac{\partial^2 \vec{P}_{NL}}{\partial t^2} \tag{2.131}$$

with

$$\frac{1}{\epsilon_0}\vec{P}_{NL} = \int \chi^{(3)}(t-\tau_1, t-\tau_2, t-\tau_3)(\vec{E}(\tau_1) \cdot \vec{E}(\tau_2))\vec{E}(\tau_3)d\tau, d\tau_2, d\tau_3.$$

$$\tag{2.132}$$

In component form, $\vec{E} = (E_x, E_y, E_z)$, and then L is the matrix operator $L =$

$$\begin{pmatrix} \frac{\partial^2}{\partial y^2} + \frac{\partial^2}{\partial z^2} - \frac{1}{c^2}\frac{\partial^2}{\partial t^2}(1 + \chi_\bullet^{(1)}) & -\frac{\partial^2}{\partial x \partial y} & -\frac{\partial^2}{\partial x \partial z} \\ -\frac{\partial^2}{\partial x \partial y} & \frac{\partial^2}{\partial x^2} + \frac{\partial^2}{\partial z^2} - \frac{1}{c^2}\frac{\partial^2}{\partial t^2}(1 + \chi_\bullet^{(1)}) & -\frac{\partial^2}{\partial y \partial z} \\ -\frac{\partial^2}{\partial x \partial z} & -\frac{\partial^2}{\partial y \partial z} & -\frac{\partial^2}{\partial x^2} + \frac{\partial^2}{\partial y^2} - \frac{1}{c^2}\frac{\partial^2}{\partial t^2}(1 + \chi_\bullet^{(1)}) \end{pmatrix}$$

$$(2.133)$$

where $1 + \chi_\bullet^{(1)}$ acting on E_x means $E_x + \int \chi^{(1)}(t - \tau)E_x(\tau)d\tau$.

We seek solutions of (2.131) as

$$\vec{E} = \mu \vec{E}_0 + \mu^2 \vec{E}_1 + \mu^3 \vec{E}_2 + \mu^4 \vec{E}_3 + \cdots \qquad (2.134)$$

where $\mu(0 < \mu \ll 1)$ measures the field amplitude and $\vec{E}_0$ has the form of a singly polarized wavepacket whose direction of polarization $\hat{e} = e_x \hat{i} + e_y \hat{j}$ is perpendicular to the direction of propagation,

$$\vec{E}_0 = \hat{e}(B(X = \mu x, Y = \mu y, Z = \mu z, T = \mu t)e^{i(kz - \omega t)} + (*)), \qquad (2.135)$$

so that all derivatives of B are also of order μ. We will choose the slow variation of B,

$$\frac{\partial B}{\partial Z} = \frac{\partial B}{\partial Z_1} + \mu \frac{\partial B}{\partial Z_2} + \cdots, \qquad (2.136)$$

that is, the coefficients $Z_j = \mu^j z$, $j = 1, 2 \ldots$ in the asymptotic expansion for $\partial B / \partial Z$ (see Section 6g) in order to suppress any terms proportional to z in $\vec{E}_1, \vec{E}_2$ that would otherwise appear.

We now work out the result of applying the operator L to the wavepacket $\vec{E}_0$. Recall that in Exercise 2.10, we saw that

$$-\frac{1}{c^2}\frac{\partial^2}{\partial t^2} - \frac{1}{c^2}\frac{\partial^2}{\partial t^2}\int \chi^{(1)}(t - \tau)d\tau \qquad (2.137)$$

applied to $\vec{E}_0$ gave

$$e^{i(kz - \omega t)}k^2\left(\omega + i\mu\frac{\partial}{\partial T}\right)B(X, Y, Z, T) + (*) \qquad (2.138)$$

where

$$k^2(\omega) = \frac{n^2(\omega)\omega^2}{c^2}, \quad n^2(\omega) = 1 + \hat{\chi}'(\omega).$$

Therefore

$$L\vec{E}_0 = e^{i(kz-\omega t)}(\mu L_0 B\hat{e} + \mu^2 L_1 B\hat{e} + \mu^3 L_2 B\hat{e} + \cdots) + (*), \qquad (2.139)$$

where

$$L_0 = \begin{pmatrix} -k^2 + k^2(\omega) & 0 & 0 \\ 0 & -k^2 + k^2(\omega) & 0 \\ 0 & 0 & k^2(\omega) \end{pmatrix} \qquad (2.140)$$

$$L_1 = \begin{pmatrix} 2ik\frac{\partial}{\partial Z_1} + 2ikk'\frac{\partial}{\partial T} & 0 & -ik\frac{\partial}{\partial X} \\ 0 & 2ik\frac{\partial}{\partial Z_1} + 2ikk'\frac{\partial}{\partial T} & -ik\frac{\partial}{\partial Y} \\ -ik\frac{\partial}{\partial X} & -ik\frac{\partial}{\partial Y} & 2ikk'\frac{\partial}{\partial T} \end{pmatrix} \qquad (2.141)$$

$$L_2 = \begin{pmatrix} -\frac{\partial^2}{\partial Y^2} + a2ik\frac{\partial}{\partial Z_2} \\ +\frac{\partial^2}{\partial Z_1^2} - (kk'' + k'^2)\frac{\partial^2}{\partial T^2} & -\frac{\partial^2}{\partial X \partial Y} & -\frac{\partial^2}{\partial X \partial Z_1} \\[4pt] -\frac{\partial^2}{\partial X \partial Y} & \begin{matrix}\frac{\partial^2}{\partial X^2} + 2ik\frac{\partial}{\partial Z_2} \\ +\frac{\partial^2}{\partial Z_1^2} - (kk'' + k'^2)\frac{\partial^2}{\partial T^2}\end{matrix} & -\frac{\partial^2}{\partial Y \partial Z_1} \\[4pt] -\frac{\partial^2}{\partial X \partial Z_1} & -\frac{\partial^2}{\partial Y \partial Z_1} & \begin{matrix}\frac{\partial^2}{\partial X^2} + \frac{\partial^2}{\partial Y^2} \\ -(kk'' + k'^2)\frac{\partial^2}{\partial T^2}\end{matrix} \end{pmatrix} \qquad (2.142)$$

In (2.140), (2.141), and (2.142),

$$k(\omega) = \frac{n(\omega)\omega}{c}, \quad k' = \frac{\partial}{\partial \omega}\left(\frac{n\omega}{c}\right), \quad k'' = \frac{\partial^2}{\partial \omega^2}\left(\frac{n\omega}{c}\right).$$

The terms $2ikk'\partial/\partial T$ and $(-kk'' - k'^2)\partial^2/\partial T^2$ come from expanding $k^2(\omega + i\mu\partial/\partial T)$ in a Taylor series in $\partial/\partial T$. We will also shortly find that

$$\frac{\partial B}{\partial Z_1} + k'\frac{\partial B}{\partial T} = 0, \qquad (2.143)$$

reflecting the fact that to leading order, the envelope travels with the group velocity, and because of this the (1.1) and (2.2) elements in L_2 simplify to

$$\frac{\partial^2}{\partial Y^2} + 2ik\frac{\partial}{\partial Z_2} - kk''\frac{\partial^2}{\partial T^2} \text{ and } \frac{\partial^2}{\partial X^2} + 2ik\frac{\partial}{\partial Z_2} - kk''\frac{\partial^2}{\partial T^2}$$

respectively.

Before we proceed, recall from Section 2e that the term in $(1/\epsilon_0 c^2)\frac{\partial^2}{\partial t^2}\vec{P}_{NL}(\vec{E}_0)$ that is proportional to $\exp(ikz - i\omega t)$ is (write $\hat{\chi}^{(3)}(-\omega, \omega, \omega)$ as $\hat{\chi}^{(3)}$)

$$-\mu^3 \frac{3\omega^2}{c^2}\hat{\chi}^{(3)}B^2 B^*\hat{e}. \tag{2.144}$$

In solving for the iterates $LE_j = F_j$, $j = 1, 2, \ldots$, we must remove from F_j any term proportional to ($\hat{i}$ or $\hat{j}$) $\exp \pm i(kz - \omega t)$ because these vectors lie in the null space of L. If F_j does contain such a term, then the corresponding solution E_j would be $(\hat{i}, \hat{j})z \exp \pm i(kz - \omega t)$ and thus render the asymptotic expansion (2.134) nonuniform.

We now solve (2.131) iteratively:

$$0(\mu) : L_0\vec{E}_0 = 0. \tag{2.145}$$

This is satisfied if $\hat{e} \cdot \hat{z} = 0$ and $k = k(\omega)$.

$0(\mu^2) : L\vec{E}_1 = -L_1\vec{E}_0$

$$= -e^{i(kz-\omega t)}\left\{\left(2ik\frac{\partial B}{\partial Z_1} + 2ikk'\frac{\partial B}{\partial T}\right)\hat{e} - ik\hat{z}\nabla B \cdot \hat{e}\right\} + (*). \tag{2.146}$$

To solve (2.146) with an $\vec{E}_1$ that does not grow like z, we must choose $\partial B/\partial Z_1$ to satisfy (2.143). Then we find

$$\vec{E}_1 = \hat{z}\frac{i}{k}\nabla B \cdot \hat{e}e^{i(kz-\omega t)} + (*). \tag{2.147}$$

Observe that $\nabla \cdot \vec{E} = \nabla \cdot (\mu\vec{E}_0 + \mu^2\vec{E}_1 + \cdots)$ is zero to order μ^3.

$0(\mu^3) : L\vec{E}_2 = -L_2\vec{E}_0 - L_1\vec{E}_1$

$$-\left(\frac{3\omega^2}{c^2}\hat{\chi}^{(3)}B^2 B^* e^{i(kz-\omega t)} + \frac{3\omega^2}{c^2}\hat{\chi}^{(3)}BB^{*2}e^{-i(kz-\omega t)}\right)\hat{e}$$

$$+ \text{ third harmonics.} \tag{2.148}$$

We now must choose $\partial B/\partial Z_2$ so that all terms on the right-hand side of (2.148) are proportional to either $\hat{i}e^{\pm i(kz-\omega t)}$ or $\hat{j}e^{\pm i(kz-\omega t)}$ are zero. Writing

$\hat{e} = e_x \hat{i} + e_y \hat{j}$, we find that the coefficient of $\hat{i} e^{i(kz-\omega t)}$ is

$$-\left(\frac{\partial^2 B}{\partial Y^2} + 2ik\frac{\partial B}{\partial Z_2} - kk''\frac{\partial^2 B}{\partial T^2}\right)e_x + \frac{\partial^2 B}{\partial X \partial Y}e_y$$

$$-\left(\frac{\partial^2 B}{\partial X^2}e_x + \frac{\partial^2 B}{\partial X \partial Y}e_y\right) - \frac{3\omega^2}{c^2}\hat{\chi}^{(3)}B^2 B^* e_x$$

$$= \left(-2ik\frac{\partial B}{\partial Z_2} - \nabla_\perp^2 B + kk''\frac{\partial^2 B}{\partial T^2} - \frac{3\omega^2}{c^2}\hat{\chi}^{(3)}B^2 B^*\right)e_x. \qquad (2.149)$$

Note how terms from $L_0\vec{E}_2$ and $L_1\vec{E}_1$ proportional to e_y cancel and how the extra $\partial^2 B/\partial X^2$ in the diffraction term is supplied by $L_1\vec{E}_1$. The coefficient of $\hat{j}e^{i(kz-\omega t)}$ is the same and the coefficients of $\hat{i}$, $\hat{j}e^{i(kz-\omega t)}$ are the complex conjugates. Thus

$$\frac{\partial B}{\partial Z_2} = \frac{i}{2k}\left(\frac{\partial^2 B}{\partial X^2} + \frac{\partial^2 B}{\partial Y^2}\right) - \frac{ik''}{2}\frac{\partial^2 B}{\partial T^2} + ik\frac{n_2}{n}|B|^2 B, \qquad (2.150)$$

where $2nn_2 = 3\chi^{(3)}$, so that

$$\frac{\partial B}{\partial Z} = -k'\frac{\partial B}{\partial T} + \mu\left(\frac{i}{2k}(\frac{\partial^2 B}{\partial X^2} + \frac{\partial^2 B}{\partial Y^2}) - \frac{i}{2}k''\frac{\partial^2 B}{\partial T^2} + ik\frac{n_2}{n}|B|^2 B\right). \qquad (2.151)$$

One can remove the μ by expressing (2.151) in the coordinates $\zeta = \mu Z\mu^2 z$, $\tau = T - k'Z = \mu(t - k'z)$, or one can recover (2.91) by writing $X, Y, Z, T = \mu(x, y, z, t)$ and $A = \mu B$.

In deriving (2.151), the important ingredients were, first, that L is strongly dispersive, namely $k = k(\omega)$ has the property that $k(n\omega) - nk(\omega)$ is of order one, $n = 2, 3, \ldots$ (because this factor with $n = 2$ and 3 respectively appears in the denominators when we calculate E_1 (if $\chi^{(2)} \neq 0$) and E_2), second, that the original equation (2.131) is weakly nonlinear due to either $\hat{\chi}^{(3)}$ or $|\vec{E}|$ being small, and, third, that $\vec{E}$ is almost monochromatic and therefore its envelope B is slowly varying. The fact that the NLS equation is "universal" can be seen by observing that it is simply the nonlinear dispersion relation applied to a wavepacket of the carrier wave $\exp(ik_0z - i\omega_0 t)$. The only possible obstruction occurs when a constant field is a solution of the linear operator L, as it is in (2.131), and $\chi^{(2)}$ is nonzero. In general, what we find is that the field envelope satisfies the NLS equation (2.151) with an additional term proportional to $i\mu CB$ and an additional equation for C containing slow gradients of $B B^*$.

What we do now is first derive NLS from

$$D\left(-\mu\frac{\partial}{\partial X}, -i\mu\frac{\partial}{\partial Y}, k_0 - i\mu\frac{\partial}{\partial Z_1} - i\mu^2\frac{\partial}{\partial Z_2}, \omega_0 + i\mu\frac{\partial}{\partial T}, \mu^2|B|^2\right)$$

$$B = 0, \qquad\qquad (2.152)$$

and then go back to understand which terms might be incorrectly ignored if $\chi^{(2)} \neq 0$.

Expand (2.152) in a Taylor series; D_j, $j = 1, 2, 3, 4, 5$ stands for partial derivatives of D with respect to the arguments and D_{jk}, $j, k = 1, 2, 3, 4, 5$ refers to second partial derivatives. Because we need relations between the various partial derivatives, we use the fact that the nonlinear dispersion relation

$$D(l, m, k(l, m, \omega, |A|^2), \omega, |A|^2) = 0 \qquad\qquad (2.153)$$

holds for all $l, m, \omega, |A|^2$ and expresses k in terms of these variables. Differentiate (2.153) with respect to l, m, ω, and $|A|^2$ and find

$$D_1 + D_3\frac{\partial k}{\partial l} = 0, \quad D_2 + D_3\frac{\partial k}{\partial m} = 0, \quad D_4 + D_3\frac{\partial k}{\partial \omega} = 0, \quad D_5 + \frac{\partial k}{\partial |A|^2}$$

$$D_3 = 0. \qquad\qquad (2.154)$$

Now we assume here that D depends on l, m, k through $l^2 + m^2 + k^2$ so that $\partial k/\partial l$ and $\partial k/\partial m$ are zero when evaluated on $l = m = 0$. Therefore $D_1 = D_2 = 0$ on $l = m = 0$. By differentiating the first equation in (2.154) with respect to m and setting $l = m = 0$ we find $D_{12} = 0$; also $D_{11} = -D_3(\partial^2 k/\partial l^2) = D_3(1/k)$ and $D_{22} = -D_3(\partial^2 k/\partial m^2) = D_3(1/k)$. Expanding (2.152), we find to order μ,

$$D(0, 0, k_0, \omega_0, 0)B - i\mu D_3\frac{\partial B}{\partial Z_1} + i\mu D_4\frac{\partial B}{\partial T} = 0,$$

which, using the fact that $D(0, 0, k_0, \omega_0, 0) = 0$ and the third relation in (2.154), gives us that

$$\frac{\partial B}{\partial Z_1} + \frac{\partial k}{\partial \omega}\frac{\partial B}{\partial T} = 0. \qquad\qquad (2.155)$$

Next, differentiate $D_4 + D_3(\partial k/\partial \omega) = 0$ with respect to l and use (2.154) to find $D_{41} = -D_{31}(\partial k/\partial \omega)$. Likewise, $D_{42} = -D_{32}(\partial k/\partial \omega)$. Finally, differentiate $D_4 + D_3(\partial k/\partial \omega)$ with respect to ω and find $D_{44} + D_{34}(\partial k/\partial \omega) +$

$D_{33}(\partial k/\partial\omega)^2 = -D_3(\partial^2 k/\partial\omega^2)$. Expanding (2.152) to order μ^2 and using these relationships, we obtain

$$-i\mu^2 D_3 \left(\frac{\partial B}{\partial Z_2} - \frac{i}{2k}\frac{\partial^2 B}{\partial X^2} - \frac{i}{2k}\frac{\partial^2 B}{\partial Y^2} + \frac{i}{2}\frac{\partial^2 k}{\partial\omega^2}\frac{\partial^2 B}{\partial T^2} - i\frac{\partial k}{\partial|A|^2}|B|^2 B \right),$$

(2.156)

where $\partial^2 k/\partial\omega^2$ and $\partial k/\partial|A|^2$ are evaluated at $l = m = |A|^2 = 0$. Hence we recover (2.150). It appears, then, that the NLS equation is simply the nonlinear dispersion relation for a weakly nonlinear wavepacket and its coefficients are always given by derivatives of k with respect to l, m, ω and $|A|^2$. In this sense, it is universal.

Now let us see what might go wrong with this simple derivation. For the sake of illustration we take $\vec{E} = (0, E(x, z, t), 0)$ and

$$\frac{1}{\epsilon_0}\vec{P}_{NL} = \hat{y}\left(\chi_\bullet^{(2)} EE + \chi_\bullet^{(3)} EEE\right),$$

(2.157)

where by $\chi_\bullet^{(2)} EE + \chi_\bullet^{(3)} EEE$ we mean the last two terms on the right-hand side of (2.22). The potential difficulty comes from the term

$$\mu^4 \frac{\partial^2}{\partial T^2}\left(2\hat{\chi}^{(2)}(\omega, -\omega)BB^*\right)$$

(2.158)

which appears at order μ^4 and therefore, one might think, can be ignored. The reason it cannot be ignored is that the constant state is a doubly degenerate solution of the homogeneous equation

$$\nabla^2 E - \frac{1}{c^2}\frac{\partial^2}{\partial t^2}E - \frac{1}{c^2}\frac{\partial^2}{\partial t^2}\int \chi^{(1)}(t-\tau)E(\tau)d\tau = 0$$

(2.159)

and thus there will be terms in E_3 (the coefficient of μ^4 in the expansion (2.134)) proportional to z^2, which is unacceptable. The only way to remove this secular term is to add the (almost) constant solution $C(X, Z, T)$ to E_1. One finds

$$E_1 = C(X, Z, T) + \frac{4\omega^2\hat{\chi}^{(2)}(\omega, \omega)}{c^2(4k^2(\omega) - k^2(2\omega))}B^2 e^{2ikz - 2i\omega t} + (*).$$

(2.160)

This leads to the additional terms

$$\left(-\frac{\omega^2}{c^2}\hat{\chi}^{(2)}(\omega, 0)BC - \frac{4\omega^4}{c^4}\frac{\hat{\chi}^{(2)}(2\omega, \omega)\hat{\chi}^{(2)}(\omega, \omega)}{4k^2(\omega) - k^2(2\omega)}B^2 B^*\right)e^{ikz - i\omega t} + (*)$$

(2.161)

in $(1/c^2)(\partial^2/\partial t^2)2\chi^{(2)}E_0E_1$, which appears at order μ^3 in the determination of E_2. Then the new solvability condition for B is

$$\frac{\partial B}{\partial Z} + k'\frac{\partial B}{\partial T} + \mu\left(-\frac{i}{2k}\frac{\partial^2 B}{\partial X^2} + \frac{ik''}{2}\frac{\partial^2 B}{\partial T^2} - i\,n_2\frac{\omega}{c}B^2B^*\right.$$

$$\left. - i\frac{\omega}{c}\frac{\hat{\chi}^{(3)}(\omega,0)}{n}BC\right) = 0 \qquad (2.162)$$

with

$$2nn_2 = 3\hat{\chi}^{(3)}(-\omega,\omega,\omega) + 4\frac{\omega^2}{c^2}\frac{\hat{\chi}^{(2)}(2\omega,-\omega)\hat{\chi}^{(2)}(\omega,\omega)}{4k^2(\omega) - k^2(2\omega)}. \qquad (2.163)$$

Observe that this correction to $\hat{\chi}^{(3)}$ would have been captured by the analysis of Section 2e and so is included in the nonlinear dispersion relation (2.153).

The terms involving C, however, would have been ignored. The equation for $C(X, Z, T)$ is found at order μ^4,

$$\frac{\partial^2 C}{\partial X^2} + \frac{\partial^2 C}{\partial Z^2} - \frac{n^2(0)}{c^2}\frac{\partial^2 C}{\partial T^2} = \frac{2\hat{\chi}^{(2)}(\omega,-\omega)}{c^2}\frac{\partial^2}{\partial T^2}BB^*, \qquad (2.164)$$

where $n^2(0) = 1 + \hat{\chi}(0)$. If we solve the full vector equation (2.131), we simply replace $\partial^2/\partial X^2$ by the horizontal Laplacian $\nabla_\perp^2$ in both (2.162) and (2.164).

These equations, (2.162) and (2.164), are the analog in optics of the Zakharov equations in plasma physics and the Benney–Roskes, Davey–Stewartson equations for water waves. The strong coupling to the mean or *DC* field occurs because the *DC* field is an exact solution of the linearized equation. These terms are often overlooked because even though their effect is felt at the same order as the nonlinear n_2 effect, the determination of the evolution of $C(X, Z, T)$ occurs at an order higher than the orders at which the evolution of $B(X, Z, T)$ is found.

Potentially, these relations can have an enormous impact in optics. For example, let us imagine that we are in a fiber waveguide (see Section 3c for the derivation of NLS in this case) made from a nonlinear material with a nonzero $\chi^{(2)}$. The linear dispersion relation will read $k(\omega) = \beta(\omega)\omega/c$ where $\beta(\omega)$ is the effective linear refractive index. Because BB^* is a function of $T - k'Z$ to leading order, there is the solution

$$C = \frac{2\hat{\chi}^{(2)}(\omega,-\omega)BB^*}{k'^2 - \frac{n^2(0)}{c^2}} \qquad (2.165)$$

(although in general C cannot be solved as an algebraic function of $B B^*$), and this leads to an effective n_2 in (2.162) of

$$n_2 \to n_2 + \frac{2 \hat{\chi}^{(2)}(\omega, 0) \, \hat{\chi}^{(2)}(\omega, -\omega)}{n} \frac{}{k'^2 - \frac{n^2(0)}{c^2}}. \tag{2.166}$$

Observe that even for small $\hat{\chi}^{(2)}$, the possibility of a large n_2 occurs because of the denominator $k'^2 - (n^2(0)/c^2)$, which is the difference of the square of the inverse group velocity v_g of the carrier wave with frequency ω and the square of the inverse of the phase velocity v_p of "long" electromagnetic waves,

$$\frac{1}{v_g^2} - \frac{1}{v_p^2}. \tag{2.167}$$

It is well known that there can be a strong interaction between long (C) and short (B) waves when the group velocity of the latter is the phase velocity of the former. This is simply a form of the three-wave mixing criterion (2.128) with $\omega_1 = \omega - \mu\Omega, \omega_2 = -\omega - \mu\Omega, \omega_3 = 2\mu\Omega$ whence $k_1 + k_2 + k_3 = k(\omega - \mu\Omega) - k(\omega + \mu\Omega) + k(2\mu\Omega) = -2\mu\Omega k'(\omega) + k(2\mu\Omega) + 0(\mu^3)$, which is zero to order μ^3 when $k'(\omega) = k(2\mu\Omega)/2\mu\Omega = n(2\mu\Omega)/c$. By choosing the vacuum wavelength $\lambda = 2\pi c/\omega$ appropriately, it should be possible to make the denominator in (2.166) very small and the effective n_2 large.

As far as we know, the possibility of inducing an effective large n_2 using these ideas has not yet been explored.

Chapter 3

COMMUNICATIONS IN OPTICAL FIBERS
AND NONLINEAR WAVEGUIDES

Overview

OUR GOAL IN THIS CHAPTER is to introduce the reader to the roles that the soliton of the one-dimensional nonlinear Schrödinger (NLS) equation will play in fiber optics communications and in slab waveguides.

3a. We begin with some introductory remarks.

3b. We derive in great detail the NLS equation for monomode, polarization preserving fibers. It is also clear from this derivation how to include the effects of attenuation and to calculate the higher-order correction terms such as third-order dispersion, higher-order nonlinear dispersion, and the effect of delay in the nonlinear refractive index.

3c. We discuss nonlinear fiber optics, the possibilities, the expectations, and the challenges.

3d. We review the principles of waveguides.

3e. We introduce the TE and TM modes of a nonlinear planar waveguide.

3f. The statics of nonlinear surface and guided TE waves. In this section we learn how to compute the shapes of guided TE waves that are attached and propagate parallel to an interface that separates two dielectrics with different nonlinear refractive indices.

3g. The dynamics of nonlinear surface and guided TE waves. In this section, we develop the parallel of Snell's laws (French readers read Descartes' laws) for a single interface separating nonlinear dielectrics by addressing such questions as: What happens to a light beam that strikes the interface at an angle? Does it reflect? Transmit? To help answer these questions, we introduce a powerful tool that enormously simplifies analysis and is extremely valuable in obtaining both qualitative and quantitative information. In particular, it is very easy to read off from this picture the stability properties of guided waves.

It may also be used to help design all sorts of optical devices such as scanners and switches.

3h, i. Finally we calculate the shapes of guided TE and TM waves for the case of the symmetric planar waveguide.

3a. Communications

People have always been interested in the rapid communication of intelligent information from one place to another. At first, there were runners and riders, drums and signal fires. The problem with the former modes of communication was that they were too slow. The latter suffered from a nonliteral basis and relied upon the prearranged meaning of signals. It was therefore difficult to alter the contents of or add additional information to the message. A lighted fire might mean to a man fighting a distant battle that his wife had just given birth to twins, but it would be difficult also to communicate the sad news that she had run off with the local butcher.

More sophisticated coding was introduced by Claude Chappe who invented an optical relay semaphore or visual telegraph system. Semaphore literally means bearing a sign and telegraph means distant writing. This means of information transmission consisted of a series of hilltop towers on the top of each of which a vertical timber supported a horizontal beam called the regulator, at the ends of which were affixed two indicator arms that could assume seven angular positions 45° apart. The forty-nine positions stood for alphabetic letters and symbols. Also the regulator beam could be tilted from the horizontal 45° to vertical. Towers, each equipped with up-line and down-line telescopes, were spaced five to ten kilometers apart. At a rate of three signals per minute, the transit time from Toulon to Paris, through 120 towers, was forty minutes, or an hour for a fifty-signal message, and was more than ninety times as fast as mounted couriers or any state operated postal system in service today.

As with many modern inventions, the semaphore or visual telegraph was greatly appreciated and supported by the military establishments of the various powers. It was particularly valuable in military engagements at sea where the first use of flag signals is often credited to the Duke of York, later James II of England.* Despite the success of signaling, this method

---

*Of course many accreditations are apocryphal and spurious and if the same gentleman were also to accept the often attributed blame of his screw-up in Ireland, on balance the ledger sheet would not be in his favor. The end of the reign of the House of Stuart on the throne of England became effective at the Battle of the Boyne, an event celebrated every year by bowler hatted and orange sashed Protestant Ulstermen as an act sometimes interpreted as one of deliberate provocation by their

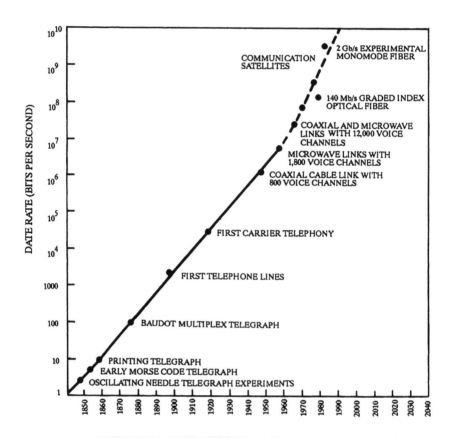

HISTORY OF IMPORTANT DEVELOPMENTS IN COMMUNICATIONS

FIGURE 3.1   Bit rates in communication systems over the last two centuries. Note the break in the curve roughly coincident with the invention of the laser.

of information was severely limited in many aspects, notably by its lack of speed and requirement for good visibility. However, it remained popular for a half-century until eventually replaced by the first step in modern telegraphy, the wire telegraph, which could send electrical signals at a much more rapid rate. Starting with the Morse code, the advance in information transmission rate, shown here in Figure 3.1, has been no less than staggering.

In a series of developments ranging from the use of wires to send signals along, wire telegraphy, to wireless or radio telegraphy, we have achieved an increase of communication rates, as measured in bits per second, by a factor

Catholic countrymen. In Dublin, immediately after the battle, James told Lady Hamilton he had lost because the cowardly Irish had run away, to which excuse the lady replied. "It would appear that your Majesty won the race." Meanwhile, back on the battlefield, the vanquished Irish taunted the victorious troops of William of Orange, "Change kings and we'll fight ye ag'in."

of one billion. The latest in this series of advances is an optical fiber system in which large amounts of information, coded as light pulses, are passed along silica fibers no thicker than a human hair. Presently most of the western world is beginning to enjoy the fruits of this communication revolution. Typically information transmission rates are of the order of 5×10^8 bits per second. Signals race along the eastern corridors of the United States, between most major Japanese, eastern Australian, and British cities, and the first transatlantic link has been established. In the layman's vernacular, this means that one can send all the information in the Encyclopaedia Brittanica from New York to London in one second. As marvelous as these advances have been, the present system still only uses a tiny fraction of the full information carrying potential of optical fibers. To see what the next step in this revolution might be, we now take a look at how fibers work and what are the main difficulties of optical transmission and the possible means by which they may be overcome. In particular we discuss how nonlinear pulses promise to revolutionize telecommunications.

The cross-section of the optical fiber is shown in Figure 3.2. The inner core consists of a special form of silica glass with very low absorption and is between 10 and 50 micrometers in diameter. This is surrounded by a glass cladding whose refractive index n_c is very close to but slightly less than n_0, the linear refractive index of the inner core. This ensures that one, the wave is guided in (its intensity is largely confined to) the inner core, and two, that the fiber only supports one (axially symmetric) mode of propagation.

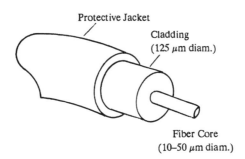

FIGURE 3.2   Optical fiber.

The basic idea of using optical fibers for communication is relatively simple. The message is coded in binary by representing a one as a pulselike modulation of a carrier wave whose wavelength is in the micrometer (10^{-6} m) range and whose frequency is in the femtosecond (10^{-15} s) range and a zero by the absence of such a pulse. The arrangement is shown in Figure 3.3. The pulses would be approximately 10–25 picoseconds (1 ps $= 10^{-12}$ s) wide and

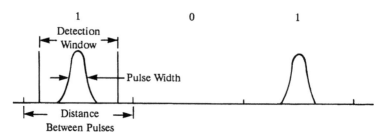

FIGURE 3.3   Soliton train.

the average distance between them would be of the order of ten times that amount. A detection-window might be of the order of seven times the pulse width. Experimental fibers have managed effective transmission rates in the gigabit range (10^9 bits/s).

The three enemies of communication in optical fibers are

- attenuation
- dispersion
- electronic switching

It is the solution of the first problem that has been principally responsible for the recent success of present-day optical fibers. Attenuation, usually denoted by γ, is measured in dB/km,

$$\gamma = -10 \log_{10} \left(\frac{\text{power after 1 km}}{\text{initial power}} \right) \tag{3.1}$$

or

$$P(z \text{ km}) = P(z = 0) \exp \left(-\frac{\gamma}{10} \ln 10 \right) z \tag{3.2}$$

where z is measured in kilometers. The rate of decrease in γ as a function of year since 1968 is 1000-fold. The main losses are due to Rayleigh scattering, absorption, and scattering due to water contamination and silica molecular bands. Ordinary glass has a $\gamma = 4000$ dB/km; light is attenuated in distances of less than a meter. Modern technology, however, has succeeded in purifying silica glass to the point where at vacuum wavelengths λ ($\lambda = 2\pi c/\omega$, ω, the light frequency) of 1.3 μm or 1.55 μm, the attenuation constant γ can be as low as 0.13 dB/km (see Figure 3.4). At this value, a signal will travel roughly 40 km before 80% of its power is lost.

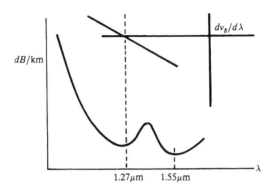

FIGURE 3.4   Schematic drawing of attenuation and dispersion as a function of vacuum wavelength.

The second enemy of optical transmission, and an increasing threat the higher the transmission rate, is dispersion. As we discussed in Chapter 2, the envelope of a wavepacket pulse of width t spreads out (doubles its width) in a distance of order $2t_0^2/k''$ where k'' is the absolute value of the dispersion. For a picosecond wide pulse and for a typical k'' of 1 to 10 ps^2/km, this distance is of the order of a kilometer or less. At the vacuum wavelength of minimum attenuation $\lambda = 1.55$ μm, k'' is approximately 20 ps^2/km, and therefore even a pulse of width 20 ps disperses in a distance of 20 km. Also the bit rate b is clearly inversely proportional to the distance between pulses, which in turn is proportional to t_0. Therefore, the higher the desired bit rate, the smaller is the choice of t_0 and the faster the dispersion. However, we have also seen that nonlinearity can act to counterbalance dispersion, and therefore the dependence of the refractive index on intensity is extremely important. The linear equation (2.78) describing the spread of a wavepacket is thus replaced, to a good approximation, by a nonlinear Schrödinger (NLS) equation, and its special solution, the soliton, is a natural candidate to be the envelope pulse and information bit along the optical fiber. It suffers no dispersion. But, as we shall see, practical considerations will require a modification of the pure soliton.

The third enemy of communication in optical fibers is electronic switching, which places a very real upper bound of about 1 gigabit per second on communication rates. To overcome loss and dispersion, the optical signal must be detected and electronically regenerated every twenty kilometers or so. Not only is this undesirable for a single channel (a single wavelength carrier wave), but it also severely inhibits the use of multiple channels (a carrier wave containing several discrete wavelengths). It is important, therefore, to avoid electronic switching and design an all-optical system. Solitons can also play a very important part in achieving this goal.

While optical fibers have become central players in the modern telecommunications revolution, their impact on the fundamental study of nonlinear optical interactions has been no less dramatic. Much of the early excitement and intense activity associated with the discovery of the fundamental nonlinear interactions of light with matter in the 1960s and early 1970s began to wane with the realization that many of these interactions could only be achieved with the most intense light pulses from Q-switched lasers. Moreover, it became difficult to discriminate between various competing nonlinear interactions. Thus stimulated light scattering such as Raman or Brillouin scattering tended to accompany filament collapse associated with critical self-focusing. The optical fiber, with its transverse mode confinement and weak losses leading to long interaction lengths, provided the ideal experimental test bed for a careful reappraisal of previously observed nonlinear optical interactions. The subject of optics is the best experimental vehicle for exploring the unique properties of the NLS equation and its various near-integrable neighbors.

In addition, the dependence of the various stimulated scattering interactions on the product of light intensity times propagation length meant that stimulated Raman and Brillouin scattering could be readily observed at intensities ranging from watts to milliwatts per square centimeter in contrast to the MW/cm^2 peak intensities required in Q-switched laser pulses. In fact, the stimulated scattering processes were found to be so efficient in optical fibers that even higher order Stokes frequency downshifted or anti-Stokes frequency upshifted waves could be observed in the laboratory with relative ease. We discuss stimulated light scattering later in Chapters 4 and 5.

Once high-quality low-loss single-mode fibers became available in the 1980s, the number of possible applications grew in a spectacular fashion. Soliton lasers, modulational instability lasers, and SBS lasers were built and fiber loops were incorporated into various coupled-cavity arrangements. Pulse compression led to optical pulses of duration of ~6 fs, or less than 3 optical cycles. Clearly the usual envelope expansions of the electric field must be highly suspect for such ultrashort pulses. Rather surprisingly, a perturbation approach incorporating higher-order group velocity dispersion appears to account for the departure from behavior expected from the NLS equation. Another novel departure was to introduce special dopants into the fiber's glass core. Erbium-doped fibers have spectacular light amplification properties and can be used as fiber lasers or as amplification stages for pulse trains in a long telecommunication length. Another extremely useful property of the single-mode fiber is that it can support two orthogonal linear polarized states of the elective field (approximate linear superpositions of these two states of polarization give us circularly or linearly polarized states of the light field). The intrinsic birefringence of the nonlinearity in glass means that relatively short fiber lengths can be used to switch the state of polarization of incident light pulses, thereby creating a polarization switch. This latter problem is described mathematically by coupled NLS-type equations.

3b. Derivation of the NLS Equation for a Light Fiber. [2k, 6j]

This is a tough section. The reader who is willing to take the NLS equation as a given can move to Section 3c. However, the serious student who hopes to make some new discoveries in optical fibers had better understand the material in this section. So, buckle up, put the head down and go to work. Make sure, however, that you have first understood Sections 2k and 6j. References [30, 31] may also be helpful.

We begin, as always, with Maxwell's equation (2.9),

$$-\nabla \times \nabla \times \vec{E} - \frac{1}{c^2} \frac{\partial^2}{\partial t^2} \vec{E} = \frac{1}{\epsilon_0 c^2} \frac{\partial^2 \vec{P}}{\partial t^2}. \tag{3.3}$$

We assume that the material is isotropic and centrosymmetric. From (2.22),

$$\frac{1}{\epsilon_0} \vec{P}(\vec{x}, t) = \int_{-\infty}^{t} \chi^{(1)}(x_\perp, t - t_1) \vec{E}(\vec{x}, t_1) dt_1$$

$$+ \int_{-\infty}^{t} \int_{-\infty}^{t} \int_{-\infty}^{t} \chi^{(3)}(x_\perp, t - t_1, t - t_2, t - t_3)(\vec{E}(t_1)$$

$$\cdot \vec{E}(t_2))\vec{E}(t_3) dt_1 dt_2 dt_3. \tag{3.4}$$

The susceptibilities $\chi^{(1)}$, $\chi^{(3)}$ depend only on the transverse coordinates $x_\perp$ (x, y) and, in most cases, only on $r = \sqrt{x^2 + y^2}$. The nonadiabatic response, namely the fact that the polarization depends on the electric field history, is important. In the linear term, the real part of the Fourier transform $\hat{\chi}^{(1)}(\omega)$ of $\chi^{(1)}(t)$ gives rise to dispersion along the fiber; the imaginary part leads to attenuation. The delay in the nonlinear term leads to a nonlinear dispersion and nonlinear absorption. The latter we ignore altogether. The former is important when we consider ultrashort pulses, wavepackets in the tens of femtosecond range, and can be treated by perturbation theory. It leads to a downshift in the frequency of the carrier wave in the fiber.

The mathematics that follows is a bit more complicated than what the reader will meet in Section 6j, principally because the electric field can no longer be treated as a scalar quantity. Therefore, we will proceed slowly in a step by step manner. Our main goal is to obtain the linearly damped NLS equation as the leading order description of pulse propagation in the fiber. A second goal is to obtain the corrections to this equation due to pumping and higher-order linear and nonlinear dispersion. We hope that the exposition will give the reader sufficient confidence to work with the relevant perturbation terms, determine their effects, and explore the effects of the many other influences that we ignore here. We emphasize that much fundamental work remains to be done before solitary wave pulses become the standard means of communication in optical fibers.

We derive the equation in three stages. First, we do the linear problem, obtain the transverse structure of the electric field, the dispersion relation giving the mode propagation constant $\beta(k = \beta(\omega)\omega/c)$ as function of frequency and develop some useful identities analogous to equations (6.264) and (6.266). Second, we show how to build a wavepacket, a linear superposition of carrier waves with a small band (order $\epsilon, 0 < \epsilon \ll 1$) of frequencies, and compute the polarization $\vec{P}$ for a wavepacket. Third, we put these ingredients together, and, expanding the electric field vector

$$\vec{E} = \epsilon \vec{E}_0 + \epsilon^2 \vec{E}_1 + \epsilon^3 \vec{E}_2 + \cdots \qquad (3.5)$$

where $\vec{E}_0$ is a wavepacket, we show how the solvability conditions for $\vec{E}_1$ and $\vec{E}_2$ give the NLS equation for the envelope $A(z, t)$ of the wavepacket. The solvability condition for $\vec{E}_3$ gives the correction terms. Fourth, we discuss the influences of effects we ignore in stages one to three.

Stage One: The Linear Problem; the Transverse Structure of $\vec{E}_0$ and the Dispersion Relation

Consider

$$\nabla^2 \vec{E} - \nabla(\nabla \cdot \vec{E}) - \frac{1}{c^2} \frac{\partial^2}{\partial t^2} \left(\vec{E} + \int_{-\infty}^{t} \chi^{(1)}(x_\perp, t - t_1) \vec{E}(t_1) \, dt \right) = 0 \quad (3.6)$$

The Fourier transform of (3.6) is

$$\left(\nabla^2 + \frac{n^2(x_\perp, \omega)\omega^2}{c^2} \right) \hat{E} - \nabla(\nabla \cdot \hat{E}) = 0 \qquad (3.7)$$

where

$$\hat{E}(\vec{x}, \omega) = \int_{-\infty}^{\infty} \vec{E}(\vec{x}, t)e^{i\omega t} \, dt, \quad \vec{E}(x, t) = \frac{1}{2\pi} \int_{-\infty}^{\infty} \hat{E}(\vec{x}, \omega)e^{-i\omega t} \, d\omega.$$

We take the refractive index $n^2(x_\perp, \omega) = 1 + \mathrm{Re}\hat{\chi}^{(1)}(x_\perp, \omega)$ to be real. We will take account of its imaginary part in stage four. Because $\chi^{(1)}(t)$ is real, $\hat{\chi}^{(1)*}(\omega) = \hat{\chi}^{(1)}(-\omega)$ and therefore $\mathrm{Re}\hat{\chi}^{(1)}(x_\perp, \omega)$ is an even function of ω. We take $n(x_\perp, \omega)$ to be axisymmetric, that is, $n(x_\perp, \omega) = n(r = \sqrt{x^2 + y^2}, \omega)$. Its graph is shown in Figure 3.5. The transition at the boundaries of core and cladding and air may be sharp. Equation (3.7) can be written

$$\tilde{L}\left(\frac{\partial}{\partial x}, \frac{\partial}{\partial y}, \frac{\partial}{\partial z}, -i\omega\right)\hat{E}(x_\perp, z, \omega) = 0 \tag{3.8}$$

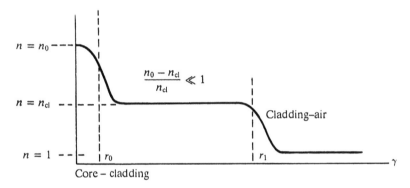

FIGURE 3.5   Refractive index in a glass fiber.

where $\tilde{L}$ is the real, symmetric operator

$$\tilde{L} = \begin{pmatrix} \dfrac{\partial^2}{\partial y^2} + \dfrac{\partial^2}{\partial z^2} + \dfrac{n^2\omega^2}{c^2} & -\dfrac{\partial^2}{\partial x \partial y} & -\dfrac{\partial^2}{\partial x \partial z} \\[2mm] -\dfrac{\partial^2}{\partial x \partial y} & \dfrac{\partial^2}{\partial x^2} + \dfrac{\partial^2}{\partial z^2} + \dfrac{n^2\omega^2}{c^2} & -\dfrac{\partial^2}{\partial y \partial z} \\[2mm] -\dfrac{\partial^2}{\partial x \partial z} & -\dfrac{\partial^2}{\partial y \partial z} & \dfrac{\partial^2}{\partial x^2} + \dfrac{\partial^2}{\partial y^2} + \dfrac{n^2\omega^2}{c^2} \end{pmatrix}, \tag{3.9}$$

and $\hat{E}$ is a column vector. We seek bounded solutions of (3.8) that decay to zero as $x^2 + y^2 \to \infty$. Accordingly, we write

$$\hat{E}(x_\perp, z, \omega) = e^{ikz}U(x_\perp, k, \omega). \tag{3.10}$$

The vector $U(x_\perp, k, \omega)$ satisfies

$$L\left(\frac{\partial}{\partial x}, \frac{\partial}{\partial y}, ik, -i\omega\right) U$$

$$\equiv \begin{pmatrix} \frac{\partial^2}{\partial y^2} - k^2 + \frac{n^2\omega^2}{c^2} - \frac{\partial^2}{\partial x\partial y} & -ik\frac{\partial}{\partial x} \\ -\frac{\partial^2}{\partial x\partial y} & \frac{\partial^2}{\partial x^2} - k^2 + \frac{n^2\omega^2}{c^2} & -ik\frac{\partial}{\partial y} \\ -ik\frac{\partial}{\partial x} & -ik\frac{\partial}{\partial y} & \frac{\partial^2}{\partial x^2} + \frac{\partial^2}{\partial y^2} + \frac{n^2\omega^2}{c^2} \end{pmatrix} U = 0$$

$$(3.11)$$

and k as a function of ω is determined as a spectral parameter. For a finite discrete set of values $k = \{k_j(\omega)\}_{j=1}^N$, there exist nontrivial solutions $U_j(x_\perp, k, \omega)$ of (3.11) which decay to zero as $x^2 + y^2 \to \infty$. We now assume that the fiber is monomode, namely that the refractive index mismatch, $n_0^2 - n_{cl}^2$, between core and cladding is sufficiently small so that there is only one confined mode solution $U(x_\perp, \omega)$ supported by the fiber (see Section 6f). What we mean by this, when the refractive index is axisymmetric, is that only the lowest mode in terms of radial structure is supported. The circular symmetry means that there is a one-parameter family of such modes, parameterized by the direction of polarization, that is, the ratio of the strengths in the transverse fields in, say, the x and y directions. For simplicity, we will therefore make a further assumption, that the fiber is polarization preserving, that is, if we choose a particular direction of polarization for the field at the beginning of the fiber, that polarization will persist. We will reexamine this assumption in stage four later. The weak guiding assumption also means that $\partial/\partial x$ and $\partial/\partial y$ are of the order $\sqrt{n_0 - n_{cl}}$ so that the third (z) component U_3 of U is much smaller than the transverse components U_1, U_2.

So, for the present, we assume a monomode, polarization preserving fiber which, for a given frequency ω, can support a transversely confined propagating mode with dispersion relation

$$k = k(\omega) = \frac{\beta(\omega)\omega}{c} \tag{3.12}$$

obtained as the eigenvalue in solving (3.11). We observe that, since $n(x_\perp, \omega)$ is even in ω, solutions of (3.11) are independent of the sign of ω. We also observe that if $U(x_\perp, k(\omega), \omega)$ is a solution of (3.11), so is $U^*(x_\perp, -k(\omega), -\omega)$. Since (3.12) is an odd function of ω (β is even in ω), we can write $U^*(x_\perp, -k(\omega), -\omega)$ as $U^*(x_\perp, k(-\omega), -\omega)$. For a right-going, monochromatic wavetrain with frequency ω_0, the Fourier transform of the electric field vector is

$$\hat{E}(x_\perp, z, \omega) = A\delta(\omega - \omega_0)U(x_\perp, k(\omega), \omega)\, e^{ik(\omega_0)z}$$

$$+ A^*\delta(-\omega - \omega_0)U^*(x_\perp, k(-\omega), -\omega)\, e^{ik(-\omega_0)z} \qquad (3.13)$$

where $k_0 = k(\omega_0)$ and $k(-\omega_0) = -k(\omega_0)$.

Assuming that we know U (and expressions for it can be found in reference [3], also Section 6f), a useful expression for the dispersion relation can be conveniently obtained by multiplying (3.11) across by the row vector $U^{*T} = (U_1^*, U_2^*, U_3^*)$ where $U = \mathrm{col}(U_1 U_2 U_3)$. We define the inner product of two complex vectors f and g to be

$$\langle f, g \rangle = \int_{-\infty}^{\infty} (f_1^* g_1 + f_2^* g_2 + f_3^* g_3)\, dx\, dy. \qquad (3.14)$$

Then

$$D([U], k, \omega) = \langle U, LU \rangle = 0 \qquad (3.15)$$

and the reader can show (by integration by parts) that $\langle U, LU \rangle = \langle LU, U \rangle$ (because of this property, we say L is self-adjoint with respect to this inner product (3.14)) and that $\langle U, LU \rangle$ is real. Without loss of generality we can take U_1, U_2 real, U_3 imaginary and then it is clear that every term in $\langle U, LU \rangle$ is real. Equation (3.15) is analogous to (6.259). Just as in that case, the first variation of the dispersion relation with respect to U is zero when U is a solution of (3.11).

EXERCISE 3.1

Show that

$$\lim_{\epsilon \to 0} \frac{D([U + \epsilon \delta U], k, \omega) - D([U], k, \omega)}{\epsilon} = 2\langle \delta U, LU \rangle, \qquad (3.16)$$

which is zero when U satisfies (3.11). $\qquad\square$

This exercise has important consequences. Since (3.15), the dispersion relation that gives k as a function of ω, holds for all ω, we may differentiate it with respect to ω:

$$0 = \frac{dD}{d\omega}([U], k(\omega), \omega) = 2\langle \frac{\partial U}{\partial \omega}, LU \rangle + \langle U, \frac{dL}{d\omega}U \rangle$$

$$= \langle U, (ik'L_3 - iL_4)U \rangle \qquad (3.17)$$

since the first term is zero by (3.15) and

$$\frac{dL}{d\omega} = \frac{\partial L}{\partial(ik)} \frac{d(ik)}{d\omega} + \frac{\partial L}{\partial(-i\omega)} \frac{d(-i\omega)}{d\omega}.$$

We denote $\partial L/\partial(ik)$ as L_3 and $\partial L/\partial(-i\omega)$ as L_4. From (3.17), we have

$$\langle U, L_4 U \rangle = k' \langle U, L_3 U \rangle \tag{3.18}$$

or

$$k' \langle U, L_3 U \rangle = +2 \, i \langle U, \frac{n(x_\perp, \omega)\omega}{c} \left(\frac{n(x_\perp, \omega)\omega}{c} \right)' U \rangle. \tag{3.19}$$

where f' means $\partial f/\partial \omega$.

$$L_4 = \frac{\partial}{\partial(-i\omega)} \left(\frac{n^2(-\omega)^2}{c^2} \right)$$

$$= i \frac{\partial}{\partial \omega} \left(\frac{n^2 \omega^2}{c^2} \right)$$

$$= 2 \, i \frac{n\omega}{c} \left(\frac{n\omega}{c} \right)'$$

Differentiating (3.17) again with respect to ω leads to (subscripts of L mean partial derivatives with respect to the denoted argument)

$$0 = \frac{d^2 D}{d\omega^2} = \left\langle \frac{\partial U}{\partial \omega}, (ik'L_3 - iL_4)U \right\rangle + \left\langle U, (ik'L_3 - iL_4) \frac{\partial U}{\partial \omega} \right\rangle$$

$$+ \langle U, (ik''L_3 - k'^2 L_{33} + 2k'L_{34} - L_{44})U \rangle$$

and a little integration by parts in the second term shows that it is the same as the first and therefore

$$\left\langle U, (k'^2 L_{33} - 2k'L_{34} + L_{44})U - 2\,i \, (k'L_3 - L_4)\frac{\partial U}{\partial \omega} \right\rangle = ik'' \langle U, L_3 U \rangle$$

$$= -2k''/k' \left\langle U, \left(\frac{n\omega}{c} \right) \left(\frac{n\omega}{c} \right)' U \right\rangle \tag{3.20}$$

The expressions (3.18) and (3.20) will be very useful in helping us write what at first sight appear to be very complicated expressions for the coefficients in the NLS equation in a simple and universal form and in particular the latter will be helpful in the case where we use a fairly crude approximation to the transverse shape of the wavepacket.

Before we go on to wavepackets, one remark on the dispersion relation is important. The nonzero dispersion ($\partial^2 k/\partial \omega^2 \neq 0$) in the propagation direction depends on two effects. First, there is a contribution because the real part of the susceptibility depends on ω. This is called material dispersion. Second, even if $\hat{\chi}^{(1)}(\omega)$ were independent of ω, the mode propagation constant $\beta(k = \beta \omega/c)$ still depends on ω as there is no way to remove ω^2 from (3.7), through a rescaling of the radius coordinate r since n^2 depends on r. This is called waveguide dispersion.

Stage Two: The Representation of Wavepackets and the Convolution Integrals Appearing in (3.4)

A wavepacket is the linear superposition of wavetrains with frequencies ω lying in an $\epsilon (0 < \epsilon \ll 1)$ neighborhood of ω_0, namely

$$\vec{E}(x_\perp, z, t) = \frac{1}{2\pi} \int_{-\infty}^{\infty} \hat{E}(x_\perp, z, \omega)e^{-i\omega t}d\omega \qquad (3.21)$$

where $\hat{E}(x_\perp, z, \omega)$ is supported (a mathematical expression for "is nonzero") on two intervals of width ϵ at ω_0 and $-\omega_0$ (see Figure 3.6).

We write

$$\hat{E}(x_\perp, z, \omega) = U(x_\perp, \omega)\hat{A}(z, \omega - \omega_0)e^{ik_0 z}$$

$$+ U^*(x_\perp, -\omega)\hat{A}^*(z, -\omega - \omega_0)e^{-ik_0 z} \qquad (3.22)$$

where $\hat{A}(\omega - \omega_0)(\hat{A}^*(-\omega - \omega_0))$ is supported only in an ϵ neighborhood around $\omega - \omega_0 = 0(-\omega - \omega_0 = 0)$. Note that if $\hat{A}$ is $A\delta(\omega - \omega_0)$, we recover the wavetrain solution (3.13). Note also that $\hat{E}(x_\perp, z, -\omega) = \hat{E}^*(x_\perp, z, \omega)$,

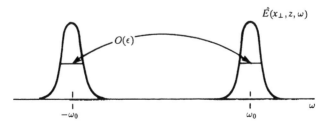

FIGURE 3.6   Graph of $|\hat{E}(x_\perp, z, \omega)|$ for a wavepacket.

thus assuring that $\vec{E}(x_\perp, z, t)$ is real. Why do we include a z-dependence in $\hat{A}(z, \omega - \omega_0)$? The reason is that when we study the propagation of the wavepacket, different frequency components travel at different speeds and so $\hat{A}(z, \omega - \omega_0)$ changes with z. We will discover how shortly.

We now convert (3.22) into our equivalent expression for the electric field $\vec{E}(x_\perp, z, t)$. Since $\hat{A}(z, \omega - \omega_0)(\hat{A}^*(z, -\omega - \omega_0))$ is supported only near $\omega - \omega_0 = 0(-\omega - \omega_0 = 0)$, we expand $U(x_\perp, \omega)(U^*(x_\perp, -\omega))$ in a Taylor series about $\omega_0(-\omega_0)$ and obtain

$$\vec{E}(x_\perp, z, t) =$$

$$= \sum_0^\infty \frac{i}{n!} \frac{d^n U}{d\omega_0^n}(x_\perp, \omega_0) e^{ik_0 z - i\omega_0 t}$$

$$\times \frac{1}{2\pi} \int_{-\infty}^\infty (\omega - \omega_0)^n \hat{A}(z, \omega - \omega_0) e^{-i(\omega - \omega_0)t} d\omega$$

$$+ \sum_0^\infty \frac{i}{n!} \frac{d^n U^*}{d\omega_0^n}(x_\perp, \omega_0) e^{-ik_0 z + i\omega_0 t}$$

$$\times \frac{1}{2\pi} \int_{-\infty}^\infty (-\omega - \omega_0)^n \hat{A}^*(z, -\omega - \omega_0) e^{-i(\omega + \omega_0)t} d\omega$$

$$= U\left(x_\perp, \omega_0 + i\frac{\partial}{\partial t}\right) A(z, t) e^{ik_0 z - i\omega_0 t}$$

$$+ U^*\left(x_\perp, \omega_0 - i\frac{\partial}{\partial t}\right) A^*(z, t) e^{-ik_0 z + i\omega_0 t} \qquad (3.23)$$

where

$$A(z, t) = \frac{1}{2\pi} \int_{-\infty}^\infty \hat{A}(\Omega) e^{-i\Omega t} d\Omega. \qquad (3.24)$$

Observe that if we choose to introduce a wavepacket description of the real electric field, the transverse shape $U(x_\perp, \omega)$ becomes an operator acting on the slowly varying envelope $A(z, t)$. Because $\left(i\frac{\partial}{\partial t}\right)^n A = 0(\epsilon^n)$, we can expand

$$U\left(x_\perp, \omega_0 + i\frac{\partial}{\partial t}\right) A(z, t) = \sum_0^\infty \frac{1}{n!} \frac{d^n U}{d\omega_0^n}(x_\perp, \omega_0) \left(i\frac{\partial}{\partial t}\right)^n A(z, t), \qquad (3.25)$$

and only the first three or four terms in (3.25) will be relevant (depending on how far we take the ϵ expansion (3.5)).

This description also allows us to develop simple expressions for the convolution operators

$$\int_{-\infty}^{\infty} \chi^{(1)}(x_\perp, t - t_1)\vec{E}(x_\perp, z, t_1)dt \tag{3.26}$$

and

$$\frac{1}{c^2}\frac{\partial^2}{\partial t^2}\int_{-\infty}^{\infty} \chi^{(3)}(x_\perp, t - t_1, t - t_2, t - t_3)$$

$$(\vec{E}(x_\perp, z, t_1) \cdot \vec{E}(x_\perp, z, t_2))\vec{E}(x_\perp, z, t_3)dt_1 dt_2 dt_3 \tag{3.27}$$

for wavepackets. The $(1/c^2)(\partial^2/\partial t^2)$ acting on (3.26) is contained in L. To obtain these expressions we first take the Fourier transform, then use the fact that the transform $\hat{E}(x_\perp, z, \omega)$ is localized about ω_0 and $-\omega_0$, expand the transform of multiplying factors like $\hat{\chi}^{(1)}(x_\perp, \omega)$, $\hat{\chi}^{(3)}(\omega_1, \omega_2, \omega_3)$ and $U(x_\perp, \omega)$ in a Taylor series about $\pm\omega_0$, and finally take the inverse transform using (3.24). Applying this operation to (3.26), we find

$$\text{F.T.} \int_{-\infty}^{\infty} \chi^{(1)}(x_\perp, t - t_1)\vec{E}(x_\perp, z, t_1)dt_1 = \hat{\chi}^{(1)}(x_\perp, \omega)\hat{E}(x_\perp, z, \omega)$$

$$= \hat{\chi}^{(1)}(x_\perp, \omega)U(x_\perp, \omega)\hat{A}(\omega - \omega_0)e^{ik_0 z}$$

$$+ \hat{\chi}^{(1)}(x_\perp, \omega)U^*(x_\perp, -\omega)\hat{A}^*(-\omega - \omega_0)e^{-ik_0 z} \tag{3.28}$$

$$= \sum_{0}^{\infty}\frac{1}{n!}e^{ik_0 z - \omega_0 t}\frac{d^n}{d\omega_0^n}\left(\hat{\chi}^{(1)}(x_\perp, \omega_0)U(x_\perp, \omega_0)\right)(\omega - \omega_0)^n \hat{A}(\omega - \omega_0)e^{i\omega_0 t}$$

$$+ \sum_{0}^{\infty}\frac{1}{n!}e^{-ik_0 z + i\omega_0 t}\frac{d^n}{d\omega_0^n}\left(\hat{\chi}^{(1)}(x_\perp, \omega_0)U^*(x_\perp, \omega_0)\right)(-\omega - \omega_0)^n$$

$$\times \hat{A}^*(-\omega - \omega_0)e^{-i\omega_0 t} \tag{3.29}$$

where we have ignored the imaginary part of $\hat{\chi}^{(1)}(x_\perp, \omega)$. Remember the real part is even in ω. We take account of the imaginary part of $\hat{\chi}^{(1)}(x_\perp, \omega)$ in stage four. Now take the inverse Fourier transform of (3.29) to obtain

$$
\int_{-\infty}^{\infty} \chi^{(1)}(x_\perp, t - t_1)\vec{E}(x_\perp, z, t_1)dt_1
$$

$$
= \hat{\chi}^{(1)}\left(x_\perp, \omega_0 + i\frac{\partial}{\partial t}\right) U\left(x_\perp, \omega_0 + i\frac{\partial}{\partial t}\right) A(z, t)e^{i(k_0 z - \omega_0 t)}
$$

$$
+ \hat{\chi}^{(1)}\left(x_\perp, \omega_0 - i\frac{\partial}{\partial t}\right) U^*\left(x_\perp, \omega_0 - i\frac{\partial}{\partial t}\right) A^*(z, t)e^{-i(k_0 z - \omega_0 t)} \quad (3.30)
$$

The evaluation of (3.27) is more complicated, not because of deeper ideas but simply because there are more terms to keep track of. However, we are only interested in writing down the leading contribution at this stage, so in fact the answer we get is extremely easy. Nevertheless we will write the expression in such a way that it is clear to the reader how to obtain the subsequent terms in the expansion.

Again, we begin by taking the Fourier Transform (multiply (3.27) by $e^{i\omega t}$ and integrate over t)

$$
\text{F.T.} \frac{1}{c^2}\frac{\partial^2}{\partial t^2} \int \chi^{(3)}(x_\perp, t - t_1, t - t_2, t - t_3)
$$

$$
\times (\vec{E}(x_\perp, z, t_1) \cdot \vec{E}(x_\perp, z, t_2))\vec{E}(x_\perp, z, t_3)dt_1 dt_2 dt_3
$$

$$
= -\frac{\omega^2}{(2\pi)^3 c^2} \int \hat{\chi}^{(3)}(x_\perp, \omega_1, \omega_2, \omega_3)(\hat{E}(x_\perp, z, \omega_1) \cdot \hat{E}(x_\perp, z, \omega_2))
$$

$$
\times \hat{E}(x_\perp, z, \omega_3)e^{i(\omega - \omega_1 - \omega_2 - \omega_3)t} dt d\omega_1 d\omega_2 d\omega_3. \quad (3.31)
$$

At this point, we could integrate over t to obtain $2\pi\delta(\omega - \omega_1 - \omega_2 - \omega_3)$. Therefore the factor ω^2 can be replaced by $(\omega_1 + \omega_2 + \omega_3)^2$. Substitute,

$$
E(x_\perp, z, \omega) = U(x_\perp, \omega)\hat{A}(z, \omega - \omega_0)e^{ik_0 z}
$$

$$
+ U^*(x_\perp, -\omega)\hat{A}^*(z, -\omega - \omega_0)e^{-ik_0 z}
$$

and then expand all terms such as

$$- \frac{(\omega_1 + \omega_2 + \omega_3)^2}{c^2} \hat{\chi}^{(3)}(x_\perp, \omega_1, \omega_2, \omega_3)(U(x_\perp, \omega_1) \cdot U(x_\perp, \omega_2))$$

$$\times U^*(x_\perp, -\omega_3)e^{ik_0 z}\hat{A}(\omega_1 - \omega_0)\hat{A}(\omega_2 - \omega_0)\hat{A}^*(-\omega_3 - \omega_0) \quad (3.32)$$

in multivariable Taylor series about the zero argument values for which the triple product of $\hat{A}$ and $\hat{A}^*$ is nonzero. In the terms displayed, that is (3.32), we expand about $\omega_1 = \omega_0, \omega_2 = \omega_0, \omega_3 = -\omega_0$. When we take the inverse transform, to leading order the resulting expression contains terms $A^3, A^2 A^*, AA^{*2}, A^{*3}$ with coefficients $\hat{\chi}^{(3)}(U \cdot U)U^*$ evaluated at $\pm\omega_0$. Next, there are terms that are order ϵ smaller, involving derivatives $\partial A^3/\partial t, (\partial/\partial t)A^2 A^*, A(\partial/\partial t)|A|^2, A^*(\partial/\partial t)A^2, (\partial/\partial t)AA^{*2}, (\partial/\partial t)A^{*3}$ whose coefficients are products of partial derivatives of $\hat{\chi}^{(3)}(U \cdot U)U^*$ with respect to the various arguments. After this, there are terms of order ϵ^2 with second derivatives, and so on. Since A and A^* are slowly varying $((1/A)(\partial/\partial t)A = O(\epsilon))$, to leading order we require only the terms without derivatives. There are four types, terms proportional to $A^3 e^{3ik_0 z}$ and $A^{*3}e^{-3ik_0 z}$, which to leading order we do not need to evaluate. In fact, all we need are the terms proportional to $\hat{A}^2\hat{A}^*e^{ik_0 z}$ and $\hat{A}\hat{A}^{*2}e^{-ik_0 z}$, because these are the terms that give rise to a cumulant (resonant) long-distance effect. The coefficient of the latter is simply obtained by taking the complex conjugate of the former. The terms proportional to $e^{ik_0 z}$ are

$$- \frac{(\omega_1 + \omega_2 + \omega_3)^2}{c^2} \hat{\chi}^{(3)}(x_\perp, \omega_1, \omega_2, \omega_3)$$

$$\Big\{ (U(x_\perp, \omega_1) \cdot U(x_\perp, \omega_2))$$

$$U^*(x_\perp, -\omega_3)\hat{A}(z, \omega_1 - \omega_0)\hat{A}(z, \omega_2 - \omega_0)\hat{A}^*(-\omega_3 - \omega_0)$$

$$+ (U(x_\perp, \omega_1) \cdot U^*(x_\perp, -\omega_2))$$

$$U(x_\perp, \omega_3)\hat{A}(\omega_1 - \omega_0)\hat{A}^*(-\omega_2 - \omega_0)\hat{A}(\omega_3 - \omega_0)$$

$$+ (U^*(x_\perp, -\omega_1) \cdot U^*(x_\perp, \omega_2))$$

$$U(x_\perp, \omega_3)\hat{A}^*(-\omega_1 - \omega_0)\hat{A}(\omega_2 - \omega_0)\hat{A}(\omega_3 - \omega_0) \Big\}. \quad (3.33)$$

Now write $e^{i(\omega - \omega_1 - \omega_2 - \omega_3)t}$ in (3.31) as

$$e^{-i(\omega_1 - \omega_0)t} \cdot e^{-i(\omega_2 - \omega_0)t} \cdot e^{-i(\omega_3 + \omega_0)t} \cdot e^{-i\omega_0 t},$$

expand all coefficients in the manner already explained, and use the integrations in ω_1, ω_2, and ω_3 to convert $\hat{A}(z, \omega_1 - \omega_0)$ into $A(z, t)$. To leading order, we obtain that the Fourier transform of the term involving $\exp(ik_0z - i\omega_0t)$ in (3.31) is

$$
-\frac{\omega_0^2}{c^2} \left\{ \hat{\chi}^{(3)}(x_\perp, \omega_0, \omega_0, -\omega_0)(U(x_\perp, \omega_0) \cdot U(x_\perp, \omega_0))U^*(x_\perp, \omega_0) \right.
$$

$$
+ \hat{\chi}^{(3)}(x_\perp, \omega_0, -\omega_0, \omega_0)(U(x_\perp, \omega_0) \cdot U^*(x_\perp, \omega_0))U(x_\perp, \omega_0)
$$

$$
\left. + \hat{\chi}^{(3)}(x_\perp, \omega_0, -\omega_0, \omega_0)(U^*(x_\perp, \omega_0) \cdot U(x_\perp, \omega_0))U(x_\perp, \omega_0) \right\}
$$

$$
\times A^2 A^* e^{i(k_0z - \omega_0t)}. \tag{3.34}
$$

The coefficient of $AA^{*2}e^{-i(k_0z - \omega_0t)}$ is the complex conjugate of the expression in braces in (3.34). Again, we remind the reader that we have ignored the imaginary part of $\hat{\chi}^{(3)}(x_\perp, \omega_1, \omega_2, \omega_3)$, which, although small, is present and gives rise to an intensity dependent absorption of the wavepacket. We are now ready to proceed to the derivation of the NLS equation.

Stage Three: The Derivations of NLS

We are now in position to derive NLS for a monomode, polarization preserving fiber. Write

$$
\vec{E}(x_\perp, z, t) = \sum_0^\infty \epsilon^{n+1} \vec{E}_n(x_\perp, z, t, \epsilon) \tag{3.35}
$$

where $\vec{E}_0$ is the wavepacket

$$
\vec{E}_0(x_\perp, z, t, \epsilon) = U\left(x_\perp, \omega_0 + i\frac{\partial}{\partial t}\right) A(z, t)e^{i(k_0z - \omega_0t)}
$$

$$
+ U^*\left(x_\perp, \omega_0 - i\frac{\partial}{\partial t}\right) A^*(z, t)e^{-i(k_0z - \omega_0t)}. \tag{3.36}
$$

We take $A(z, t)$ to be slowly varying in time t and distance z along the fiber and introduce the slow scales

$$
T = \epsilon t, \quad Z = \epsilon z. \tag{3.37}
$$

The solvability conditions for $\vec{E}_1, \vec{E}_2, \ldots$ determine $\partial A/\partial Z$ and its complex conjugate $\partial A^*/\partial Z$ as a function of $\partial A/\partial T, \partial^2 A/\partial T^2, A^2 A^*, (\partial/\partial T)A^2 A^*$, $\partial^3 A/\partial T^3$, and so on. Write ($Z_n = \epsilon^n z = \epsilon^{n-1} Z$)

$$\frac{\partial A}{\partial Z} = \frac{\partial A}{\partial Z_1} + \epsilon \frac{\partial A}{\partial Z_2} + \epsilon^2 \frac{\partial A}{\partial Z_3} + \cdots \quad (3.38)$$

for bookkeeping convenience. Observe that

$$\vec{E}_0(x_\perp, z, t, \epsilon) =$$

$$\left(U(x_\perp, \omega_0) A(Z, T) + \epsilon \frac{dU}{d\omega_0} i \frac{\partial A}{\partial T} + \cdots \right) e^{i(k_0 z - i\omega_0 t)} + (*) \quad (3.39)$$

has an order ϵ contribution.

We have chosen $U e^{ik_0 z}$ so that (3.3) is satisfied to order ϵ. At order ϵ^2, we have

$$\nabla^2 \vec{E}_1 - \nabla(\nabla \cdot \vec{E}_1) - \frac{1}{c^2} \frac{\partial^2}{\partial t^2} \left(\vec{E}_1 + \int \chi^{(1)}(t - t_1) \vec{E}(t_1) dt_1 \right) = \vec{F}_1 \quad (3.40)$$

where $\vec{F}_1$ is made up of terms that arise from the slow dependence of the envelope $A(Z, T)$ on position and time and from the order ϵ contribution to $\vec{E}_0$ in (3.39). $\vec{F}_1$ is equal to the negative of all the order ϵ^2 terms that occur when we apply

$$\nabla^2 - \nabla(\nabla \cdot - \frac{1}{c^2} \frac{\partial^2}{\partial t^2} \left(1 + \int dt_1 \chi^{(1)}(t - t_1) \right)$$

to $\epsilon \vec{E}_0$, which is simply the order ϵ^2 term in the expansion

$$L \left(\frac{\partial}{\partial x}, \frac{\partial}{\partial y}, ik_0 + \epsilon \frac{\partial}{\partial Z}, -i\omega_0 + \epsilon \frac{\partial}{\partial T} \right) \epsilon U \left(x_\perp, \omega_0 + i\epsilon \frac{\partial}{\partial T} \right)$$

$$\times A(Z, T) e^{i(k_0 z - \omega_0 t)} + L \left(\frac{\partial}{\partial x}, \frac{\partial}{\partial y}, -ik_0 + \epsilon \frac{\partial}{\partial Z}, i\omega_0 + \epsilon \frac{\partial}{\partial T} \right)$$

$$\times \epsilon U^* \left(x_\perp, \omega_0 - i\epsilon \frac{\partial}{\partial T} \right) A^*(Z, T) e^{-i(k_0 z - \omega_0 t)} \quad (3.41)$$

We obtain

$$
\vec{F}_1 = \left(-iL\left(\frac{\partial}{\partial x}, \frac{\partial}{\partial y}, ik_0, -i\omega_0\right)\frac{dU}{d\omega_0}(x_\perp, \omega_0)\frac{\partial A}{\partial T}\right.
$$

$$
\left. -L_3 U(x_\perp, \omega_0)\frac{\partial A}{\partial Z_1} - L_4 U(x_\perp, \omega_0)\frac{\partial A}{\partial T}\right) e^{i(k_0 z - \omega_0 t)} \tag{3.42}
$$

$$
+ (*).
$$

But, differentiating (3.11) with respect to ω, and evaluating the result at $\omega = \omega_0$, we get ($k_0' = \partial k/\partial \omega(\omega_0)$),

$$
L\frac{dU}{d\omega} = (-ik_0' L_3 + iL_4)U(x_\perp, \omega_0) \tag{3.43}
$$

and have

$$
\vec{F} = -L_3 U(x_\perp, \omega_0)\left(\frac{\partial A}{\partial Z_1} + k_0'\frac{\partial A}{\partial T}\right) e^{i(k_0 z - \omega_0 t)} + (*). \tag{3.44}
$$

Remember that L_3 means the operator L differentiated with respect to the third argument and evaluated at ω_0. How do we solve (3.40)? Write

$$
\vec{E}(x_\perp, z, t) = U_1\, e^{i(k_0 z - \omega_0 t)} + U_1^* e^{-i(k_0 z - \omega_0 t)}
$$

and obtain

$$
LU_1 = -L_3 U(x_\perp, \omega_0)\left(\frac{\partial A}{\partial Z_1} + k_0'\frac{\partial A}{\partial T}\right). \tag{3.45}
$$

Now the self-adjoint operator L has a nontrivial null space (that is, $LU_1 = 0$ has nontrivial solutions, e.g., $U_1 = $ constant U) and therefore the Fredholm alternative theorem, which guarantees that (3.45) has a solution, demands that the right-hand side of (3.45) is orthogonal to $U(x_\perp, \omega_0)$, namely that

$$
\langle U(x_\perp, \omega_0), -L_3 U(x_\perp, \omega_0)\rangle \left(\frac{\partial A}{\partial Z_1} + k_0'\frac{\partial A}{\partial T}\right) = 0. \tag{3.46}
$$

But, using (3.19),

$$
\langle U(x_\perp, \omega_0), -L_3 U(x_\perp, \omega_0)\rangle = -\frac{2\,i\omega_0}{k_0' c}\langle U, n\left(\frac{n\omega_0}{c}\right)' U\rangle
$$

is nonzero, and therefore we must choose

$$\frac{\partial A}{\partial Z_1} + k_0' \frac{\partial A}{\partial T} = 0, \tag{3.47}$$

the familiar result that, on distances of the order of the wavepacket width, the envelope $A(Z, T)$ moves with the group velocity of the underlying carrier wave. The solution to (3.45) then is $U_1 = 0$, because there is no point in adding in another constant times U.

Digression: The reader may want to omit this remark on a first pass. However, one might ask: What happens if we had taken $\vec{E}_0$ to be a slowly varying function $A(Z, T)$ times the "leading order" behavior $U(x_\perp, \omega_0)e^{ik_0 z - i\omega t}$ of the wavetrain centered on ω_0? In that case, the first term on the right-hand side of (3.42) is absent, and the solvability condition for U would read

$$\langle U, L_3 U \rangle \frac{\partial A}{\partial Z_1} + \langle U, L_4 U \rangle \frac{\partial A}{\partial T} = 0. \tag{3.48}$$

But, from the identity (3.18),

$$\langle U, L_4 U \rangle = k' \langle U, L_3 U \rangle,$$

and so again we recover (3.47). This now means

$$\vec{F}_1 = (k_0' L_3 - L_4) U \frac{\partial A}{\partial T}$$

$$= (-ik_0' L_3 + iL_4) U \left(i \frac{\partial A}{\partial T} \right)$$

$$= L \frac{dU}{d\omega} \left(i \frac{\partial A}{\partial T} \right).$$

Thus

$$U_1 = \frac{dU}{d\omega_0} \cdot i \frac{\partial A}{\partial T},$$

and therefore

$$\epsilon \vec{E}_0 + \epsilon^2 \vec{E}_1 = \epsilon \left(U(x_\perp, \omega_0) A(Z, T) + \epsilon \frac{dU}{d\omega_0} i \frac{\partial A}{\partial T} \right) e^{i(k_0 z - \omega_0 t)} + (*)$$

$$= \epsilon \left(U \left(x_\perp, \omega_0 + i\epsilon \frac{\partial}{\partial T} \right) A(Z, T) e^{i(k_0 z - \omega_0 t)} + (*) \right) + O(\epsilon^3),$$

which is what we have already. End of digression.

We now proceed to the next level, order ϵ^3, at which we solve for $\vec{E}_2(x_\perp, z, t)$,

$$\nabla^2 \vec{E}_2 - \nabla(\nabla \cdot \vec{E}_2) - \frac{1}{c^2}\frac{\partial^2}{\partial t^2}\left(\vec{E}_2 + \int_{-\infty}^{\infty} \chi^{(1)}(t - t_1)\vec{E}_2(t_1)dt_1\right) = \vec{F}_2$$

(3.49)

where $\vec{F}_2$ contains

1. the negative of those terms proportional to $e^{\pm i(k_0 z - \omega_0 t)}$ arising from the order ϵ^3 contribution of (3.41),

2. the nonlinear terms proportional to $A^2 A^* e^{i(k_0 z - \omega_0 t)}$ and $AA^{*2}e^{-i(k_0 z - \omega_0 t)}$ arising from the nonlinear polarization (3.27),

3. the terms proportional to $A^3 e^{3i(k_0 z - \omega_0 t)}$ and $A^{*3}e^{-3i(k_0 z - \omega_0 t)}$ arising from the nonlinear polarization (3.27).

We have already computed most of (2) in (3.34). We do not need to compute (3), at least to this order, because a term in $\vec{F}_2$ proportional to $e^{3i(k_0 z - \omega_0 t)}$ belongs to the range of the operator L and leads to a bounded solution $\vec{E}_2$. We would need to find this contribution if we were taking the expansion to order ϵ^5. Specifically, we may write

$$\vec{E}_2 = U_2^{(3)}(x_\perp)e^{3i(k_0 z - \omega_0 t)} + (*) + U_2^{(1)}(x_\perp)e^{i(k_0 z - \omega_0 t)} + (*)$$

(3.50)

and obtain

$$L\left(\frac{\partial}{\partial x}, \frac{\partial}{\partial y}, 3\,ik_0, -3\,i\omega_0\right)U_2^{(3)} = F_{2,3}$$

(3.51)

$$L\left(\frac{\partial}{\partial x}, \frac{\partial}{\partial y}, ik_0, -i\omega_0\right)U_2^{(1)} = F_{2,1}$$

(3.52)

where $F_{2,\pm 3}$ is the vector multiplying $e^{\pm 3i(k_0 z - \omega_0 t)}$ in $\vec{F}_2$ and $F_{2,\pm 1}$ is the vector multiplying $e^{\pm i(k_0 z - \omega t)}$. The important thing to note is that there are no nontrivial solutions of the homogeneous equation

$$L\left(\frac{\partial}{\partial x}, \frac{\partial}{\partial y}, 3ik_0, -3i\omega_0\right)U_2^{(3)} = 0$$

because the wavenumber $k(3\omega_0)$ corresponding to frequency $3\omega_0$ is not equal to $3k(\omega_0) = 3k_0$ precisely because there is dispersion in the longitudinal direction. (Note: if $k = \beta\omega/c$, β constant, then $k(3\omega) = 3k(\omega)$ and all terms in $\vec{F}_2$ would give rise to a nontrivial solvability condition. If this were the case, the NLS equation would *not* be the appropriate equation.)

On the other hand, equation (3.52) has a nontrivial solvability condition, namely $F_{2,1}$ must be orthogonal to the solution U:

$$\langle U, F_{2,1} \rangle = 0. \tag{3.53}$$

We now compute the part of $F_{2,1}$ arising from part (1), namely, the negative of the factor of $e^{ik_0 z - i\omega_0 t}$ in the order ϵ^3 contribution of (3.41). We find this to be:

$$\left(-L_3\frac{\partial A}{\partial Z_2} - \frac{1}{2}L_{33}\frac{\partial^2 A}{\partial Z_1^2} - L_{34}\frac{\partial^2 A}{\partial Z_1 \partial T} - \frac{1}{2}L_{44}\frac{\partial^2 A}{\partial T^2} \right) U(x_\perp, \omega_0)$$

$$+ i\left(L_3\frac{\partial}{\partial Z_1} - L_4\frac{\partial}{\partial T} \right) \cdot \frac{\partial A}{\partial T}\frac{dU}{d\omega_0} + \frac{1}{2}L\frac{d^2 U}{d\omega_0^2}\frac{\partial^2 A}{\partial T^2}$$

$$= \left(-L_3\frac{\partial A}{\partial Z_2} - \left(\frac{1}{2}L_{33}k_0'^2 - L_{34}k_0' + \frac{1}{2}L_{44}\right)\frac{\partial^2 A}{\partial T^2} \right) U(x_\perp, \omega_0)$$

$$+ i(k_0'L_3 - L_4)\frac{\partial^2 A}{\partial T^2}\frac{dU}{d\omega_0} + \frac{1}{2}L\frac{d^2 U}{d\omega_0^2}\frac{\partial^2 A}{\partial T^2}$$

using (3.47). All partial derivatives are evaluated at $l = m = 0$, $k = k_0$, $\omega = \omega_0$. Therefore, in all its glory, the solvability condition, including the contributions of (1) and (2) (the latter are calculated in (3.36)), is

$$0 = -\langle U, L_3 U\rangle\frac{\partial A}{\partial Z_2} - \left\langle U, \left(\frac{1}{2}L_{33}k_0'^2 - L_{34}k_0' + \frac{1}{2}L_{44}\right)U\right\rangle\frac{\partial^2 A}{\partial T^2}$$

$$+ \left\langle U, (ik_0'L_3 - iL_4)\frac{dU}{d\omega_0}\right\rangle\frac{\partial^2 A}{\partial T^2} + \frac{1}{2}\left\langle U, L\frac{d^2 U}{d\omega_0^2}\right\rangle\frac{\partial^2 A}{\partial T^2}$$

$$- \frac{\omega_0^2}{c^2}\langle U(\hat{\chi}_3^{(3)}(U \cdot U)U^* + \hat{\chi}_2^{(3)}(U \cdot U^*)U + \hat{\chi}_1^{(3)}(U^* \cdot U)U))A^2 A^*$$

$$\tag{3.54}$$

where $\hat{\chi}_3^{(3)} = \hat{\chi}^{(3)}(x_\perp, \omega_0, \omega_0, -\omega_0)$ and $\hat{\chi}_1^{(3)}$ and $\hat{\chi}_2^{(3)}$ are similarly defined. Because of symmetry, all are in fact equal and we simply write each of them as $\hat{\chi}^{(3)}$. At the moment (3.54) looks like an awful mess! But now comes a miracle. Take the derivative of (3.43) and find

$$L\frac{d^2 U}{d\omega_0^2} = (-2\, ik_0' L_3 + 2\, i L_4)\frac{dU}{d\omega_0} - ik_0'' L_3 U + (k_0'^2 L_{33} - 2k_0' L_{34} + L_{44})U$$

so that all the terms in (3.54) that are linear in A and its derivatives collapse to

$$-\langle U, L_3 U\rangle \left(\frac{\partial A}{\partial Z_2} + \frac{ik_0''}{2}\frac{\partial^2 A}{\partial T^2}\right).$$

Next, use identity (3.18) to calculate

$$\langle U, L_3 U\rangle = \frac{1}{k_0'}\langle U, L_4 U\rangle,$$

and define

$$2n_2 = \frac{k_0'\langle U, \chi^{(3)}(U\cdot U)U^* + 2\hat{\chi}^{(3)}(U^*\cdot U)U\rangle}{\langle U, n(\frac{n\omega_0}{c})'U\rangle} \tag{3.55}$$

and we obtain

$$\frac{\partial A}{\partial Z_2} + \frac{ik_0''}{2}\frac{\partial^2 A}{\partial T^2} - i\frac{\omega_0}{c}n_2 A^2 A^* = 0. \tag{3.56}$$

Now that we have both $\partial A/\partial Z_1$ and $\partial A/\partial Z_2$, we may write

$$\frac{\partial A}{\partial Z} = \frac{\partial A}{\partial Z_1} + \epsilon\frac{\partial A}{\partial Z_2} = -k_0'\frac{\partial A}{\partial T} + \epsilon\left(\frac{-ik_0''}{2}\frac{\partial^2 A}{\partial T^2} + \frac{i\omega_0}{c}n_2|A|^2 A\right)$$

and introducing the coordinates

$$\tau = T - k_0' Z = \epsilon(t - k_0' z)$$

$$\zeta = \epsilon \quad Z = \epsilon^2 z,$$

we have

$$\frac{\partial A}{\partial\zeta} + \frac{ik_0''}{2}\frac{\partial^2 A}{\partial\tau^2} - i\frac{\omega_0}{c}n_2|A|^2 A = 0, \tag{3.57}$$

the canonical and universal NLS equation.

Note that if the wavepacket were not confined in the transverse direction, $\hat{\chi}^{(1)}$, n, and $\hat{\chi}^{(3)}$ would be independent of $x_\perp$, $U \propto e^{ilx+imy}$, $k = n(\omega)\omega/c$, and $2n_0 n_2 = 3\hat{\chi}^{(3)}$ exactly as we obtained in Section 2k.

Here endeth the derivation of NLS!

Remark: If we had taken the approximation

$$\vec{E}_0 = U(x_\perp, \omega_0) A e^{ik_0 z - i\omega_0 t} + (*)$$

instead of (3.36), we would not have the $d^2 U / d\omega_0^2$ term in (3.54). However, as we have already indicated in the previous digression, $\vec{E}_1$ would be nonzero and these terms would give rise to the terms proportional to $dU/d\omega_0$ in (3.54). In order that the solvability condition (3.54) simplifies to (3.56), we must use the identity (3.20). After applying the solvability condition, we would find that

$$U_2^{(1)} = -\frac{1}{2} \frac{d^2 U}{d\omega_0} \frac{\partial^2 A}{\partial T^2}$$

and so $\epsilon \vec{E}_0 + \epsilon^2 \vec{E}_1 + \epsilon^3 \vec{E}_2 + \cdots$ becomes (3.36) plus $O(\epsilon^3)$ third order harmonics.

Stage Four: The Terms We Have Ignored

Now that we have established that the NLS equation describes the propagation of the envelope $A(z, t)$ of the electric field in a monomode, polarization preserving fiber over distances proportional to the square of the packet width (or square of the inverse of the spectral width), we turn next to the corrections that should be considered to bring the equation into closer contact with reality. We will discuss the actual sizes of the coefficients in (3.57) and the applicability of a corrected version of (3.57) to the propagation of light pulses in a fiber in the next section.

The corrections to (3.57) are

1. Attenuation, linear and nonlinear
2. Higher-order linear dispersion
3. Nonlinear dispersion and an effective Raman scattering due to delay in $\chi^{(3)}$
4. Other effects, such as external forcing and fiber irregularities

Attenuation

As we will see in the next section, attenuation is the largest correction and, *at values of the pulse width and power currently used*, is not only comparable to the dispersion and nonlinear terms in (3.57) but may even be larger than them. The linear attenuation comes from the imaginary part of $\hat{\chi}^{(1)}(x_\perp, \omega)$, which we will call $\hat{\chi}_I^{(1)}(x_\perp, \omega)$. To add this contribution, we go back to

$$-\frac{1}{c^2}\frac{\partial^2}{\partial t^2}\epsilon\int_{-\infty}^{\infty}\chi^{(1)}(t-t_1)\vec{E}_0(x_\perp, z, t_1)dt_1$$

appearing on the left-hand side of the governing equation (3.3) and (3.6), which is equal to

$$\frac{\epsilon}{c^2}\left(\omega_0 + i\,\epsilon\frac{\partial}{\partial T}\right)^2\hat{\chi}^{(1)}\left(x_\perp, \omega_0 + i\,\epsilon\frac{\partial}{\partial T}\right)U\left(x_\perp, \omega_0 + i\epsilon\frac{\partial}{\partial T}\right)$$

$$\times A(Z, T)e^{i(k_0 z - \omega_0 t)}$$

plus its complex conjugate. We now include, to leading order in ϵ, the term

$$i(\omega_0^2/c^2)\,\epsilon\,\hat{\chi}_I^{(1)}(x_\perp, \omega_0)Ae^{i(k_0 z - \omega_0 t)}$$

plus its complex conjugate in our calculations. We will assume it is of order ϵ^3 and therefore include its negative as part of $\vec{F}_2$ in (3.49) and $F_{2,1}$ in (3.52). Hence we add to the solvability condition (3.54) the additional term

$$-i\frac{\omega_0^2}{c^2\epsilon^2}\langle U, \hat{\chi}_I^{(1)}(x_\perp, \omega_0)U\rangle\, A.$$

Define (by analogy with (3.55))

$$2n_I = \frac{k_0'}{\epsilon^2}\frac{\langle U, \hat{\chi}_I^{(1)}(x_\perp, \omega_0)U\rangle}{\langle U, n(n\omega_0/c)'U\rangle}, \tag{3.58}$$

and this leads to the additional term

$$n_I\frac{\omega_0}{c}\,A$$

in (3.57), whence

$$\frac{\partial A}{\partial \zeta} + \frac{ik_0''}{2}\frac{\partial^2 A}{\partial \tau^2} - i\frac{\omega_0}{c}n_2|A|^2 A + n_I\frac{\omega_0}{c}A = 0. \tag{3.59}$$

The fact that $n_I\frac{\omega_0}{c}A$ may be larger than either dispersion or nonlinearity does not invalidate the envelope equation (3.59). It does, however, make it harden to solve since the exact soliton of the NLS equation is no longer the leading order approximation to the solution of (3.59). Note that if we take $\vec{E}$ to be monochromatic and linear (i.e. $\partial A/\partial \tau = n_2 = 0$), $A = A_0 e^{-n_I(\omega_0/c)\zeta}$, which is exactly consistent with what we get if we write

$$k = \frac{(n(\omega_0) + i\epsilon^2 n_I)\omega_0}{c}. \tag{3.60}$$

Note also that if n is independent of $x_\perp$, $2\epsilon^2 nn_I = \hat{\chi}_I^{(1)}$. The nonlinear attenuation correction can be calculated in a similar manner by looking for the contributions from the imaginary part of $\hat{\chi}^{(3)}$, but it is very small and we will ignore it here.

Higher-Order Linear Dispersion

It has become clear that the linear part of NLS is generated by expanding the dispersion relation $D(k, \omega) = 0$ or $k = k(\omega)$ in a Taylor series as

$$k\left(\omega_0 + i\ \epsilon \frac{\partial}{\partial T}\right) A = k(\omega_0)A + i\ \epsilon k_0' \frac{\partial A}{\partial T} - \frac{\epsilon^2 k_0''}{2} \frac{\partial^2 A}{\partial T^2} + \cdots. \tag{3.61}$$

The next term in the series is $-i\epsilon^3(k_0'''/6)(\partial^3 A/\partial T^3)$, and this gives rise to the higher-order correction $\epsilon(k_0'''/6)(\partial^3 A/\partial T^3)$ in (3.57) and (3.58). Remember, $\vec{E} = \epsilon \vec{E}_0$, and so this higher-order dispersion term appears at the $o(\epsilon^4)$ level.

Nonlinear Dispersion and Effects of Delay in $\chi^{(3)}$

Go back to (3.33). Recall that we expand (3.33) in a multivariable Taylor expansion about those values $\omega_1, \omega_2, \omega_3$, which make zero the arguments of the multiplying factors containing triple products of $\hat{A}$ and $\hat{A}^*$. The first term is (3.34). The next term consists of all terms linear in $\pm\omega_1 - \omega_0, \pm\omega_2 - \omega_0, \pm\omega_3 - \omega_0$, each of which when multiplied by the appropriate $\hat{A}(\omega_j - \omega_0)$, $j = 1, 2, 3$ or $\hat{A}^*(-\omega_j - \omega_0)$, $j = 1, 2, 3$ gives rise, after the inverse Fourier transform is taken, to terms proportional to $\epsilon A^2(Z, T)\partial A/\partial T(Z, T)$ and $\epsilon A(Z, T)A^*(Z, T)\partial A/\partial T(Z, T)$.

Now go to the order ϵ^4 level in the equation (3.3),

$$\nabla^2 \vec{E}_3 - \nabla(\nabla \cdot \vec{E}_3) - \frac{1}{c}\frac{\partial^2}{\partial t^2}\left(\vec{E}_3 + \int_{-\infty}^{\infty} \chi^{(1)}(t - t_1)\ \vec{E}_3(t_1)dt_1\right) = \vec{F}_3. \tag{3.62}$$

$\vec{F}_3$ contains the nonlinear terms we have first discussed as well as linear terms discussed in the two previous subsections that give rise to attenuation and higher-order linear dispersion. When the solvability condition is applied to (3.62), we obtain

$$\frac{\partial A}{\partial Z_3} = -\frac{2\pi n_I \omega_0}{\epsilon c}A + \frac{k_0'''}{6}\frac{\partial^3 A}{\partial T^3} - \epsilon i\beta_1\frac{\partial}{\partial T}|A|^2 A - \epsilon i\beta_2 A\frac{\partial}{\partial T}|A|^2 \quad (3.63)$$

where $Z_3 = \epsilon^3 z$. Consider the last two terms. The first gives rise to a correction to nonlinear dispersion. The second can be considered as a manifestation of a delay in the nonlinear refractive index whereby the term $n_2|A|^2 A$ in (3.57) is replaced by $A \int_{-\infty}^{\infty} n_2(t - t_1)|A|^2(t_1)dt$. Using (3.38) and introducing $\zeta = \epsilon^2 z$, $\tau_1 = \epsilon(t - k_0' z)$, the perturbed NLS equation is

$$\frac{\partial A}{\partial \zeta} + i\frac{k_0''}{2}\frac{\partial^2 A}{\partial \tau_1^2} - \epsilon\frac{k_0'''}{6}\frac{\partial^3 A}{\partial \tau_1^3} - i\frac{\omega_0}{c}(n_2|A|^2 + i\, n_I)A$$

$$+ \epsilon\beta_1\frac{\partial}{\partial \tau}|A|^2 A + \epsilon i\beta_2 A\frac{\partial}{\partial \tau}|A|^2 = 0. \quad (3.64)$$

Other Effects

We mention several other effects that have so far been ignored. First, since attenuation is present and, as we will see in the following section, is important for propagation of pulses over distances of thousand of kilometers, the signal has to be reinforced at regular intervals. This makes for an effective n_{2_I} which is both negative (acts as a pump) and positive (acts as an attenuator) and which is periodic in z or ζ. Second, we have not at all considered what happens if the cross-section of the core is not everywhere a perfect circle. For example, its radius $\rho_{\text{core}} = \rho_0 + \epsilon\rho(\theta, z)$, the core-cladding boundary, may be weakly dependent on angle in a random fashion; the dependence at different values of z may be weakly correlated. This means that there is an induced linear birefringence so that the fiber is not perfectly polarization preserving. Introducing two directions of polarization,

$$\hat{E}(x_\perp, z, \omega) = U(x_\perp, \omega)(\hat{i}A + \hat{j}B)e^{ik_0 z} \quad (3.65)$$

leads to both a linear and nonlinear coupling of the two envelopes A and B and the possibility of nontrivial energy exchanges between these two modes. Third, for a sequence of solitons of sufficient power, each pulse may produce a backscattered Stokes wave through a SBS process (see Section 5f) whose step-by-step amplification due to interaction with consecutive solitons may overcome its attenuation and lead to an erosion of information content. An analysis of this effect would require the introduction of a coupling of (3.58) with an equation for the envelope of the Stokes field.

As a final remark, we point out that the NLS can be rewritten as

$$\frac{\partial A}{\partial T} + \omega_0' \frac{\partial A}{\partial Z} - i \frac{\omega_0''}{2} \epsilon \frac{\partial^2 A}{\partial Z^2} - i \frac{\omega_0}{c} \epsilon n_2 \omega_0' |A|^2 A = 0, \tag{3.66}$$

namely, as the evolution of an initial shape in Z as a function of T.

3c. Nonlinear Fiber Optics, Possibilities and Challenges, Realities and Practicalities

Two technological advances have made optical communication a reality. The first is the huge reduction in fiber losses to about 20 dB/km [32, 33]. The second is the development of super efficient erbium doped amplifiers [34] which regenerate attenuated pulses. While these practical developments took place over the two decades of the 1970's and 1980's, already theoreticians were speculating about the possibilities of schemes to improve bit flow rates. If each bit pulse had to be isolated in a window five times its width, a pulse of width 200 ps would require a window of 1 ns and at best allow a bit flow rate of 10^9 bits per second (1 gigabit). For a ten gigabits flow rate, one would need a considerably shorter pulse, of width 20 ps. But such narrow pulses will spread (and thereby lose amplitude) much more rapidly, because they contain a wider band of wavelengths and each wave moves at a different speed. Therefore, it was natural that, having solved in principle the challenge of attenuation, attention be turned to the second enemy of optical communications, dispersion.

In a ground breaking paper [35] in 1973, Hasegawa and Tappert suggested that the soliton solution of the NLS equation would be the ideal bit unit in optical fibers. It had two obvious advantages. First, in a soliton, dispersion is balanced by nonlinearity so that in a real sense, a soliton bit would be dispersionless. Nonlinearity acts to phase lock all the different wavelengths contained in the pulse and so all travel together, thereby maintaining the pulse shape. Second, solitons are normal modes of a completely integrable Hamiltonian system and therefore, in principle, one could construct multiple channel soliton streams based on neighboring but different wavelengths, say $\lambda_1 = 1.55 \ \mu m, \lambda_2 = 1.60 \ \mu m, \lambda_3 = 1.65 \ \mu m$. In the absence of nonintegrable perturbations such as attenuation and amplification, a faster moving soliton of one stream would pass through a slower moving soliton without distortion. The only interaction memory would be a phase shift which would be so small that, even over transoceanic distances, a bit would not be displaced from its detection window. Attractive as the idea was, the practicalities involved in making the soliton bit a reality provided major challenges for theoreticians and experimentalists. Foremost among these was the fact that the required amplification of attenuated pulses every 20 kms or so could adversely affect the advantages of the soliton's integrable characteristics. Indeed, as we write this updated version of the optical communications story today, in October 2002, it is a different nonlinear solution of a modified NLS equation that seems destined to be the optical communication bit for the next decade or so. But let us not get ahead of ourselves. We will now, using the NLS equation as the central theoretical model building tool, review the history and developments of the last thirty years. We will also briefly describe two means of amplification, the currently used erbium doped fibers and the use of the Raman effect. In a field where things constantly charge and where new ideas are always needed, it is a good idea to have many options.

We have seen in §3b that the electric field in a monomode, polarization preserving fiber can be represented by

$$\vec{E}(x_\perp, z, t) = \hat{e}\epsilon U\left(x_\perp, \omega_0 + i\epsilon \frac{\partial}{\partial \tau_1}\right) A(\zeta, \tau_1) e^{i(k_0 z - \omega_0 t)} + (*) + O(\epsilon^3)$$

$$(3.67)$$

where A as a function of $\zeta = \epsilon^2 z$ and $\tau_1 = \epsilon(t - k_0' z)$ satisfies the NLS equation (3.59). In the original coordinates z and retarded time $\tau = t - k_0' z$, the amplitude $B(z, \tau) = \epsilon A(\zeta, \tau_1)$ of the field satisfies

$$\frac{\partial B}{\partial z} + \frac{ik_0''}{2}\frac{\partial^2 B}{\partial \gamma^2} - i\frac{2\pi n_2}{\lambda_0}|B|^2 B = \gamma B,$$

$$(3.68)$$

where we have included for this discussion the most important correction term, the attenuation. In (3.68), $\lambda_0 = 2\pi c/\omega_0$ is the vacuum wavelength and $\gamma = -2\pi \epsilon^2 n_I/\lambda_0$. Our goal in this section is to compare the sizes of each of the terms for pulses in the 20 picosecond range with powers in the range of 50 mW (milliwatts) to 1 W. It is useful to measure time in picoseconds (ps), distance in kilometers (km) and the electric field in volts per kilometer. From the equation relating power and field amplitude

$$P = 2\epsilon_0 e\beta S_0|B^2|$$

$$(3.69)$$

($\beta = 1.5$, the linear refractive index of SiO_2, $S_0 = \langle U, U\rangle$ the effective cross section of the fiber pulse which is about $100(\mu m)^2$, $\epsilon_0 = 8.85 \times 10^{-12} CV^{-1}m^{-1}$,

$c = 3 \times 10^8$ ms^{-1}), we see that an electric field strength of 10^9 V(km)$^{-1}$ gives rise to a pulse power P of approximately 1 W (one watt). We scale the variables in (3.65) as follows:

$$z \to z_0 z, \; \tau \to t_0 \tau, \; B \to \sqrt{\frac{P}{2\epsilon_0 c\beta \; S_0}} B \qquad (3.70)$$

whence (3.68) reads

$$\frac{\partial B}{\partial z} + \frac{ik_0'' z_0}{2t_0^2} \frac{\partial^2 B}{\partial z^2} - i \frac{\pi n_2 P z_0}{\epsilon_0 c\lambda_0 \beta \; S_0} |B|^2 B = \gamma z_0 B \qquad (3.71)$$

We can define three length scales associated with attenuation/amplification, dispersion and nonlinearity which we call L_γ, L_D and L_{NL} respectively.

Attenuation

As we have already mentioned, modern fiber manufacturing techniques have greatly reduced the losses in glass fibers. Today one can achieve values of

$$\tilde{\gamma} = -10 \log_{10} \frac{\text{power } (z = 1 \text{ km})}{\text{power } (z = 0)}$$

measured in dB/km of about 0.13 at a vacuum length of $\lambda_0 = 1.55$ μm. This means that

$$P(z = 1 \text{ km}) = B(z = 0) \exp\left(-\frac{2\tilde{\gamma}}{20} \ln_e 10\right)$$

so that γ in (3.68) is $-\tilde{\gamma}/20 \ln_e 10$ or -0.02(km)$^{-1}$. This means the amplitude of a *linear wavetrain* would decay by a factor e^{-1} in a distance of 50 km. We call $|\gamma|^{-1}$ the attenuation length and denote it by $L_\gamma = |\gamma|^{-1}$ km. On the other hand, we will see from soliton perturbation theory that the amplitude of a soliton of (3.68) decays as $e^{2\gamma z}$ so that its effective attenuation distance is $\frac{1}{2}L_\gamma = 25$ km. However, this last result presupposes that the dominant balance in (3.68) is the dispersion and nonlinear terms and that attenuation can be treated as a small perturbation. Whatever the exact attenuation distance is, it is necessary that, as a consequence of attenuation, pulses are periodically amplified. In current practice this is done every 20-25 kms by inserting a short strip of erbium doped amplifier. Therefore, we should think of γ in (3.65) and (3.71) as being a periodic function of z with period of about 20 kms. For most of this distance γ is negative and equal to 0.02 (km)$^{-1}$. See Fig. (3.7).

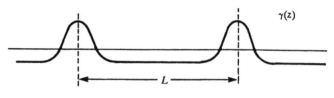

$$\gamma(z)$$

FIGURE 3.7

Dispersion

For a pulse of width t_0 (Figure 3.8), the size of the dispersion term in (3.68) is $(|k_0''|/2t_0^2)[B]$ where we use $[B]$ to denote the amplitude of the electric field. The coefficient $(|k_0''|/2t_0^2) = L_D^{-1}$ also has the dimension of inverse length. For $k = \beta(\omega)\omega/c$, a little calculation will show that

$$v_g^{-1} = \frac{\partial k}{\partial \omega} = \frac{1}{c}\left(\beta - \lambda\frac{d\beta}{d\lambda}\right)$$

and

$$\frac{\partial^2 k}{\partial \omega^2} = \frac{\lambda}{2\pi c^2}\left(\lambda^2\frac{d^2\beta}{d\lambda^2}\right)$$

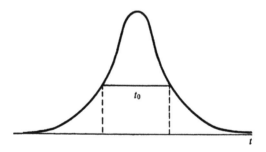

$$t_0$$

FIGURE 3.8

where β, the effective mode propagation constant, is expressed as a function of vacuum wavelength $\lambda = 2\pi c/\omega$. For normal fibers, the dimensionless quantity $\lambda^2(d^2\beta/d\lambda^2)$ has a zero (the zero dispersion point) for glass at $\lambda_1 = 1.27$ μm. Dispersion, the dependence of β on λ, arises for two reasons. The first is material dispersion where the dependence of β on ω or λ comes about because of the frequency dependence of the refractive indices n_0 and n_c of the core and cladding respectively. The second is waveguide dispersion and arises because in a waveguide, the effective refractive index β is a function of ω even though n_0 and n_c are constants. Its also depends on the core radius r_0 and refractive index mismatch $n_0 - n_c$. Thus the dimensionless dispersion $\lambda^2(d^2\beta/d\lambda^2)$ consists of two terms ($\lambda^2(d^2\beta/d\lambda^2)$ material $+\lambda^2(d^2\beta/d\lambda^2)$ waveguide). As λ increases, the former passes through zero from positive (normal) to negative (anomalous) values at approximately $\lambda = 1.27\ \mu$m in silica glass. The latter (waveguide) contribution is always positive and therefore can be used to shift the point of zero dispersion to higher vacuum wavelengths. For example, by choosing the core radius between 2 and 3 μm and the refractive index mismatch $n_0 - n_c/n_0$ greater then 0.004, the zero dispersion point can be shifted to the neighborhood of wavelengths of 1.55 μm, where the attenuation losses are minimum. As we shall shortly see, this fact has been exploited. Near $\lambda = \lambda_1$,

$$\lambda^2 \frac{d^2\beta}{d\lambda^2} = -5.3 \times 10^{-2}\left(1 - \frac{\lambda_1}{\lambda}\right)$$

Therefore, for λ about 1.55 μm, $c = 3 \times 10^{-7}$ km ps^{-1},

$$k'' = -\frac{1.55 \times 5.3}{2\pi \times 0.9} 10^2 \left(1 - \frac{\lambda_1}{\lambda}\right) \simeq -20 \text{ ps}^2/\text{km}.$$

Therefore,

$$L_D = \frac{2t_0^2}{k_0''} \simeq \frac{1}{10}t_0^2 \text{ kms}.$$

where t_0, the pulse width, is measured in picoseconds. For a pulse width of $t_0 = 20$ ps, $L_D \simeq 40$ kms. Note that L_D increases as the square of pulse width (narrow pulses have a broader frequency spread and therefore the individual waves separate more quickly) and increases as the dispersion $|k_0''|$ decreases. The fact that L_y and L_D are similar means that the attenuation/amplification perturbation is, for $t_0 = 20$ ps and $k_0'' = -20$ ps(km)$^{-1}$, an order one effect.

Nonlinearity

From (3.71), the nonlinear length

$$L_{NL} = \frac{\epsilon_0 c\beta\lambda_0 S_0}{\pi n_2 P}. \tag{3.72}$$

For $\epsilon_0 = 8.85 \times 10^{-9} \mathrm{CV}^{-1}(\mathrm{km})^{-1}$, $c = 3 \times 10^5 \mathrm{km}\ \mathrm{s}^{-1}$, $\beta = 1.5$, $\lambda_0 = 1.55 \times 10^{-9}$ km, $S_0 = 10^{-16}(\mathrm{km})^2$, $n_2 = 1.2 \times 10^{-28}(\mathrm{km})^2 \mathrm{V}^{-2}$, we see that $L_{NL} \sim L_D \simeq 40$ kms for pulses with power P of about 40 mW (or electric field amplitude of $\frac{1}{5}10^9 \mathrm{V}(\mathrm{km})^{-1}$). Increasing the power or nonlinear refractive index n_2 strengthens nonlinearity by shortening L_{NL}.

Exercise 3.2: The key to obtaining a low-power, short-width pulse is a high n_2. We invite the reader to explore the possibilities introduced in Section 2k by rederiving (2.162), (2.164) in the context of a waveguide fiber made of a material with a nonzero $\chi^{(2)}$.

Exercise 3.3: Although the soliton is not an exact solution of (3.65) because of the presence of attenuation and amplification, it is nevertheless a useful exercise to calculate the averages over a wave period $\overline{P}_s$ of the power

$$P_s = \int (\vec{E} \times \vec{H}) \cdot \hat{z}\, dS$$

associated with a soliton. For $\vec{E} = (0, E_y, 0)$, $\vec{H}(\beta\ \epsilon_0 c E_y, 0, 0)$ and $E_y(x_\perp, z, t) \simeq U(x_\perp, \omega) B(z, t) e^{ik(\omega)z - i\omega t} + (*)$ where $B(z, t)$ is the soliton solution of (3.68) with $\gamma = 0$, namely

$$B(z, t) = B_0 \operatorname{sech}\left(\sqrt{\frac{2\pi n_2}{\lambda_0 |k_0''|}} B_0(t - k'(\omega)z)\right) \exp i\frac{\pi n_2}{\lambda_0} B_0^2 z.$$

We find

$$\overline{P}_s = 2\epsilon_0 c\beta\ S_0 B_0^2 \operatorname{sech}^2 \sqrt{\frac{2\pi n_2}{\lambda_0 |k_0''|}} B_0(t - k'(\omega)z)$$

where $S_0 = \langle U, U \rangle = \int |U|^2(\vec{x}_\perp, \omega)(x_\perp) d\vec{x}_\perp$ is the effective cross section. The peak average power is $2\epsilon_0 c\beta\ S_0 B_0^2$ is equivalent to what we called P in (3.69). We have noted that an electric field B_0 of $\frac{1}{5}10^9$ V(km)$^{-1}$ corresponds to a peak power of 40 mW.

But the problem with the soliton bit is that, for $L_D \simeq L_{NL} \simeq 40$ kms, $L_\gamma \simeq 20$ kms, all terms in (3.68) and (3.71) are of equal importance and soliton theory does not apply. Since it was impractical to increase L_γ, attention turned to another possibility. What if, theoreticians and experimentalists reasoned, one were to consider another possibility. Instead of wishing that the attenuation term in (3.68) was the smallest effect, why not arrange for it to be the largest?

The means to do this was the dispersion shifted fiber [36] whereby the zero λ_1 of dispersion k'' was shifted to the region of $\lambda = 1.55\ \mu$m where attenuation is lowest. In that case, $L_D = \frac{2t_0^2}{|k_0''|}$ could be made to increase by at least a factor of 10, to the range of 400–500 kms. To go lower could be counter productive because if the nonlinear length had to follow suit, the pulse power would decrease to 4 mW and it was important to keep the signal to noise ratio well above unity.

How could this idea help? Let γ in (3.71) be a periodic function of z (see Figure 3.7) of period about 20–25 kms namely, $\gamma(z) = \gamma - \gamma L_\gamma \sum_n \delta(z - nL_\gamma)$, $\gamma < 0$. Choose z_0 to be $L_D \simeq 400 - -500$ kms. Remove the forcing term in (3.71) by setting

$$B(z, \tau) = C(z, \tau) \exp\left(L_D \int^z \gamma(z) dz\right)$$

whence

$$\frac{\partial C}{\partial z} + i(\operatorname{sgn} k_0'')\frac{\partial^2 C}{\partial \tau^2} - i\frac{L_D}{L_{NL}}\exp\left(2L_D\int^z \gamma(z)dz\right)C^2 C^* = 0 \qquad (3.73)$$

The coefficient sgn k_0'', the sign of k_0'', is negative for wavelengths of interest. Now, seen on the dispersion (and nonlinear) length scale L_D, the exponent of the exponential in the coefficient of the nonlinear term is rapidly varying. The idea was to replace it with its average over the attenuation-amplification cycle. Even though the average of $\gamma(z)$ over $0 \le z \le L_\gamma$ is zero, the average of the exponential is not. Imagine the following experiment. Begin with a soliton at $z = 0$ just after a short segment of erbium doped amplifier. Let it run from $z = 0$ to $z = L_\gamma$ being attenuated at the constant rate γ. Its amplitude B, constant for a pure soliton, will decay as $\exp(2\gamma z)$ (the decay rate is twice that of a linear pulse because it broadens as its amplitude diminishes; see Chapter 6, page 392). Now give a sudden boost of energy over a short interval. The overall power input over the short interval is $\exp(2|\gamma|L_\gamma)$ which exactly cancels the power lost over the attenuation segment. The smaller and broader soliton pulse will absorb most of this power but some will also go into a radiation. Imagine that the radiation is lost. What is the fixed point (the shape at the end of the nth such segment, n large) of such a map? It is the soliton solution of (3.74) with the exponential replaced by its average. This turns out to be a very good approximation and brings the constant coefficient NLS equation with all its integrability and soliton properties back into play. The 1990 experimental study of Mollenauer et al [36] confirmed that the dispersion shifted fiber worked. Additional references of interest are [37, 38, 39]

Andante, allegro! But there were problems. First, the reader would be correct to question the validity of the averaging step. There are possibilities of resonances resulting from the fact that higher harmonics of the attenuation-amplification period can match the dispersion-nonlinear length scale. But this problem is a small one. The more serious problem is the fact that the erbium doped amplifier introduces broad band noise due to spontaneous emission which can affect the soliton phase and thereby its speed. It can, after distances of the order of 5000 kms, be knocked out of its detection window as a result of these random disturbances. This Elgin-Gordon-Haus [40, 41] effect was remedied by the use of a sliding filter [42, 43], invented by Mollenauer and Mamyshev, to which the soliton would adjust adiabatically but which would suppress all other noise. It seemed finally that with the help of dispersion shifted fibers and sliding filters, soliton transmission would become a reality.

The next challenge was the use of multiple channels on the same fiber. Operating around the wavelength $\lambda = 1.55\mu m$, the hope was to use 16 channels whose center wavelengths ranged from $1.20\ \mu m$ to $1.95\ \mu m$. The problem with multichannels is that, because of the amplification at 20 km intervals, the collisions between the soliton bits of different channels are not solitonic. To address and overcome this difficulty, the dispersion designed fiber was proposed. The idea is that the dispersion in each 20 km segment of the fiber would be designed so that at every point z the nonlinear coefficient $\exp(2Ld \int_0^z \gamma(z)dz)$ in (3.74) would be balanced by dispersion. Choose the dispersion k'' along the fiber segment to be a function of z as well as ω. Then (3.74) becomes (assume $k'' < 0$)

$$\frac{\partial C}{\partial z} - id(z)\frac{\partial^2 C}{\partial z^2} - i\frac{L_D}{L_{NL}}\exp(2L_D d\int_0^z \gamma(z)dz)C^2C^4 = 0 \qquad (3.74)$$

where $d(z) = \frac{|k''(\omega,z)|}{|k_0''(\omega)|}$. Now choose $d(z)$ to match $\exp(2Ld\int_0^z \gamma(z)dz)$ at every z. Then rescaling z to $y(z)$ when

$$y'(z) = d(z) = \frac{L_D}{L_{NL}}\exp(2L_D d\int_0^z \gamma(z)dz)$$

gives us

$$\frac{\partial C}{\partial y} - i\frac{\partial^2 C}{\partial z^2} - 2iC^2C = 0,$$

the exact focusing NLS equation. This brilliant innovation worked extraordinarily well even with using a fairly crude piecewise constant approximation for $d(z)$. In 1996, Mollenauer, Mamyshev and Newbelt [44] demonstrated the soliton multichannel transmission on 6 channels at the rate of 70 Gbits per channel over transoceanic distances.

The only problem with this wonderful idea is that its implementation required replacing all existing fibers with dispersion designed fibers. Indeed, many companies were geared up to do exactly this when another innovation arose, one which could use existing fiber networks. The new idea, called dispersion management, relies on the observation that the linear paraxial approximation to the envelope equation

$$\frac{\partial B}{\partial z} + \frac{ik''}{2}\frac{\partial^2 B}{\partial \tau^2} = 0$$

has the property that a pulse propagated over a distance z_1 spreads out from the shape $B(-z_1, \tau)$ to $B(0, \tau)$. Now if one changes the sign of k'' and then propagates $B(0, \tau)$ a further distance z_1, one finds $B(z_1, \tau) = B(-z_1, \tau)$. The mathematical reason for this is that if $B(z, \tau)$ satisfies $\frac{\partial B}{\partial z} + \frac{ik''}{2}\frac{\partial^2 B}{\partial \tau^2} = 0$ over a distance z_1 spreads out from the shape $B(-z_1, \tau)$ to $B(0, \tau)$. Now if one changes the sign of k'' and then propagates $B(0, \tau)$ a further distance z_1, one finds $B(z_1, \tau) = B(-z_1, \tau)$. The mathematical reason for this is that if $B(z, \tau)$ satisfies $\frac{\partial B}{\partial z} + i\frac{k''}{2}\frac{\partial^2 B}{\partial \tau^2} = 0$, then $B(-z, \tau)$ satisfies $\frac{\partial B}{\partial z} - i\frac{k''}{2}\frac{\partial^2 B}{\partial \tau^2} = 0$. (See §2g on phase conjugation.) The physical reason is that the relative speeds of the different waves making up the pulse $B(-z_1, \tau)$ depend on the sign and size of $k''(z)$. Each wave runs at its own speed for the distance z_1. Then at $z = 0$, the middle of the fiber segment, the dispersion sign is changed. Now the relative speeds of the different waves exactly reverses. After a further distance z_1, the original pulse is exactly restored. The advantage is that, because of dispersion, over most of the cycle, the amplitude of the pulse is low. Nonlinearity is less important. Therefore pulses from different channels pass through each other with little distortion. The probability of collision when the pulses are in the very short amplification segment of the 20 km attenuation-amplification cycle is small.

The equation describing this behavior is (3.74) with $\exp(2L_D \int_0^z \gamma(z)dz)$ being replaced by its average value and by the coefficient of $\frac{\partial^2 C}{\partial \tau^2}$ being $a(z)$ which switches from $+1$ (regular dispersion) to -1 (anomalous dispersion) approximately half way through the 20 km cycle. There is a slight anomalous bias (namely the average of $a(z)$ over the cycle is slightly negative) in order to balance the nonlinearity. Nonlinearity is important to maintain the pulse power at a level to overcome the noise. The solution of

$$\frac{\partial C}{\partial z} + ia(z)\frac{\partial^2 C}{\partial \tau} - i\frac{L_D}{L_{NL}}\langle\exp 2L_D \int^z \gamma(z)dz\rangle C^2 C^* = 0 \qquad (3.75)$$

is no longer the solution given by

$$C(z, \tau) = C_0 \operatorname{sech}\left(\sqrt{\frac{b}{-2a}} C_0 \tau\right) \exp\left(\frac{ib}{2} C_0^2 z\right)$$

for the case where $a(z) = a$ is constant and $b = \frac{L_D}{L_{NL}}\langle \exp 2L_D \int_0^z \gamma(z)dz \rangle$. Instead it has a more Gaussian like shape with oscillatory tails (See Figure 3.9). This alternating dispersion arrangement does not require a replacement of all existing optical lines. Instead, one can put an extra length of fiber with higher regular dispersion at each repeater station where amplification occurs. A pulse would run through 20–25 kms of anomalous dispersion fiber, with $k_0'' = -20$ ps/km say, and then through one fifth that distance in a coiled fiber of dispersion $k_0'' = 100$ ps/km at the repeater station. If it turns out that the dispersion managed fiber arrangement enjoys widespread use, then no doubt new optical fiber will be designed with this feature built in.

In 1999, an experiment by Le Guen et al [45], demonstrated terrabit (10^{12} bits per second) error free transmission using a dispersion managed arrangement using existing standard fibers. It would appear that for at least the next decade or so that it is this arrangement and the solution of (3.75) which will be the arrangement of practical choice. But in a field where there have been so many new developments and new ideas, who can really predict the future?

Stay tuned!   We emphasize that whereas the choice of communication bit has changed from linear pulse to soliton to nonlinear pulse of (3.75), the governing equation, the NLS equation (3.68), has remained virtually the same.

Erbium Doped Fibers

Developments in introducing dopants such as erbium ions into fiber cores has meant that the remarkable broad frequency band amplification properties of these materials can be exploited in a soliton communications network. Such erbium-doped fibers have been used as gain media for lasers themselves. Chains of lumped erbium fiber amplifiers of length 2.2OA m have been spliced into existing optical fiber at 25-km intervals and pumped by a semiconductor laser source. This avoids, among other thins, the undesirable phenomenon of pulse energy oscillation associated with distributed Raman amplifiers which we describe in the next section. Such a soliton communication system with a bit rate of 2.4 gigabits/s has been experimentally realized over a propagation distance of 12,000 km! [36].

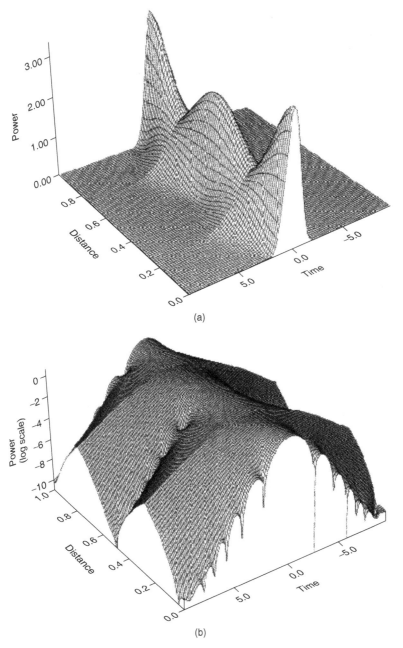

(a)

(b)

FIGURE 3.9 (a) Power versus distance and time of a dispersion managed solution. (b) Power (log scale) versus distance and time. (c) Intensity as function of time at fixed z. Note the amplitude oscillates from positive in the center, to negative, to positive and so on in the wings. The wiggles in the wings are very small. The scale is in powers of 10^{-1}. The envelope is more Gaussian in shape than $sech^2$.

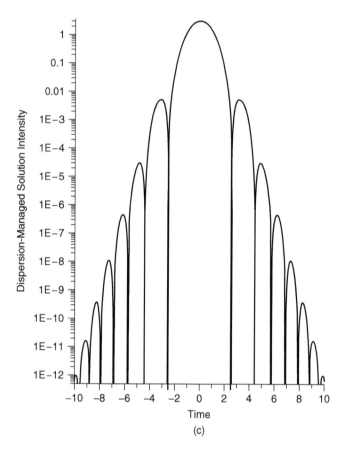

FIGURE 3.9   (*continued*)

Raman Boosting: An Alternative Means of Pulse Amplification

The Raman effect, discussed in Sections 4g and 5e for discrete level atomic and molecular systems, can be viewed as either detrimental or beneficial for pulse propagation in optical fibers. This apparent paradox can be resolved from an inspection of the Raman gain spectrum as measured for silica glass making up the fiber core. We see that the Raman spectrum is extremely broad, extending all the way down to zero frequency shift relative to the incident optical wave. This is in sharp contrast to the situation in the atomic or molecular systems discussed in Sections 4g and 5e, where typically the Raman line is sharp and well separated from the incident wave optical frequency. One important consequence of gain close to the soliton pulse carrier frequency is that the high-frequency components of the pulse spectrum can act as pump waves to drive the lower-frequency components. The result is called the soliton "self-frequency shift", a perturbation whose effect is captured by the $\epsilon i \beta_2 A \frac{\partial}{\partial \tau} |A|^2$

term in (3.64). The magnitude of this shift is a function of pulse width with shorter optical pulses undergoing larger shifts due to the fact that these broader frequency spectra (Fourier transforms of the pulse width) overlap with more of the Raman gain spectrum shown in Figure 3.10. For example, a 260 fs (10^{-15} s) soliton pulse over a fiber length of 392 m is experimentally observed to be frequency downshifted by 4% of the optical carrier frequency. The perturbation responsible for the retarded contribution above can also split an $N = 2$ soliton pulse into separate components. Fortunately, for soliton communication systems with pulse width of several tens of picoseconds, the soliton self-frequency shift can be safely ignored.

A beneficial aspect of the Raman effect is that it can be exploited as a boost to the decaying soliton resulting from attenuation. While a soliton propagating over a 40-km distance can lose more than 80% of its energy due to losses, a wak pump laser if injected into both ends of the fiber at a frequency corresponding to the maximum of the Raman gain curve shown above (i.e., upshifted from the soliton frequency by $\sim$400 cm^{-1} $\equiv 1.2 \times 10^{1}2$ s^{-1}) can restore the soliton pulse's energy. The mechanism is identical to that causing the soliton self-frequency shift, but now the energy is being drained from an independent higher-frequency source rather than from the soliton pulse spectrum itself. This mechanism is elaborated on further in Sections 4g and 5e. A schematic of a soliton-based communication systems is sketched in Figure 3.10 showing a Raman boost applied in both directions through a directional coupler at regular intervals of $L = 40$ km. The pump laser that replenishes the soliton pulse energy is a semiconductor laser diode. This boost mechanism has been demonstrated experimentally over tens of kilometer distances. An undesirable aspect of Raman compensation is that the pump energy from the diode laser is absorbed as it converts into soliton pulse energy with

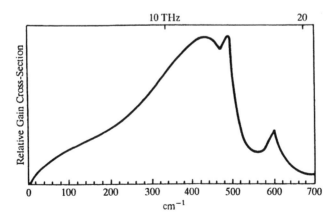

FIGURE 3.10   Relative Raman gain versus frequency difference between pump and signal.

distance. As a result, the pulse energy oscillates in z following the oscillation in the Raman gain itself, as shown in Figure 3.11.

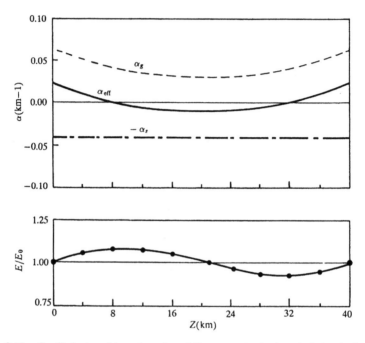

FIGURE 3.11   Coefficients of loss $(-\alpha_g)$ and Raman gain (α_g) and their algebraic sum (α_{eff}) for an amplification period $L = 40$ km.

For an additional reference on fiber optics, the reader may also consult Agrawal [46].

3d.   Review of Waveguiding Principles and The Potential Applications of Waveguides for Nonlinear Materials

The enormous information carrying capacity of optical fibers poses many important and difficult questions. How can we cram all of this information, which might be wavelength or time multiplexed, into a single optical channel, and how can we separate each bit train, encoded say as solitary pulses, at the output end for further processing?

Multiplexing is a technique whereby different information channels (telephone conversations, faxes, data transmission on computer links, etc.) are intermingled in order to increase the information carrying capacity of a communications link. A critical part of any high-speed communications link is switching networks, which in contrast to optical fibers with repeaters many hundreds of kilometers long, are local miniaturized stations possibly just a few hundred micrometers in length. The challenge then is to separate out by some means ultrahigh-speed information trains encoded as optical pulses over extremely short distances. The directional coupler, which consists of two light guides fabricated side by side, provides a means of redirecting, combining, or separating individual channels of information. Directional couplers are phase switches, and existing devices rely on the first order electro-optic effect whereby an applied voltage modifies the propagation constant of the waveguided mode. The schematic of this device in Figure 3.12 shows light coupled in through the input channel, which would be the output end of a fiber spliced to the coupler, while the second channel can receive an applied electrical voltage from an external source. Many such directional couplers may be placed end to end to build sophisticated switching networks capable of separating tens or hundreds of channels. The basic problem with this device, as with the fiber repeater, is the electrical component. This again limits switching speed, and an all-optical alternative is an obviously desirable solution.

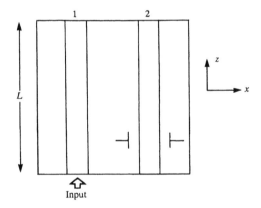

FIGURE 3.12   Directional coupler.

Optical nonlinearity in waveguide media affords the possibility of achieving all of these switching actions in an all-optical integrated structure, avoiding the need for electrical control. The idea is simple and already familiar to the reader! The switching action now relies on the fact that the intrinsic $\chi^{(3)}$ nonlinear response of dielectric materials induces an intensity dependent change in the mode propagation constant through the nonlinear refractive index change described in Chapter 2. In weakly nonlinear materials the switching action is essentially the same as for the electrical case described above, and the mode profiles of the individual waveguide channels remain unperturbed as the energy transfers from one channel to the other. Mathematically such switching action is simple, being described by the kind of coupled mode analysis treated in Section 2i. Strongly nonlinear interactions have stimulated considerable excitement [47, 48, 49], promising novel nonlinear switching effects. The important theoretical observation here is that the underlying "linear" mode profile of the guide becomes increasingly distorted as the light intensity is increased. Moreover, various permutations and combinations of nonlinear dielectric slabs making up planar waveguides offer a rich variety of bifurcating solutions whereby the shapes of the underlying nonlinear modes undergo profound changes as a function of light intensity. The simple mathematical description of coupled mode analysis now has to be abandoned, and we are led to study the solutions of nonlinear evolution equations of the NLS form where both the linear and nonlinear coefficients may be functions of the coordinate transverse to the direction of light propagation, and both may change discontinuously as a function of this coordinate. The goal of the remainder of this chapter is to enumerate some of these nonlinear surfaces and guided wave solutions, investigate their stability, and provide an equivalent particle prescription that allows one to study sophisticated nonlinear switching effects via a simple Newtonian dynamical model. The soliton will reemerge as the robust nonlinear structure that allows for such a simple picture. Now, however, the soliton shape corresponds to a self-trapped light channel in space, and the balance that maintains its invariant "sech" shape comes from nonlinear self-focusing and linear diffraction.

A word of caution is needed at this stage! Whereas the theoretical analysis that follows provides some very nice and interesting phenomenology in nonlinear science, the practical realization of many of these highly nonlinear devices awaits the development of suitable nonlinear materials. The subject is therefore in its infancy, and, perhaps not surprisingly, some of the most promising materials are glasses specially doped with semiconductor materials. Doping increases absorption losses, but this may not pose a significant problem as the propagation lengths are so short.

A thin-film planar waveguide consists of three layers of different dielectric materials with the central guiding layer having a refractive index greater than both outer layers. This sandwich structure is usually grown by depositing a thin film of dielectric material on a glass substrate. The upper layer is often just air, but a further layer of dielectric material may be deposited on top of the thin film, and this is referred to as the cladding. The principle of light confinement in such a structure is exactly the same as that for an optical fiber with light rays undergoing total internal reflection at the high-low index interface. This interface need not be sharp, and graded index structures in which the refractive index makes a continuous transition from one value to another are common; in this case light rays are gradually bent as they move toward the lower index region as discussed earlier. We will see in Section 6e that if the guiding layer with refractive index n_0 and width $2L$ is sandwiched between two (thick) layers of refractive index n_1, the number of modes the waveguide can support depends on the waveguide parameter

$$V = 2\pi \frac{2L}{\lambda} \sqrt{n_0^2 - n_1^2} \tag{3.76}$$

where λ is the vacuum wavelength of the light and $k = 2\pi c/\omega$. For the purpose of the discussion in these sections, V is taken to be small so that only *one* confined mode can be supported by the waveguide. This weak guidance assumption makes the analysis of fibers and waveguides much simpler as it means that the fields can be assumed to be essentially transverse.

3e. Transverse Electric (TE) and Transverse Magnetic (TM) Modes of a Planar Waveguide. [6e]

An essential difference between an optical fiber and a planar waveguide, from the point of view of ease of theoretical analysis, lies in their differing geometries. The cylindrical symmetry of the fiber ensures that different electric and magnetic field components are coupled in Maxwell's equations, leading in general to relatively complicated hybrid modes. However, under the weak guidance assumption, applicable to most commercial fibers, the fields are essentially transverse, and the so-called linearly polarized modes are solutions to a scalar wave equation. The result is that pulse propagation in optical fibers can be described by an NLS equation, as discussed earlier. Thin-film planar waveguides, on the other hand, are fundamentally different in that the planar geometry leads, without appealing to the weak guidance assumption, to a natural decoupling of the Cartesian field components that satisfy Maxwell's equations into separate sets describing transverse electric (TE) and transverse magnetic (TM) waves. Figure 3.13 illustrates each case for an infinite plane wave incident at an angle to the interface separating two semi-infinite dielectric media. (See also references [40–42].)

The interface located at $x = 0$ extends to $+\infty$ and $-\infty$ in the y-direction, and the plane of incidence is defined as the x, z plane. In (a) the electric field vector is polarized at right angles to the plane of incidence (in the y-direction), that is, $\vec{E} = (0, E_y, 0)$ whereas in (b) the magnetic field vector is polarized at right angles to the plane of incidence ($\vec{H} = (0, H_y, 0)$). The two sets of equations describing TE and TM excitation are derived by assuming that all of

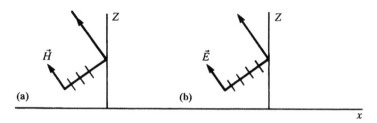

FIGURE 3.13   Picture of TE and TM modes.

the time dependence is contained in the term $e^{-i\omega t}$ and substituting $\partial/\partial y = 0$
in Maxwell's equations:

TE (*s*-polarization): $\vec{E} = (0, E_y, 0)$, $\vec{H} = (H_x, 0, H_z)$

$$\frac{\partial}{\partial x} H_z - \frac{\partial}{\partial z} H_x = i\omega D_y$$

$$\frac{\partial}{\partial x} E_y = -i\omega\mu H_z \qquad (3.77)$$

$$\frac{\partial}{\partial z} E_y = i\omega\mu H_x$$

TM (*p*-polarization): $\vec{H} = (0, H_y, 0)$, $\vec{E} = (E_x, 0, E_z)$

$$\frac{\partial}{\partial x} E_z - \frac{\partial}{\partial z} E_x = -i\omega\mu H_y$$

$$\frac{\partial}{\partial x} H_y = -i\omega D_z \qquad (3.78)$$

$$\frac{\partial}{\partial z} H_y = i\omega D_x$$

Because the constitutive relation between the electric displacement vector $\vec{D}$
and the electric field $\vec{E}$ is in general a tensor relation, the analysis of TM
waves is particularly difficult. In the last section of this chapter, we derive
one particular example of a TM surface wave. Initially, however, we address
the much simpler problem of computing TE nonlinear surface waves.

We saw in Chapter 2 that the equation for the envelope of a TE wave

$$\vec{E} = \hat{y}\ F(x, z)e^{i\beta(\omega/c)z - i\omega t} + (*) \qquad (3.79)$$

in a nonlinear dielectric with refractive index $n^2(x, |F|^2)$ is given by

$$2i\beta \frac{\omega}{c} \frac{\partial F}{\partial z} + \frac{\partial^2 F}{\partial x^2} - (\beta^2 - n^2(x, |F|^2)) \frac{\omega^2}{c^2} F = 0. \qquad (3.80)$$

In deriving (3.80) from the original equation for the propagation of the electric field (2.10), we assume that the neglected term $\partial^2 F / \partial z^2$ is much smaller than $2i\beta(\omega/c)(\partial F/\partial z)$, which in turn is much smaller than $\beta^2(\omega^2/c^2)F$ and of the same order as the diffraction term $\partial^2 F/\partial x^2$ and the term $(\beta^2 - n^2)(\omega^2/c^2)F$, reflecting the difference between the mode propagation constant β (yet to be determined) and the refractive index. In particular, therefore, if $n^2 = n_0^2 + \alpha_0|F|^2$ in one medium, say, and $n^2 = n_1^2 + \alpha_1|F|^2$ in another, the requirement that $\beta^2 - n^2$ is small means that both $n_0^2 - n_1^2$ and the product of the nonlinear refractive indices α_0 and α_1 with the intensity $|F|^2$ must be small and of equal orders of magnitude. In what follows, we take the small parameter in the theory to be

$$\Delta = n_0^2 - n_1^2. \qquad (3.81)$$

At this stage it is convenient also to nondimensionalize the variables x and z by c/ω; namely we let $x' = (\omega/c)x, z' = (\omega/c)z$ and, dropping the primes, obtain

$$2 i\beta \frac{\partial F}{\partial z} + \frac{\partial^2 F}{\partial x^2} - (\beta^2 - n^2 \left(\frac{c}{\omega}x, |F|^2 \right) F = 0. \qquad (3.82)$$

3f. Nonlinear Surface and Guided TE Waves: Statics. [6d]

Let us begin by recalling how an intensity dependent refractive index with $(dn/dI)(I = |F|^2)$ positive can lead to a localized wavepacket. We seek stationary (z-independent) solutions $F(x)$ to equation (3.80) with $n^2 = n_0^2 + \alpha_0|F|^2, \alpha_0 > 0$,

$$\frac{d^2 F}{dx^2} - (\beta^2 - n_0^2 - \alpha_0|F|^2)F = 0 \qquad (3.83)$$

where F tends to zero as $x \to \pm\infty$. There is no loss of generality in assuming F to be real, although any solution may be multiplied by the constant factor $e^{i\phi}$. It is useful, before writing down the exact solution, to analyze (3.83), with F real, using phase plane methods. Multiply (3.83) by dF/dx, integrate with respect to x, use the fact that F tends to zero at $x = \pm\infty$ to set the integration constant to zero, and find

$$\frac{dF}{dx} = \pm F\sqrt{\beta^2 - n_0^2 - \tfrac{1}{2}\alpha_0 F^2}. \tag{3.84}$$

The graph of (3.84) is shown in Figure 3.14. As x transverses from $-\infty$ to ∞, the solution $F(x)$ follows the homoclinic orbit $ABCD$, which joins the saddle point A to itself, exiting on the unstable manifold and returning on the stable manifold in the direction shown. We have used the $+$ sign in (3.84) for the portion ABC and the $-$ sign for the portion CDA. By making a hyperbolic secant substitution, we can integrate (3.84) explicitly to find

$$F(x) = \sqrt{\frac{2(\beta^2 - n_0^2)}{\alpha_0}} \ \mathrm{sech}\sqrt{\beta^2 - n_0^2}(x - \overline{x}). \tag{3.85}$$

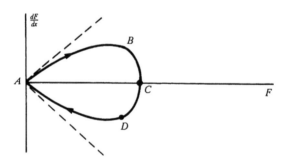

FIGURE 3.14   Homoclinic orbit consisting of the unstable and stable directions of the saddle point A. The dashed lines represent the lines $dF/dx = \pm\sqrt{\beta^2 - n_0^2}F$. It is the nonlinear term, when $\alpha_0 > 0$, that overcomes the exponential growth from $x = -\infty$ and allows the solution to join up with the exponentially decaying solution as $x \to \infty$.

There is no restriction on $\overline{x}$, the center of the focused layer. The power

$$P = \int_{-\infty}^{\infty} FF^* dx \tag{3.86}$$

in the wavepacket is $4/\alpha_0\sqrt{\beta^2 - n_0^2}$. We define

$$\eta_0 = \tfrac{1}{2}\sqrt{\beta^2 - n_0^2} \tag{3.87}$$

whence the power $P = 8\eta_0/\alpha_0$, and the effective refractive index or propagation constant β is given by

$$\beta = \sqrt{n_0^2 + \frac{\alpha_0^2}{16}P^2} = \sqrt{n_0^2 + 4\eta_0^2} > n_0, \tag{3.88}$$

whose graph is shown in Figure 3.15. Observe that β must be greater than n_0 and that there is a stationary solution for all values of P. Observe also that if we had simply looked for solution $\vec{E}(x, z) = \hat{y}F(x, z)\exp(in_0(\omega/c)z - i\omega t)$, then the effective propagation wavenumber would have been $n_0 + 2\eta_0^2/n_0$ which is (3.88) to order $\left(\frac{\eta_0}{n_0}\right)^4$.

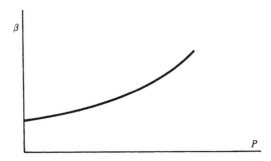

Figure 3.15   Effective refractive index as a function of wavepacket power P.

The z-dependent localized wavepacket corresponding to (3.85) has the form

$$F(x, z) = 2\eta_0\sqrt{\frac{2}{\alpha_0}} \operatorname{sech} 2\eta_0\left(x - v\frac{z}{2\beta} - \bar{x}\right) e^{iv/2x - i\left(\frac{v^2}{4} + \beta^2 - n_0^2 - 4\eta_0^2\right)\frac{z}{2\beta}}$$

$$\tag{3.89}$$

with power $P = 8\eta_0/\alpha_0$. This is simply the one-soliton solution of equation (3.80). Because we have allowed z-dependency, we are no longer required to take $\beta^2 - n_0^2 - 4\eta_0^2 = 0$. However note that the z-independent solutions requires $v = 0$ and $4\eta_0^2 = \beta^2 - n_0^2$. The corresponding electric field $\vec{E}$ is given by

$$\vec{E}(x, z) = \hat{y}\sqrt{\frac{2}{\alpha_0}}2\eta_0 \operatorname{sech} 2\eta_0\left(x - v\frac{z}{2\beta} - \bar{x}\right)$$

$$\times \exp iz\left(\beta - \frac{v^2}{4\beta} - \frac{\beta^2 - n_0^2 - 4\eta_0^2}{2\beta}\right) + \frac{iv}{2}x - i\omega t + (*). \tag{3.90}$$

Also notice that the direction of the phase of the incoming wave is approximately (the terms v^2, $\beta^2 - n_0^2$ and n_0^2 are small compared to β^2)

$$\beta\left(\frac{v}{2\beta}, 0, 1\right),\tag{3.91}$$

which is also the direction, $\frac{dx}{dz} = \frac{v}{2\beta}$, taken by the maximum of the argument of the hyperbolic secant. Therefore, whereas (3.85) represents the envelope of a confined light beam travelling parallel to the z-axis, equation (3.90) represents a confined light beam making an angle $\tan^{-1} v/2\beta$ with the z direction.

What we have learned, then, is that in a nonlinear dielectric with a focusing ($\alpha_0 > 0$) nonlinearity, a light beam can be naturally confined through the balance of diffraction and nonlinear focusing. We will shortly be asking what happens when such a wavepacket or light beam impinges on an interface at $x = 0$ dividing two dielectric materials with refractive indices $n_0^2 + \alpha_0|F|^2$ and $n_1^2 + \alpha_1|F|^2$. First, however, we seek stationary (i.e., z-independent) solutions which correspond to light parallel to the z-axis.

Accordingly, we seek solutions to (3.83) for the envelope $F(x)$ of a nonlinear TE wave that is both independent of z and localized at the interface $x = 0$. The envelope $F(x)$ satisfies

$$\frac{d^2 F}{dx^2} - \left(\beta^2 - n^2\left(\frac{cx}{\omega}, |F|^2\right)\right) F = 0\tag{3.92}$$

where

(a) the refractive index on both sides of the interface has a Kerr dependence on the field intensity, that is, $n^2 = n_0^2 + \alpha_0|F|^2$, $x < 0$ and $n^2 = n_1^2 + \alpha_1|F|^2$, $x > 0$

(b) F and dF/dx are continuous at the interface $x = 0$ reflecting the continuity of the electric and magnetic fields tangent to the interface

(c) $F(x)$ decays exponentially as $x \rightarrow \pm\infty$. There is no loss of generality in taking F to be real.

We observe that in the limit where α_0 and α_1 both tend to zero, equation (3.92) with the boundary conditions (a) and (c) has no nontrivial (i.e., $F \neq 0$) solution. Why? In order that the solution decay exponentially for all x, β^2 must be greater than both n_0^2 and n_1^2. However, in that case, we cannot ensure the continuity of both F and dF/dx at $x = 0$ unless $F \equiv 0$. The nonlinear problem (3.92), however, has nontrivial solutions. We will write them down explicitly in a moment, but first let us again use a phase plane analysis. As before, multiply equation (3.92) by dF/dx, integrate with respect to x, and apply the boundary condition at x equals either plus or minus infinity to find

$$\frac{dF}{dx} = \pm F\sqrt{(\beta^2 - n_j^2) - \frac{\alpha_j}{2}F^2}, \quad j = 0, 1 \tag{3.93}$$

when $j = 0$ or 1 depending on whether x is less than or greater than zero. These curves are graphed in Figure 3.16 for $x < 0$ (dashed line) and $x > 0$ (solid line), where it has been assumed that $0 \leq \alpha_0 < \alpha_1$ and $n_0 > n_1 > 0$.

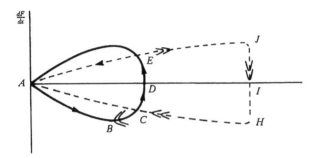

FIGURE 3.16   Graph of equation (3.88), dF/dx vs. F. One solution is marked by the single arrow *ABCDEA*. The other *AJIHCBA* is marked by the double arrow.

EXERCISE 3.2

Given that $n_0 > n_1$, use the arguments below to show that unless $\alpha_1 > \alpha_0 \geq 0$, there can be no nontrivial solution. □

Remark: Since the field F must tend to zero as $x \to \pm\infty$ and since at infinity the term $1/2\alpha_j F^2$, $j = 0, 1$ is negligible compared to $\beta^2 - n_j^2$, $j = 0, 1$, the propagation constant must be greater than both n_0 and n_1.

Observe that we can construct a solution as follows. Begin at A and follow the solid line homoclinic orbit *ABCDEA* in the counterclockwise direction until you reach E. In the portion *ABCD*, we use the negative sign in (3.93), in *DE* the positive sign. At E, where F and dF/dx for both the solid and dashed curves are equal, switch onto the dashed line back to A. Note that in order that an intersection occurs, D must be to the left of I, which means

$$\frac{\beta^2 - n_1^2}{\alpha_1} < \frac{\beta^2 - n_0^2}{\alpha_0},$$

which in precisely the condition that the power associated with this solution is defined (the argument of the square root in the up coming equation (3.101) is positive). In Figure 3.17, we draw this solution F vs. x. If we had continued along the solid line to A, we would have obtained a solution completely symmetric about D that would correspond to a local pulse in the right-hand medium given by (3.85) with subscript 0 replaced by 1.

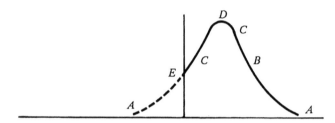

FIGURE 3.17   Graph of F vs. x for $ABCDA$. This solution is stable.

We can also construct a second solution, shown in Figure 3.18 and with the double arrow in Figure 3.16. Beginning at A (now corresponding to $x = -\infty$), follow the dashed line AJ (the branch of (3.93) with positive slope) through I and H around to C (at which F and dF/dx of both the dashed and solid curves are equal), and then follow the solid line CBA back to A.

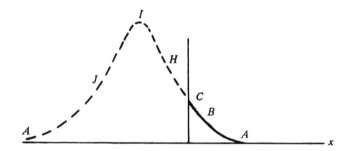

FIGURE 3.18   Graph of F vs. x for $AJIHCA$. This solution is unstable.

We will see in the next section that the first solution is stable, the second unstable in the sense that if we add back in the terms $2i\beta(\partial F/\partial z)$ and ask if a small perturbation to either solution amplifies as z increases, we will find that the first solution is neutrally stable (the perturbed solution simply oscillates about the z-independent one) whereas the second solution is unstable (the perturbation grows exponentially in z) and the wavepacket leaves the interface. *Thus stable nonlinear surface waves attach themselves to the side of the interface with smaller linear and larger nonlinear refractive index.*

EXERCISE 3.3

Can you construct a heuristic and intuitive argument for the stability of *ABCDA* and of *AJIHEA*? (*Hint*: Use the idea that light beams move toward regions of increased refractive index.) ☐

We now write these solutions down explicitly. The analysis is straightforward. In order to solve (3.93), make a hyperbolic secant substitution for F. We find the first solution is

$$
F^+(x) = \begin{cases} \sqrt{\dfrac{2(\beta^2 - n_0^2)}{\alpha_0}} \; \mathrm{sech}\sqrt{\beta^2 - n_0^2}(x - x_0), & x < 0 \\[3mm] \sqrt{\dfrac{2(\beta^2 - n_1^2)}{\alpha_1}} \; \mathrm{sech}\sqrt{\beta^2 - n_1^2}(x - x_1), & x > 0 \end{cases} \tag{3.94}
$$

where in order to satisfy the boundary condition (F^+ and dF^+/dx continuous) at $x = 0$,

$$
x_j = \frac{1}{2\sqrt{\beta^2 - n_j^2}} \ln \frac{1 + r_j\sqrt{1 - \mu^2}}{1 - r_j\sqrt{1 - \mu^2}}, \qquad j = 0, 1 \tag{3.95}
$$

and

$$
r_0 = \sqrt{\frac{(\beta^2 - n_1^2)}{(\beta^2 - n_0^2)}}, \qquad r_1 = 1, \qquad \mu^2 = \frac{n_0^2 - n_1^2}{\left(1 - \frac{\alpha_0}{\alpha_1}\right)(\beta^2 - n_1^2)}.
$$

For a reason that will shortly be clear, carry out the following exercise.

EXERCISE 3.4

Define η_0 by the relation

$$
\beta^2 = n_0^2 + 4\eta_0^2 \tag{3.96}
$$

and show from (3.95) that

$$
\mathrm{sech}^2 2\eta_0 x_0 = \frac{(n_0^2 - n_1^2)\frac{\alpha_0}{\alpha_1}}{\left(1 - \frac{\alpha_0}{\alpha_1}\right) 4\eta_0^2}. \tag{3.97}
$$

Notice that for a solution x_0, the right-hand side of (3.97) must be less than unity. We will meet the relation (3.97) again soon. Show that the condition that the right-hand side of (3.97) is less than unity is precisely the condition that $\mu^2 < 1$ and that the point D in Figure 3.15 lies to the left of the point I. □

Observe that for the existence of this solution, we require $n_0^2 > n_1^2$, $\alpha_0 < \alpha_1$, $\beta^2 > n_0^2$, n_1^2. The second solution $F^-(x)$ is the dual of $F^+(x)$, found simply by noting that if x_0, x_1 are chosen so that $F^+(x)$ satisfies the boundary conditions at $x = 0$, then $-x_0, -x_1$ also define a solution with the same properties

$$
F^-(x) = \begin{cases}
\sqrt{\dfrac{2(\beta^2 - n_0^2)}{\alpha_0}}\ \operatorname{sech}\sqrt{\beta^2 - n_0^2}(x + x_0), & x < 0 \\[4mm]
\sqrt{\dfrac{2(\beta^2 - n_1^2)}{\alpha_1}}\ \operatorname{sech}\sqrt{\beta^2 - n_1^2}(x + x_1), & x > 0.
\end{cases}
\tag{3.98}
$$

The condition that D lies to the left of I in Figure 3.15 means that the amplitude of $F^+(x)$ is less than that of $F^-(x)$.

There is an interesting limit solution when $\alpha_0 = 0$ whence the refractive index of the left-hand medium is independent of the field intensity. For $x > 0$.

$$
\lim_{\alpha_0 \to 0} F^+(x) = \sqrt{\frac{2(\beta^2 - n_1^2)}{\alpha_1}}\ \operatorname{sech}\left(\sqrt{\beta^2 - n_1^2}(x - x_1)\right)
$$

$$
\frac{\omega}{c}x_1 = \frac{1}{2\sqrt{\beta^2 - n_1^2}}\ \ln \frac{\sqrt{\beta^2 - n_1^2} + \sqrt{\beta^2 - n_0^2}}{\sqrt{\beta^2 - n_1^2} - \sqrt{\beta^2 - n_0^2}}
\tag{3.99}
$$

and for $x < 0$,

$$
\lim_{\alpha_0 \to 0} F^+(x) = \sqrt{\frac{2(n_0^2 - n_1^2)}{\alpha_1}}\ \exp(\sqrt{\beta^2 - n_0^2}\,x).
\tag{3.100}
$$

EXERCISE 3.5

It takes a little effort to show that

$$
\lim_{\alpha_0 \to 0} \frac{1}{\sqrt{\alpha_0}}\ \exp(-\sqrt{\beta^2 - n_0^2}\,x_0) = \sqrt{\frac{n_0^2 - n_1^2}{4\alpha_1(\beta^2 - n_0^2)}}.
$$
  □

There is no corresponding limit of the $F^-(x)$ solution because if $\alpha_0 = 0$ and there is no nonlinearity in the left-hand medium, it is impossible to localize the pulse there.

We now calculate the powers P^+ and P^- in these two solutions:

$$P^{\pm} = \int_{-\infty}^{\infty} \left(F^{\pm}(x)\right)^2 dx$$

$$= 2\left[\frac{\sqrt{\beta^2 - n_0^2}}{\alpha_0} + \frac{\sqrt{\beta^2 - n_1^2}}{\alpha_1} \pm \left(\frac{1}{\alpha_1} - \frac{1}{\alpha_0}\right)\sqrt{\beta^2 - n_1^2 - \frac{n_0^2 - n_1^2}{1 - \frac{\alpha_0}{\alpha_1}}}\right].$$

$$(3.101)$$

The positivity of the expression under the square root is equivalent to $\mu^2 < 1$ and the fact that D lies to the left of I in Figure 3.15.

EXERCISE 3.6

Show in the limit, $n_1 \to n_0, \alpha_1 \to \alpha_0,$

$$P^{\pm} = \frac{4}{\alpha_0}\sqrt{\beta^2 - n_0^2} = \frac{8\eta_0}{\alpha_0}$$

using (3.96) as the definition of η_0. □

EXERCISE 3.7

Show that if we define η_0 by (3.96), the radical

$$\beta^2 - n_1^2 - \frac{n_0^2 - n_1^2}{1 - \frac{\alpha_0}{\alpha_1}} = 4\eta_0^2\left(1 - \frac{\frac{\alpha_0}{\alpha_1}(n_0^2 - n_1^2)}{4\eta_0^2(1 - \frac{\alpha_0}{\alpha_1})}\right).$$

□

The existence of stationary solutions requires the parameter

$$\frac{\frac{\alpha_0}{\alpha_1}(n_0^2 - n_1^2)}{4\eta_0^2(1 - \frac{\alpha_0}{\alpha_1})}$$

to be less than one. Therefore, stationary solutions are only achieved when the effective refractive index β is sufficiently large.

In Figure 3.19, we graph the relations (3.95) and (3.101), giving values of $\bar{x}$ and β as a function of power P. For $F^+(x)$, $\bar{x} = x_1$ and $P = P^+$ whereas

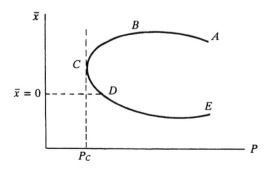

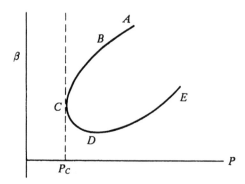

FIGURE 3.19   Plots of $\bar{x}$ and β as a function of P.

for $F^-(x)$, $\bar{x} = x_0$ and $P = P^-$. The point D corresponds to a value of β where the radical in (3.101) is zero, namely when

$$\beta^2 = n_1^2 + \frac{n_0^2 - n_1^2}{1 - \frac{\alpha_0}{\alpha_1}}.$$

This is also the point at which $\mu^2 = 1$ and the last value x_1 for which the pulse $F^+(x)$ has a maximum in the right-hand medium. The point C corresponds to the configuration that requires minimum power. The branch $ABCD$ corresponds to $F^+(x)$, DE to $F^-(x)$. The branch ABC is stable, the branch CDE is unstable. In particular, this means that when the maximum of $F^+(x)$ gets close to the interface (x_1 lies between the points C and D), the surface wave is unstable. The graph β and $\bar{x}$ vs. P in the limit $\alpha_0 \to 0$ are shown in Figure 3.20.

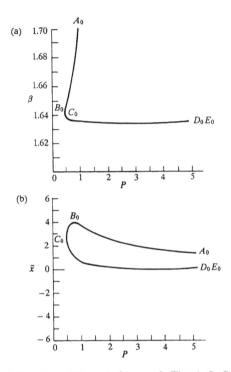

FIGURE 3.20   Plots of β vs. P and $\bar{x}$ vs. P for $\alpha = 0$. The $A_0 B_0 C_0$ branch is stable and $C_0 D_0 E_0$ branch is unstable.

3g.   Nonlinear Surface Waves at a Single Interface: Dynamics. [6k]

In this subsection, we return to the z-dependent (which we call the dynamic) problem and attempt to answer questions such as (see also [50–54]:

- Why does the stationary solution attached to an interface between two different dielectrics require a minimum power when a stationary solution in a single medium does not?

- Are the stationary solutions, shown in Figures 3.17 and 3.18, and represented by the various branches in Figure 3.19 stable or unstable?

- What happens when a TE light beam or wavepacket, which initially has the structure given by equation (3.90) with $v > 0$, impinges on the interface from the left-hand medium? How much of the beam is reflected or transmitted?

- If transmitted, how does the shape and angle of the beam change? Because of the different nonlinear refractive indices ($\alpha_1 > \alpha_0$), a light beam that crosses the interface from the weaker to the stronger nonlinear medium has more power than it needs to retain its single soliton shape and breaks up into one or more solitons plus some radiation in the right-hand medium.

- What happens to a soliton that is initiated in the right-hand medium (formula (3.90) with subscript 0 replaced by 1 and v negative) when it impinges on the interface at $x = 0$?

The reader will recognize the answers of these questions as being analogous to Snell's laws at an interface between linear dielectrics. Our approach will be to treat the wavepacket or light beam as a particle whose position is given by $\overline{x}(z)$ where z is treated as the "time" variable. The change in refractive index at the interface gives rise to a perturbative potential $U(\overline{x})$ for this particle, and we will find an equation of motion for the particle to be

$$\frac{d^2\overline{x}}{dz^2} = -\frac{\partial U}{\partial \overline{x}}. \tag{3.102}$$

where the potential

$$U\left(2\eta_0\overline{x}, \frac{n_0^2 - n_1^2}{4\eta_0^2(1 - \frac{\alpha_0}{\alpha_1})}\right)$$

depends on the initial power in the wavepacket and the ratio $(n_0^2 - n_1^2)/(\alpha_1 - \alpha_0)$ of the differences of linear and nonlinear refractive indices. The initial velocity of the particle corresponds to the angle of incidence of the beam at the interface. The approach is represented schematically in Figure 3.21. It is clear that if the light beam–particle equivalence obtains, it is extremely simple to calculate exactly what happens to the light beam by referring to the particle in the potential well picture. The behavior of the light beam can then be readily inferred from the shape of the potential and the initial velocity (angle of incidence) of the particle (light beam).

We begin by choosing convenient nondimensional rescalings for the dependent and independent variables. Our equation is

$$2i\beta\frac{\partial F}{\partial z} + \frac{\partial^2 F}{\partial x^2} - \left(\beta^2 - n^2\left(\frac{cx}{\omega}, |F|^2\right)\right)F = 0 \tag{3.103}$$

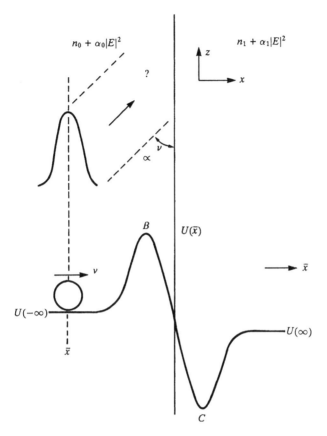

FIGURE 3.21

where $n^2(cx/\omega, |F|^2) = n_0^2 + \alpha_0|F|^2$, $x < 0$ and $n_1^2 + \alpha_1|F|^2$, $x > 0$. We will imagine that the initial condition on (3.103) is the single soliton (3.89) that approaches the interface from the left medium. (The assumption of a single soliton as initial condition is not particular; we know from the general theory of Chapter 6 that any initial pulse will decompose into a set of solitons and radiation. We simply treat their interactions with the interface separately. The case of an initial pulse that is a bound double soliton must be treated separately but can be handled in a similar way.) Since we are including z-dependence in F, the choice of β is quite arbitrary as long as $\beta^2 - n_0^2$ and $\beta^2 - n_1^2$ are both small enough so that the assumption that $\partial^2 F/\partial z^2 \ll \beta\frac{\omega}{c}\frac{\partial F}{\partial z} \ll \beta^2\frac{\omega^2}{c^2}F$ remains valid. This means, of course, that we are assuming that the jump in linear refractive index $n_0 - n_1$ is also small (compared to $n_0 + n_1$). Here for the moment we will leave β to be a definite but as yet unspecified constant.

Recall that x and z have already been nondimensionalized by ω/c; that is, they are measured in units of length for which $c/\omega = 1$. We write

$$x' = \sqrt{\Delta}x$$

$$z' = \Delta\frac{z}{2\beta} \tag{3.104}$$

$$A(x', z') = \sqrt{\frac{\alpha_0}{2\Delta}}\, F(x, z)\exp i(\beta^2 - n_0^2)\frac{z}{2\beta}$$

where $\Delta = n_0^2 - n_1^2$ and find

$$i\frac{\partial A}{\partial z'} + \frac{\partial^2 A}{\partial x'^2} + 2|A|^2 A = V(x')A,$$

$$V(x') = \begin{cases} 0, & x' < 0 \\ 1 - 2(\alpha^{-1} - 1)|A|^2, & x' > 0. \end{cases} \tag{3.105}$$

With $z = 0$, we have

$$A(x', 0) = \frac{2\eta_0}{\sqrt{\Delta}}\,\mathrm{sech}\,\frac{2\eta_0}{\sqrt{\Delta}}(x' - x_0')\exp i\frac{v_0}{2\sqrt{\Delta}}x'.$$

Therefore the six parameters $n_0, n_1, \alpha_0, \alpha_1$, the power $P = 8\eta_0/\alpha_0$, and angle of incidence v_0 (proportional to the $\sin\psi_i$) of the incoming wavepacket are effectively reduced to three:

$$\eta_0' = \frac{\eta_0}{\sqrt{\Delta}},\ \text{the nondimensional and rescaled power}$$

$$v_0' = \frac{v_0}{\sqrt{\Delta}},\ \text{the nondimensional and rescaled angle of incidence}$$

$$\alpha = \frac{\alpha_0}{\alpha_1},\ \text{the ratio of the nonlinear refractive indices.} \tag{3.106}$$

In what follows, we will drop the primes on x, z, η_0, and v. Define

$$p(z) = \int_{-\infty}^{\infty} AA^* dx, \quad \bar{x}(z) = \frac{1}{p}\int_{-\infty}^{\infty} xAA^* dx,$$

$$v(z) = \frac{i}{p}\int_{-\infty}^{\infty}\left(A\frac{\partial A^*}{\partial x} - A^*\frac{\partial A}{dx}\right)dx \tag{3.107}$$

and find from (3.105) that

$$\frac{dp}{dz} = 0 \tag{3.108}$$

$$\frac{d\overline{x}}{dz} = v \tag{3.109}$$

$$\frac{dv}{dz} = -\frac{2}{p}\int_{-\infty}^{\infty} \frac{\partial V}{\partial x} A A^* dx \tag{3.110}$$

where

$$V(x) = 0, \ x < 0 \ \text{and} \ 1 - 2(\alpha^{-1} - 1)|A|^2, \ x > 0. \tag{3.111}$$

So far, things are exact. We now introduce an approximation, namely that the intensity $|A|^2$ of the field moves collectively and therefore is a function of a single coordinate $x - \overline{x}(z)$. This means we can replace the derivative $\partial/\partial x$ of the integrand in (3.110) by a derivative $-(\partial/\partial\overline{x})$. What this means physically is that the light beam does not break up and that its principal response to the perturbation is to reshape or change its angle with respect to the interface. Of course, this assumption is not perfect, but we shall see how good it is shortly.

Specifically, from (3.111),

$$\frac{\partial V}{\partial x} = \delta(x)\left(1 - 2(\alpha^{-1} - 1)|A|^2\right) + 2H(x)(1 - \alpha^{-1})\frac{\partial|A|^2}{\partial x} \tag{3.112}$$

where $\delta(x)$ and $H(x)$ are the Dirac delta function and Heaviside functions, respectively. Then equation (3.110) becomes

$$\frac{d^2\overline{x}}{dz^2} = -\frac{1}{p}M(|A|^2) \tag{3.113}$$

where

$$M(|A|^2) = 2|A|^2 + 2(1 - \alpha^{-1})|A|^4. \tag{3.114}$$

Define

$$U_L(\overline{x}) = \frac{1}{p}\int^{\overline{x}} M(|A(s)|^2)ds \tag{3.115}$$

and (3.113) is

$$\frac{d^2\overline{x}}{dz^2} = -\frac{\partial U_L(\overline{x})}{\partial\overline{x}}, \tag{3.116}$$

which is Newton's law for the motion of a particle at position $\bar{x}$ in a conservative field with potential $U_L(\bar{x})$. So far we have only assumed that $|A|^2$ is a function of $x - \bar{x}(z)$. The idea is that if the wavepacket remains principally in the left medium, only a very small part of its leading edge crosses the interface and therefore the perturbative potential $V(x)$ is small. The principal effect of the perturbation is to change the position (location) and velocity (direction) of the wavepacket (light beam). Only a very small amount is converted into other solution components. (Later on, when we deal with the case of a soliton crossing the interface, we will calculate the amount lost to radiation. For now, ignore it. In most, but not all, circumstances, it is negligible, a fact that can be readily verified by direct numerical experiments on the full governing equation (3.105).)

We take $A(x, z)$ to have soliton form

$$A(x, z) = 2\eta(z)\,\mathrm{sech}\,2\eta(z)(x - \bar{x}(z))\exp\left(\frac{iv(z)x}{2} + 2i\sigma(z)\right) \qquad (3.117)$$

where

$$\frac{d\sigma}{dz} = -\frac{v^2}{8} + 2\eta^2.$$

However, from (3.108), $\eta = \frac{1}{4}p$ is constant and therefore equal to η_0, its initial value. (Recall that we have dropped the prime on η_0 and that $\eta_0' = \Delta^{-1/2}\eta_0$; likewise, we have omitted writing the prime on v and $v_0' = v'(z = 0) = \Delta^{-1/2}v_0$.) Using (3.115) and (3.117), we now calculate $U_L(\bar{x})$ and find

$$U_L(2\eta_0\bar{x}, \alpha S_0) = (1 - (\alpha S_0)^{-1})\tanh 2\eta_0\bar{x} + (3\alpha S_0)^{-1}\tanh^3 2\eta_0\bar{x} \quad (3.118)$$

where the parameter

$$\alpha S_0 = \frac{\alpha}{4\eta_0^2(1 - \alpha)}. \qquad (3.119)$$

In terms of the original variables

$$\eta_0 \to \frac{\eta_0}{\sqrt{\Delta}},$$

$$\bar{x} \to \sqrt{\Delta}\bar{x},$$

$$2\eta_0\bar{x} \to 2\eta_0\bar{x},$$

the parameter αS_0 is precisely equal to the right-hand side of (3.97). We stress that U_L is defined only for $\bar{x} < 0$. The equilibrium solutions occur at the place where U_L is constant (i.e., $\partial U_L/\partial \bar{x} = 0$), which is

$$\operatorname{sech}^2 2\eta_0 \bar{x} = \alpha S_0, \tag{3.120}$$

so that the position at which the particle is at rest (the corresponding light beam travels parallel to the interface) is given by the negative root of (3.120). This is precisely what we had obtained in (3.97). Further, we observe that, for $\alpha S_0 > 1$, $U_L(x)$ is monotonically increasing from $-\infty$ to 0 and therefore there are no equilibria. For $\alpha S_0 < 1$, which when written in terms of the original power $P = 8\eta_0/\alpha_0$ of the incident light beam is

$$\frac{\alpha_0^2 P^2}{16} > \frac{n_0^2 - n_1^2}{1 - \alpha}\alpha, \tag{3.121}$$

namely, the power has to overcome the ratio of linear to nonlinear refractive index mismatches, $U_L(\bar{x})$ achieves a single maximum and therefore the stationary solution is unstable. Therefore one cannot have a nonlinear TE surface wave attached to the stronger linear refractive index, weaker nonlinear refractive index side of an interface.

In Figure 3.22, we draw U_L for two values of αS_0. We can read off from these graphs what happens as long as the incident beam angle of incidence (the initial velocity v_0 of the particle) is such that its initial kinetic energy $\frac{1}{2}v_0^2$ (achieved at some point $z = 0, \bar{x} = \bar{x}_0 < 0$ say) is insufficient to overcome the maximum increase in potential energy namely, the maximum value for $(U_L(2\eta_0 \bar{x}) - U_L(2\eta_0 \bar{x}_0))$ when $\bar{x} < 0$. From (3.116), we have

$$\tfrac{1}{2}v^2 - \tfrac{1}{2}v_0^2 = U_L(2\eta_0 \bar{x}_0, \alpha S_0) - U_L(2\eta_0 \bar{x}, \alpha S_0) \tag{3.122}$$

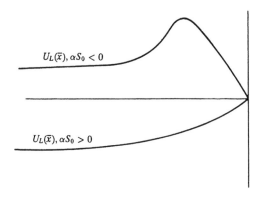

$U_L(\bar{x}), \alpha S_0 < 0$

$U_L(\bar{x}), \alpha S_0 > 0$

FIGURE 3.22

and therefore, if $\frac{1}{2}v_0^2 < \max_{\bar{x} \leq 0}(U_L(2\eta\bar{x}_0, \alpha S_0) - U_L(2\eta_0\bar{x}, \alpha S_0))$, the particle (light beam) is reflected and at $\bar{x} = \bar{x}_0$ has the same velocity magnitude $|v_0|$ as the incoming particle. The corresponding reflected light beam has its asymptotic angle of reflection equal to its angle of incidence. Also since its power is a constant of the motion, the coefficient of reflection is unity.

On the other hand, if the initial angle of incidence is sufficiently large, the light beam and the particle cross the interface. We reach, therefore, a somewhat unexpected conclusion. In the parameter ranges for which this perturbation theory holds (we will discuss this point later), a light beam is either totally reflected or totally transmitted! We will show how to treat the case of a transmitted beam in a moment. First, however, we develop the dual picture of what happens when a light beam is confined to the right-hand medium.

This time we define

$$A(x', z') = \sqrt{\frac{\alpha_1}{2\Delta}} F(x, z) \exp i(\beta^2 - n_1^2)\frac{z}{2\beta} \tag{3.123}$$

and then $A(x', z')$ satisfies

$$i\frac{\partial A}{\partial z'} + \frac{\partial^2 A}{\partial x'^2} + 2|A|^2 A = V(x')A \tag{3.124}$$

with

$$V(x') = \begin{cases} -1 - 2(\alpha - 1)|A|^2, & x' < 0 \\ 0, & x' > 0 \end{cases}$$

The initial shape of $F(x, z)$ is taken to be

$$\sqrt{\frac{2}{\alpha_1}} 2\eta \operatorname{sech} 2\eta_1 \left(x - \frac{v}{2\beta}z - x_1 \right)$$

$$\times \exp\left(\frac{iv\,x}{2\beta} - i\left(\frac{v^2}{4} + \beta^2 - n_1^2 - 4\eta_1^2 \right)\frac{z}{2\beta} \right)$$

so that, far from the interface, $A(x', z')$ has the form

$$\frac{2\eta_1}{\sqrt{\Delta}} \operatorname{sech} \frac{2\eta_1}{\sqrt{\Delta}}\left(x' - \frac{v}{2\beta\sqrt{\Delta}}z' - x_1' \right) \exp\left(\frac{iv}{2\Delta}x' - i(\frac{v^2}{4\Delta}) - \frac{4\eta_1^2}{\Delta} \right)\frac{z'}{2\beta}.$$

As before, we call $\eta_1/\sqrt{\Delta} = \eta_1', v/\sqrt{\Delta} = v'$ and drop the primes on η_1', v_1', x', z' and assume $A(x, z)$ to have the single soliton form

$$A(x, z) = 2\eta \operatorname{sech} 2\eta(x - \overline{x}(z)) \exp\left(i\frac{v(2)}{2}x + 2i\sigma\right) \tag{3.125}$$

where

$$\frac{d\sigma}{dz} = -\frac{v^2}{8} + 2\eta^2. \tag{3.126}$$

Conservation of power gives us $\eta = \eta_1$, a constant, and exactly as before we obtain

$$\frac{d^2\overline{x}}{dz^2} = -\frac{\partial U_R}{\partial \overline{x}}(2\eta_1\overline{x}, \ S_1) \tag{3.127}$$

where, for $\overline{x} > 0$,

$$U_R(2\eta_1\overline{x}, S_1) = (1 - S_1^{-1}) \tanh 2\eta_1\overline{x} + (3S_1)^{-1} \tanh^3 2\eta_1\overline{x}. \tag{3.128}$$

In Figure 3.23, we graph U_R for three choices of values for the parameter

$$S_1 = \frac{1}{4\eta_1^2(1 - \alpha)}. \tag{3.129}$$

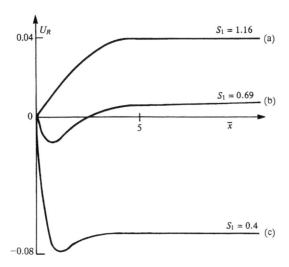

FIGURE 3.23   Graph of $U_R(\overline{x})$ for three cases. The first one is for $S_1 \geq 1$ and the second and third are for $S_1 < 1$, where there is always a minimum for some $\overline{x} > 0$. In (b), $U(0) < U(\infty)$ and the opposite is true in (c).

From the shape of U_R, we draw the following conclusions. For $S_1 > 1$, U_R is monotonically increasing. There are no critical points. A particle (light beam) beginning on the right at $\bar{x} = \bar{x}_1$ where $v = v_1$ with either a negative velocity (angle of incidence) $v_1 < 0$ or with positive velocity such that $\frac{1}{2}v_1^2 < U_R(\infty) - U_R(2\eta_1\bar{x},\ S_2)$ always crosses the interface. On the other hand, for $S_1 < 1$, there is always a minimum of U_R situated at the positive root of

$$\text{sech}^2 2\eta_1\bar{x} = S_1. \tag{3.130}$$

A particle at this point remains there. Likewise a light beam running in a direction parallel to the interface at this point remains stable. This is a strong (nonlinear) stability result because from U_R we can calculate the finite amount of the perturbation required to knock the particle (light beam) out of its minimum (confined) state.

EXERCISE 3.8

Define η by $\beta^2 - n_i^2 = 4\eta_i^2 = 4\eta_i'^2\Delta$ and show that x_1, as defined by (3.95), satisfies

$$\text{sech}^2 2\eta_1 x_1 = S_1 = \frac{\Delta}{4\eta_i^2(1-\alpha)}. \tag{3.131}$$

$\square$

We next discuss what happens if a particle crosses the interface from left to right. Suppose the field $F(x, z)$ has modulus

$$|F(x, z)| = \sqrt{\frac{2}{\alpha_0}} 2\eta_0 \,\text{sech}\, 2\eta_0(x - \bar{x})$$

and velocity v as it crosses the interface. We will assume that if v, the velocity of the particle or the angle of incidence of the light beam, at the interface is finite and not close to zero, the soliton crosses quickly and retains its shape during the crossing. As it meets the new nonlinear medium, however, it dissociates into several soliton components, because the envelope $A(x, z)$ in the right-hand medium is no longer a soliton as its amplitude is $1/\sqrt{\alpha}$ times its inverse width, namely,

$$\frac{2\eta_0}{\sqrt{\alpha}}\text{sech}2\eta_0(x - \bar{x})\exp\left(\frac{iv}{2}x + 2\sigma\right), \quad \frac{d\bar{x}}{dz} = v. \tag{3.132}$$

In Chapter 6, we discuss the calculation for computing the soliton and radiation components of (3.132). We find that the number N of solitons produced in the right-hand medium is given by the integer in the interval

$$\left(\frac{1}{\sqrt{\alpha}} - \frac{1}{2}, \quad \frac{1}{\sqrt{\alpha}} + \frac{1}{2} \right). \tag{3.133}$$

The amplitudes are

$$\eta_{1r} = 2\eta_0 \left(\frac{1}{\sqrt{\alpha}} - r + \frac{1}{2} \right), \quad r = 1 \ldots N \tag{3.134}$$

whereas all the initial velocities are the same and equal to v. Each new soliton, however, interacts with the interface differently, and therefore the velocities quickly become unequal and each can be treated by a single particle analysis. Observe that for $\alpha > \frac{4}{9}$, $N = 1$, and only one soliton is produced on the right.

How much radiation is produced? We can calculate this from the trace formula given in Section 6k,

$$\int F F^* dx = \frac{8}{\alpha_1} \sum_{r=1}^{N} \eta_{1r} - \frac{4}{\pi \alpha_1} \int_{-\infty}^{\infty} \ln(1 - |R(k)|^2) dk. \tag{3.135}$$

The left-hand side is $8\eta_0/\alpha_0$, and the first term on the right-hand side can be calculated from (3.132) and for $N = 1$ is $8\eta_1/\alpha_1$ with η_1 given by (3.134). Therefore the amount of energy that goes into radiation is given by

$$-\frac{4}{\pi \alpha_1} \int_{-\infty}^{\infty} \ln(1 - |R(k)|^2) \, dk = \frac{8\eta_0}{\alpha_0}(1 - \sqrt{\alpha})^2. \tag{3.136}$$

Observe that as $\alpha \to 1$, the amount is negligible. Even in the worst possible case $\alpha = \frac{4}{9}$, the loss to radiation is only 11%.

On the other hand, if a soliton of modulus

$$|F(x, z)| = \sqrt{\frac{2\Delta}{\alpha_1}} 2\eta_1 \operatorname{sech} 2\eta_1 (x - \overline{x}(z)), \quad \frac{d\overline{x}}{dz} = v$$

crosses from right to left, the initial amplitude $A(x, z)$ in the left-hand medium is

$$2\eta_1 \sqrt{\alpha} \operatorname{sech} 2\eta_1 (x - \overline{x}(z)) \exp \left(\frac{iv}{2} x + 2i\sigma \right)$$

and the number of solitons in the left-hand medium lies in the interval

$$\left(\sqrt{\alpha} - \tfrac{1}{2}, \quad \sqrt{\alpha} + \tfrac{1}{2}\right). \tag{3.137}$$

Since $\alpha < 1$, N is at most one and will be equal to one if $\alpha > \tfrac{1}{4}$. Its amplitude is

$$\eta_0 = 2\eta_1(\sqrt{\alpha} - \tfrac{1}{2}),$$

and again using the trace formula

$$\frac{8\eta_1}{\alpha_1} = \frac{16\eta_1}{\alpha_0}(\sqrt{\alpha} - \tfrac{1}{2}) - \frac{4}{\pi\alpha_0}\int_0^\infty \ln(1 - |R(k)|^2)dk, \tag{3.138}$$

the amount converted to radiation is

$$\frac{8\eta_1}{\alpha_1}\frac{(1 - \sqrt{\alpha})^2}{\alpha},$$

which is again small if α is close to unity.

In Figures 3.24, 3.25 we show pictures of light beam which crosses left to right (Figure 3.24) and right to left (Figure 3.25). These pictures were produced by solving the full equations numerically. The results are very close to those predicted by the theory.

We are now in a position to describe how to handle the general case where the light beam may cross the interface more than once. For simplicity of presentation, we take $\alpha > \tfrac{4}{9}$ so that a single soliton on one side gives rise to only one on the other. The rule is to form a composite potential

$$U(\overline{x}) = H(\overline{x})U_R(2\eta_1\overline{x}, S_1) + (1 - H(\overline{x}))U_L(2\eta_0\overline{x}, \alpha S_0). \tag{3.139}$$

In (3.139), $\eta_0 = \alpha_0 P/8\sqrt{\Delta}$ and $\eta_1 = \alpha_1 P/8\sqrt{\Delta}$, $x, \overline{x}$, and z are measured in the scaled coordinates given by (3.104), and P is the power $\int F F^* dx$ (x is the unscaled coordinate) in the original wavepacket.

In Figure 3.26 we show the composite potential $U(\overline{x})$ in several different cases. In Figures 3.27 and 3.28, we show comparisons of the trajectories obtained by theory and numerical experiment. As you can see, the predicted behavior of the beam is very close to that which is observed.

Modifications of these ideas have been used to describe the behavior of beams in dielectrics when carrier diffusion is important [45] and the interaction of two beams at an interface [46].

Why does the whole theory work, and how might it be useful? The key point is that for sufficiently large power in the input as measured by $\eta_0/\sqrt{\Delta}$,

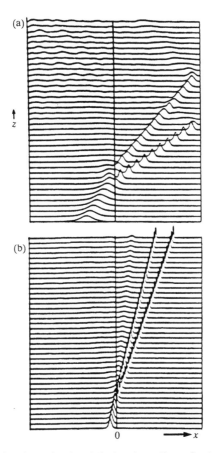

FIGURE 3.24 Soliton break-up in the right-hand medium. In (a), an incoming packet breaks into three as it crosses the interface, with the smallest one being reflected back to the left medium where it eventually disperses into radiation. In (b) the angle of incidence is bigger so that no wavepacket is reflected. In both cases, the amplitudes of each soliton are very close to the theoretical predictions. Here $P = 4$, $v_0 = 0.4$ in (a) and 1.0 in (b), $\Delta = 0.1$ and $\alpha = 0.125$.

the soliton does indeed stay intact and only a fraction of this power is reflected. However, we find numerically that as the parameter $\eta_0/\sqrt{\Delta}$ drops below 0.05, the amount of power that is reflected as radiation is about the same as the amount of power that stays in the weak beam, which is better reflected or transmitted depending on the angle of incidence. For more details, the reader should consult references [43, 44].

For moderate to large powers, the theory is very good indeed and captures most of the observed behavior, quantitatively when the linear and nonlinear mismatches are small and qualitatively even when they are large. Even in the latter case, the agreement with the results obtained by integrating the

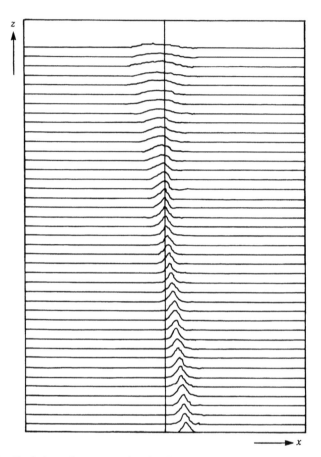

FIGURE 3.25   Evolution of a wavepacket that initially is in the right-hand medium. Once it crosses the interface, it loses its soliton shape as predicted by the theory. Here $\tau = z/2\beta$. The parameters are the same as those of Figure 3.23 except that the wavepacket is initially in the right-hand medium and $v_0 = -1.0$.

full partial differential equation is very good. The reason for this is that the perturbation not only depends on Δ and $\alpha_0 - \alpha_1$ but also on the amount by which the beam overlaps the interface. For much of its trajectory in z, this amount is very small.

Because the equivalent particle motion is conservative, the question of when and how beams can be trapped remains. For example, when can the beam initiated at an angle in the left-hand dielectric emerge as a beam parallel to the interface in the right? The resolution of this requires us to take account of the small amount of radiation reflected at the interface and manner in which this phase shifts the center of the soliton beam. Look at the third figure in Figure 3.26. A beam that has sufficient angle (velocity) to make it over the

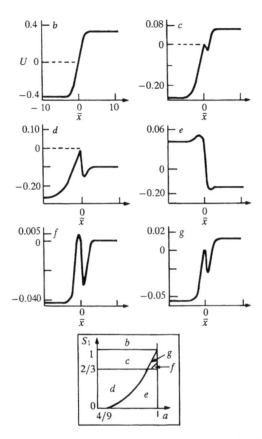

FIGURE 3.26 Graphs of $U(\overline{x})$ for regions of different behavior in (S_1, α) plane where $\alpha > \frac{4}{9}$. Observe that there are not stable critical points for negative $\overline{x}$. In (c) and (d), the maximum is always a corner at $\overline{x} = 0$. In (e), (f), and (g), there is a maximum for $\overline{x} < 0$. In (d), (e), and (f) an equivalent particle that enters from the left with enough energy to overcome the maximum will travel to $\overline{x} = \infty$ and not return. On the other hand in (b), (c), and (g) the particle (or the wavepacket) may enter the right medium, remain there for awhile, and eventually return to the left medium.

maximum in the region $\overline{x} < 0$ also makes it above this maximum on the other side of the minimum and therefore is transmitted or eventually reflected. On the other hand, if its effective starting point (due to a phase shift with the radiation) is to the right of the maximum, then conceivably it may not have sufficient angle (kinetic energy) to escape the potential well. In this case the beam is trapped.

In summary, then, the equivalent particle theory allows us to learn a lot with relatively little work and should become a standard tool in the analysis and design of nonlinear waveguides.

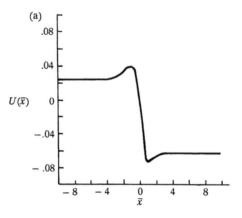

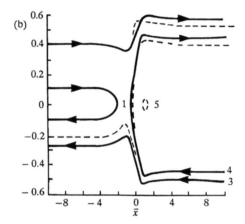

FIGURE 3.27   Potential and phase plane for values of the parameters $\alpha = 0.75$, $\Delta = 0.1$, and $P = 2$. We show all typical trajectories. Observe that the agreement is better when the wavepacket does not cross the interface, which in this case are the oscillatory trajectory in the right medium and the trajectory that comes from $-\infty$ and is reflected before crossing the interface. In all other cases the agreement is very good.

3h. Nonlinear Guided and Surface TE Waves in a Symmetric Planar Waveguide

Consider the geometry shown in Figure 3.29, where the square of the refractive indices in the regions $x > L$ and $x < -L$ is $n_1^2 + \alpha_1 |F|^2$ and n_0^2 in the strip $-L < x < L$. We take $n_0 > n_1$, $\alpha_0 < \alpha_1$, and the electric field to be

$$\vec{E}(x, z) = \hat{y}(F(x, z) \exp(i\beta \frac{\omega}{c} z - i\omega t) + (*)). \qquad (3.140)$$

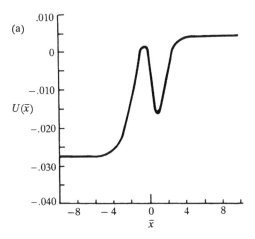

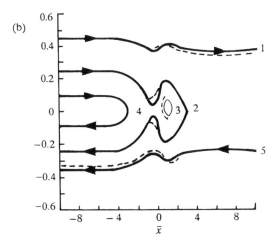

FIGURE 3.28 Potential and phase plane for values of the parameters $\alpha = 0.75$, $\Delta = 0.1$, and $P = 1.52$. The difference with the previous case of Figure 3.27 is that now we can have a wavepacket come from $-\infty$, cross the interface, and be reflected ending up at $-\infty$ (trajectory 2), which is the opposite of trajectory 4 of Figure 3.27. Again the agreement is good in all cases but better when the wavepacket does not cross the interface.

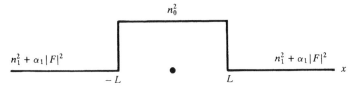

FIGURE 3.29

In Section 6e, we work out the various confined solutions, TE_n, $n = 0, 1, 2, \ldots$ when the nonlinear refractive indices are zero. Here we examine the dependence of the existence and stability of these solutions as the power is increased. The thin-film planar waveguide geometry sketched in Figure 3.29 differs fundamentally from the previous single interface problem in that linear guided waves exist even when $\alpha_0 = 0$. This linear guiding effect results in light confinement predominantly within the thin-film layer with evanescent waves decaying into outer substrate or cladding layers. The guiding mechanism is precisely that discussed earlier for a fiber. The choice of material constants in Figure 3.29 is clearly not unique, and various permutations of linear refractive indices and nonlinear coefficients may be considered. The particular symmetric single-mode case depicted in Figure 3.29 has the merit of containing within it much of the bifurcation behavior associated with other cases.

It is clear that the method of solution follows precisely that outlined for the single interface with the exception that the linear film allows a solution without any zeros (TE_0) or with one zero (TE_1) for a single mode waveguide. Solutions for $F(x)$ in each medium are

$$\underline{TE_0} \quad q^2 = \beta^2 - n_1^2, \quad r^2 = n_0^2 - \beta^2, \quad \Gamma^2 = \beta^2 - n_0^2$$

$$F(x) = \begin{cases} \sqrt{\frac{2}{\alpha_1}}\, q \operatorname{sech} q(x - x_2), & x < -L \\ A \cos r(x - L_0), & \beta < n_0, \quad |x| < L \\ A \cosh \Gamma(x - L_0), & \beta > n_0 \\ \sqrt{\frac{2}{\alpha_1}}\, q \operatorname{sech} q(x - x_1), & x > L \end{cases}$$

$$\underline{TE_1} \quad q^2 = \beta^2 - n_1^2, \quad r^2 = n_0^2 - \beta^2, \quad \Gamma^2 = \beta^2 - n_0^2$$

$$F(x) = \begin{cases} \sqrt{\frac{2}{\alpha_1}}\, q \operatorname{sech} q(x - x_2), & x < -L \\ A \sin r(x - L_0), & \beta < n_0, \quad |x| < L \\ A \sinh \Gamma(x - L_0), & \beta > n_0 \\ \sqrt{\frac{2}{\alpha_1}}\, q \operatorname{sech} q(x - x_1), & x > L \end{cases}$$

The constants A and L_0 appearing in the solutions in the linear region are determined from the boundary conditions at $x = \pm L$. The solution structure is obviously much richer and analysis more complicated. We note the essential results for the TE_0 waves. Matching F and dF/dx at the boundaries $x = \mp L$ gives, after considerable algebraic manipulation, a complicated transcendental relation, a nonlinear dispersion relation determining the effective waveguide refractive index β:

$$\frac{q^2 - \Gamma \tanh^2 \Gamma^2 (L - L_0)}{q^2 - \Gamma^2 \tanh^2 \Gamma (L + L_0)} = \frac{\cosh^2 \Gamma (L - L_0)}{\cosh^2 \Gamma (L + L_0)}.$$

A close inspection of this transcendental equation shows that a unique solution exists for $L_0 = 0$ corresponding to symmetric modes of the layered structure (i.e., $x_1 = -x_2$). In this case the amplitude of the field in the linear film is obtained from

$$A^2 = \frac{2}{\alpha_0} \frac{q^2}{\cosh^2 \Gamma L} \left(1 - \frac{\Gamma^2}{q^2} \tanh^2 \Gamma L \right), \quad \beta > n_0$$

$$= \frac{2}{\alpha} \frac{q^2}{\cos^2 r L} \left(1 - \frac{r^2}{q^2} \tan^2 r L \right), \quad \beta < n_0.$$

In addition, a set of asymmetric modes exists that bifurcate from the branch of symmetric solutions at a critical value of the effective index β_c. Rather than list expressions for these modes, we refer the interested reader to [47] for details. Figure 3.30 shows a power vs. β characteristic with insets showing the field shapes $F(x)$ at various locations on the different branches.

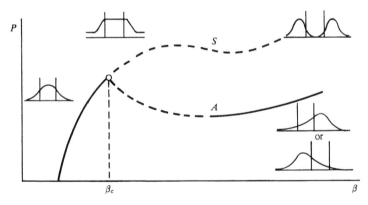

FIGURE 3.30   Sketch of the power P versus β characteristic for a symmetric thin-film planar waveguide. Shapes of the symmetric and nonsymmetric nonlinear guided modes are depicted at various locations along their respective branches.

In contrast to the single interface problem, a guided wave solution exists for all incident powers, degenerating into a linear TE$_0$ wave as $P \rightarrow 0$. The point β_c signifies a symmetry-breaking bifurcation to a doubly degenerate branch of solutions (A-branch) where the nonlinear guided wave moves gradually to the left (or right) and eventually out of the linear guide into the surrounding nonlinear layers at the minimum of the *A*-branch. As one moves further along the branch to larger β-values, the nonlinear waves become progressively more localized in the outer media and are undistinguishable from the single interface nonlinear surface waves discussed earlier. Beyond β_c the solution on the symmetric branch (*S*-branch) loses stability. This loss of stability manifests itself as an ejection of most of the energy from the vicinity of the waveguide into the surrounding nonlinear medium, followed by propagation as a self-focused channel away from the guide. The remaining energy stays localized in the thin film and acts much like a low-power (linear) guided wave. The stability of such nonlinear guided waves has only recently been addressed. The *A*-branch of solutions are unstable in the negatively sloped regions.

Breaking of symmetry in the structure leads to a power versus β characteristic below in Figure 3.31(a). A single nonlinear medium results in a much simpler characteristic (Fig. 3.31b).

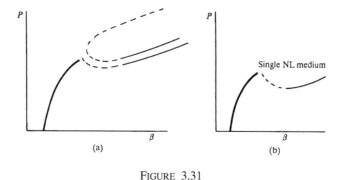

FIGURE 3.31

3i.   TM (Z-Independent) Nonlinear Surface Waves

The general solution of this problem still remains open due to complications associated with the tensor properties of the relationship between the electric displacement vector and the electric field. As an illustration, we provide a recent example where a first integral can be found if we assume a specific form for the components of the electric displacement vector [48]. Assuming a Kerr-type nonlinearity, we write the *x* and *z* diagonal components of the dielectric tensor as

$$\epsilon_{xx} = \epsilon_x + \alpha_1 E_x^2 + \alpha_2 E_z^2$$

$$\epsilon_{zz} = \epsilon_z + \alpha_1 E_z^2 + \alpha_2 E_x^2$$

and the vector electric field is

$$\vec{E}(\vec{r}, t) = \tfrac{1}{2}[\varrho_x(x)\hat{x} + i\varrho_z(x)\hat{z}]e^{i(\omega t - kz)} + (*),$$

where the transverse coordinate x is in units of $k_0 = \omega/c$. Substituting these expressions into the TM set of equations, one obtains the following equations:

$$\frac{d}{dx}\varrho_z = \frac{1}{k}[\epsilon_{xx} - k^2]\varrho_x$$

$$\frac{d}{dx}(\epsilon_{xx}\,\varrho_x) = -k\,\epsilon_{zz}\,\varrho_z.$$

We have allowed for the fact that two electric field components are $\pi/2$ out of phase TM waves (factor of i above). These equations have a first integral

$$\left(\frac{d\varrho_z}{dx}\right)^2 = (k^2 - \epsilon_x)\varrho_x^2 - \epsilon_z\,\varrho_z^2 - \alpha_z(\varrho_x\varrho_z)^2 - \frac{1}{4}(\varrho_x^4 + \varrho_z^4).$$

If we now consider an interface at $x = 0$, separating a linear $(x < 0)$ and nonlinear dielectric $(x > 0)$ medium, the bounded TM solution for $x < 0$ is the usual evanescent wave

$$\varrho_z(x) = \varrho_{0z}e^{qx}, \quad x < 0$$

where $q^2 = \beta^2 - \epsilon_s$, $\varrho_z(0)$, and ϵ_s is the linear medium dielectric constant. Continuity of D_x and ϱ_z across the interface yields the set of algebraic equations

$$\varrho_{0z} = -\frac{\epsilon_{xx}\,q}{k\,\epsilon_s}\varrho_{0x} = -\frac{(\epsilon_x + \alpha_1\varrho_0^2 + \alpha_2\varrho_{0z}^2)q}{k\,\epsilon_s}\varrho_{0x}$$

$$\frac{\alpha_1}{2}\varrho_{0z}^4 + \left[\epsilon_z\left(\frac{\epsilon_2}{q}\right)^2\right]\varrho_0^2 + \left[\left(\frac{k\,\epsilon_s}{q}\right)\varrho_{0z} - \frac{\alpha_1}{2}\varrho_{0x}^4\right] = 0,$$

which are the dispersion relations for the TM-polarized waves at the interface. Given the material parameters, the boundary values of the electric field components can be determined as a function of effective wave index k. These allow one to determine numerically the field distributions in the nonlinear medium from the original set of coupled first-order ordinary differential equations. No attempt has been made to date to study the stability properties of the TM-polarized waves.

Chapter 4

THE INTERACTION BETWEEN LIGHT AND MATTER

Overview

A DETECTABLE NONLINEAR RESPONSE BETWEEN light and matter depends on one of two key ingredients: Either there is a resonance between the light wave and some natural oscillation mode of the material, or the light is sufficiently intense. In the previous chapters, we have dealt with the latter situation, and the response of matter was incorporated into the theory through the frequency dependent linear and nonlinear susceptibilities. In this chapter, we let the matter variables themselves play active roles and derive coupled nonlinear differential equations for the electric and matter fields. Of course, matter has many, many degrees of freedom, and it would be downright exhausting and well-high impossible to attempt to write dynamical equations for all of them. However, because of the weak coupling between light and matter, and since it usually can be arranged that the electric field consists of a collection of almost monochromatic wavetrains so that its frequency content is narrowly localized around a finite number of frequencies, only those degrees of freedom that can directly or indirectly resonate with the electric field have a long-time cumulative influence.

Direct resonance can occur in isolated intervals of the electromagnetic spectrum at ultraviolet and visible frequencies (10^{15} s^{-1}), where the material oscillator is an electronic transition, infrared (10^{13} s^{-1}), where the material has vibrational modes, and the far infrared-microwave range (10^{11} s^{-1}), where there are rotational modes. These interactions are also called one-photon processes. The lower frequency modes can be excited at optical frequencies ($\sim 10^{15}$ s^{-1}) through indirect resonant processes in which the difference in frequencies and wavevectors of two light waves, called the pump and the Stokes wave respectively, matches the frequency and wavevector of one of these lower frequency modes. These three-frequency interactions are sometimes called two-photon processes. In the case where the "lower frequency" mode is an electronic transition or in the vibrational range (in which case the Stokes frequency can be of the same order of magnitude as that of the light wave), this process is called *Raman scattering*. When the lower frequency mode is an acoustic mode of the material, it is called *Brillouin scattering*.

Because of weak coupling and the monochromatic character of the electromagnetic wave, we can reduce the number of equations to a manageable number, and the resulting models give a reasonably accurate picture of observed behavior. The influence of all the other degrees of freedom must also be accounted for because, while all are weakly and nonresonantly coupled to the modes of interest, there are enough of them to have a collective effect. This effect is usually included in the analysis by the insertion of damping terms called *homogeneous broadening*, which simulate the drain of energy away from the excitations of interest to the heat bath consisting of the many other degrees of freedom including the quantum fluctuations about the classical electric field. In the limit where these attenuation processes are relatively fast, we can adiabatically eliminate the matter variables and write algebraic expressions for them in terms of the electric field. The adiabatic elimination of the matter variables returns us to the notion of a refractive index, and in fact we will gain explicit formulae for both the real and imaginary parts of the susceptibility function $\hat{\chi}(\omega)$ that help explain Figure 2.2.

The outline of this chapter is as follows.

4a. We use Schrödinger's equation to derive the Bloch equations, which describe the dynamics of the oscillators (excited atoms) of matter [2, 4]. These excited atoms are represented by the density matrix ρ, whose off-diagonal elements ρ_{jk} correspond to a polarization induced by a transition between energy levels E_j and E_k and the differences $\rho_{jj} - \rho_{kk}$ of whose diagonal elements correspond to the relative occupations of the various energy levels.

4b. We derive the Maxwell equation for the envelope of an almost monochromatic electromagnetic wave.

4c. We combine the two equations for a gas of two-level atoms and obtain the Maxwell-Bloch equations.

4d. We show how to recover the notion of susceptibility and refractive index by assuming that the loss or homogeneous broadening terms in the Bloch equations dominate the time derivatives and by adiabatically eliminating the matter variables so as to obtain the analog of equation (2.14), $1/\epsilon_0 \hat{P}(\omega) = \hat{\chi}(\omega)\hat{E}(\omega)$.

4e. We show that the mutual interaction of two identical counter-propagating waves in a two-level medium significantly alters the nature of the nonlinear interaction through the presence of a spatial grating (interference) at half the optical wavelength.

4f. We extend the analysis to the case of two distinct co- or counter-propagating waves interacting with a three-level medium where both waves excite separate quantum transitions sharing a common level.

4g. We then use the results to derive an effective two-level atom model that describes two-photon absorption and stimulated Raman scattering.

4h. We now make a significant departure from our quantum approach and introduce the reader to two important nonlinear mechanisms in liquids and solids: molecular reorientation and electrostriction. The material equations derived from a purely classical basis provide the background for the material in the following sections.

4i. We introduce a model due to Debye to describe retardation effects caused by the nonresonant inertial response of the optical material. The Maxwell–Debye equations corresponding to the cases treated in 4c and 4e are derived.

4j. We end the chapter with a short essay on the important adiabatic (Born–Oppenheimer) approximation, which underpins many of the theoretical simplifications in derivations made throughout the chapter.

4a. The Bloch Equations

The ultrahigh velocities of electrons in their orbits about nuclei, the much slower vibrational motions of the relatively massive nuclei forming chemical bonds, and the even slower rotational motions of these molecules are all quantized in matter. A near-monochromatic light source, under frequency tuning, can act as a sensitive probe of the discrete structure associated with each of these internal degrees of freedom of the atoms or molecules that constitute the material. The fundamental quantum relation between energy and frequency ($E = \hbar\omega = hc/\lambda$) and the vastly differing time scales of the various types of motion leads, in turn, to a natural partitioning of the discrete energy spectrum of matter into progressively smaller subsets of discrete quantum states associated with electronic, vibrational, and rotational degrees of freedom. This important observation is the content of the *Born–Oppenheimer approximation* (discussed in Section 4j), which allows one to reduce a mathematically intractable spectral eigenvalue problem to a set of separate spectral problems for each kind of dynamical motion. In effect, an ansatz is made that the total quantum wavefunction representing all degrees of freedom is separable into a product of individual wavefunctions, one for each degree of freedom, showing an explicit dependence on the coordinate representing that particular type of motion. The spectral eigenvalue problem defining the appropriate physical energy spectrum is defined by the effective mass appearing in the kinetic energy operator of the linear Schrödinger equation and the nature of the potential energy function. The latter is the classical Coulomb potential describing the attraction between one or more electrons and nuclei, the Morse (or anharmonic oscillator) potential representing the integral of the restoring force for the vibrational motion of the nuclei in a chemical bond, or the rigid rotor effective potential relevant for rotation about some symmetry axis. Subtleties due to weak couplings between the different degrees of freedom can introduce important physical consequences in laser spectroscopy, but these are not of direct interest here. The schematic in Figure 4.1 is meant to convey a picture of the energy spectrum of matter as it is probed on a progressively finer energy scale.

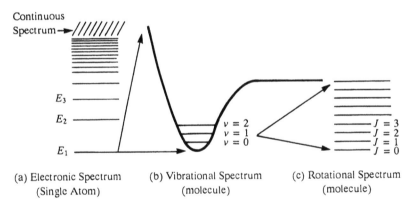

FIGURE 4.1   Energy spectrum of matter.

Mathematically, the state of matter is described by the quantum mechanical wavefunction ψ and Hamiltonian operator H whose discrete eigenvalues define the set of energy levels and whose corresponding eigenfunctions are the basis states. For the unperturbed state, we will write the Hamiltonian as H_0 and assume that the energy levels E_j are given by $\hbar\omega_j$ and the corresponding eigenstate is given by ψ_j, namely,

$$H_0\psi_j = \hbar\omega_j\psi_j, \quad j = 1, \ldots, N, \tag{4.1}$$

and the eigenstates $\{\psi_j\}$ form a complete, orthonormal basis

$$\int \psi_j^* \psi_k d\vec{R} = \delta_{jk} \qquad (4.2)$$

for the Hilbert space to which the wavefunction belongs.

The electric field is treated as a classical variable because the photon flux is large. For example, even in a He–Ne laser with the moderate flux intensity S (the Poynting vector $1/2\epsilon_0 cn|E|^2$) of a kilowatt per square meter, the photon flux $S/\hbar\omega$ is of the order of 10^{21} photons per square meter per second. Now suppose a light source containing one or several almost monochromatic wavetrains with several frequencies is applied to a material that in the unperturbed state has frequency *differences* $\omega_{jk} = \omega_j - \omega_k$ that are close to some of those of the applied electric field. Light photons are absorbed and the atoms of matter access some of their higher energy states. For example, let us imagine that the "atoms" of matter consist of an electron orbiting a nucleus, and let the electron coordinate with respect to the nucleus be denoted by $\vec{R}$. In the unperturbed ground state, the wavefunction is spherically symmetric and localized around a radius $|\vec{R}| = R_0$. In the perturbed state (see Figure 4.2), the electron cloud is distorted in the direction of the electric field and an effective electric dipole is induced in the material. Moreover, the dipole oscillates and, as is well known, a classical oscillating dipole radiates an electromagnetic wave so that we expect that the induced dipole (in reality, a very large number of such dipoles) will, in turn, act to modify the electric field. Our goal in this section is to calculate the polarization vector

$$\vec{P} = n_a\, e \int \vec{R}\psi\psi^* d\vec{R} \qquad (4.3)$$

induced by the applied electric field $\vec{E}$.* In (4.3), n_a is the volume density of atoms and e the electric charge.

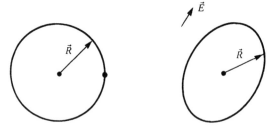

FIGURE 4.2   Response of electron cloud to electric field.

*The distortion shown in Figure 4.2 can also lead to higher-order (e.g. quadrupole) contributions to the effective polarization P. However, usually these are very small, of the order of 10^{-6} times the dipole contribution, and so we will ignore them.

The recipe for calculating $\vec{P}$ is simple and we carry it out in seven steps.

Step 1. Schrödinger's equation: We assume the wavefunction $\psi(\vec{R}, t)$ satisfies schrödinger's equation

$$\hbar \frac{\partial \psi}{\partial t} = H \psi \tag{4.4}$$

where the Hamiltonian H is the sum of the unperturbed Hamiltonian H_0 and the perturbation potential $\delta V = -e \int \vec{E} \cdot d\vec{R}$ where $\vec{E}$ is the external field. Because $\vec{E}(\vec{r}, t)$ changes very little over atomic distances as measured by $\vec{R}$, we can write

$$\delta V = -e\vec{E} \cdot \vec{R} \tag{4.5}$$

and treat $\vec{E}$ as constant over interatomic distances $|\vec{R}|$. For laser intensities of interest, the perturbation potential is usually much smaller than the potential $V_0(\vec{R})$ of the unperturbed Hamiltonian H_0, which is the Coulomb potential describing the interaction between an electron and nucleus when we are considering an electronic transition, or the Morse or rigid rotor potentials if we are considering a vibration or rotation in a chemical bond.

Step 2. The unperturbed states: We solve (4.4) by assuming that we may write

$$\psi(\vec{R}, t) = \sum_{j=1}^{N} a_j(t) \psi_j(\vec{R}) \tag{4.6}$$

where the $\psi_j(\vec{R})$ are the unperturbed states. They can be arranged to have the properties that both $\int \vec{R} \psi_j(\vec{R}) d\vec{R}$ and $\int \vec{R} \psi_j \psi_j^* d\vec{R}$ are zero. If $H = H_0$, then

$$a_j(t) = a_j(0)e^{-i\omega_j t}. \tag{4.7}$$

Step 3. The polarization: The polarization per atom is given by

$$\vec{P}_{\text{atom}} = e \int \vec{R} \psi \psi^* d\vec{R} = \sum_{jk} \rho_{jk} \vec{p}_{kj} = Tr \vec{p}\rho, \tag{4.8}$$

where the density matrix element

$$\rho_{jk} = a_j a_k^* \tag{4.9}$$

depends on time and the dipole matrix element, independent of time, is

$$\vec{p}_{jk} = e \int \vec{R} \psi_j^* \psi_k d\vec{R}. \tag{4.10}$$

The individual density matrix elements $\rho_{jk}(t)$ can be directly related to physical observables, in contrast to the probability amplitudes $a_j(t)$. Diagonal terms give the occupation probability of a particular quantum state whereas off-diagonal terms are related directly to the polarization associated with a given transition induced by the electromagnetic wave. For a particular dipole transition to occur, the dipole transition matrix element (i.e., the corresponding $\vec{p}_{jk}$) must be nonzero, and the strength of the transition is determined by the magnitude of this quantity. Certain symmetries allow us to establish a priori whether this matrix element is nonzero, but its magnitude can only be determined by explicit computation. For example, if the material is made up of atoms, the spherical symmetry of the individual atom allows one to separate the angular from radial dependence in both the Hamiltonian operator and the wavefunction. The evaluation of the integral over all space decouples into the product of radial and angular integrals. If any of these integrals are zero, then a dipole transition cannot occur between the two relevant discrete quantum states. As a special case, all diagonal terms p_{jj} are identically zero. Also, certain off-diagonal elements may be zero. The restrictions imposed by the various symmetries in the problem are called *selection rules*.

The quantity $Tr\vec{p}\rho$ is the trace of the product of the matrices $\vec{p}$ and ρ. The total polarization

$$\vec{P} = n_a \vec{P}_{\text{atom}}. \tag{4.11}$$

Our job now is to calculate $\vec{P}_{\text{atom}}$ and in particular to write dynamical equations for the density matrix elements ρ_{jk}.

Step 4. The "raw" Bloch equations: We first derive equations for the probability amplitudes $a_j(t)$ by substituting (4.6) into (4.4) with $H = H_0 - e\vec{E} \cdot \vec{R}$, using (4.1), and then multiplying the equation by ψ_k^*. We obtain, on using (4.2),

$$\frac{\partial a_k}{\partial t} = -i\omega_k a_k + \frac{i\vec{E}}{\hbar} \cdot \sum_{\ell=1}^{N} \vec{p}_{k\ell} a_\ell, \quad k = 1, \ldots, N. \tag{4.12}$$

Multiply (4.12), with k replaced by j, by a_k^*, then multiply the complex conjugate of (4.12) by a_j, add, and obtain

$$\frac{\partial \rho_{jk}}{\partial t} = -i(\omega_j - \omega_k)\rho_{jk} + \frac{i\vec{E}}{\hbar} \sum_{l=1}^{N} \vec{p}_{jl}\rho_{lk} - \frac{i\vec{E}}{\hbar} \cdot \sum_{l=1}^{N} \vec{p}_{lk}\rho_{jl}, \qquad (4.13)$$

where j, k run from 1 to N and we have used the fact that $\vec{p}_{kl}^* = \vec{p}_{lk}$. These are the "raw" Bloch equations. We call them raw because there are altogether too many of them. There are in fact $N - 1 + N(N - 1)/2$ of them, where N is the number of material degrees of freedom. We now want to reduce this number if possible to a much smaller number of equations for the dynamically relevant states.

Step 5. Reduction of the "raw" Bloch Equations: The key realization is this: We can easily solve (4.13) if we neglect the terms depending on the electric field $\vec{E}$. Let us therefore ask how big they are in relation to the terms without $\vec{E}$. The ratio is

$$\frac{Ep}{\hbar\omega} \qquad (4.14)$$

where E, p, and ω are typical sizes of the electric field, dipole matrix element, and frequency difference $\omega_{jk} = \omega_j - \omega_k$, respectively. The last corresponds to an electronic transition and is in the neighborhood of 10^{15} s^{-1}. (Notice that only frequency or energy level differences matter!) The ratio (4.14) is therefore the ratio of the (Rabi) frequency

$$\omega_{\text{Rabi}} = \frac{Ep}{\hbar} \qquad (4.15)$$

to the transition frequency ω, a ratio that is roughly between 10^{-4} and 10^{-7}. The inverse of the Rabi frequency measures the time it takes for the field to transfer energy coherently from the field to the atoms and back.

Since the magnitudes of the perturbation terms are so small, why can we not ignore them altogether? We already learned the answer to this when we considered weak nonlinear coupling of oscillators and wavetrains. If the perturbation term has frequencies that resonate with any of the natural frequencies $\omega_{jk} = \omega_j - \omega_k$ of the material, then the perturbation terms lead to secular behavior of ρ_{jk}. We would find that

$$\rho_{jk} - \rho_{jk}(0)e^{-i\omega_{jk}t} \propto \frac{Ep}{\hbar}te^{-i\omega_{jk}t}, \qquad (4.16)$$

so that after times of the order of the inverse Rabi frequency, the cumulative effect of the resonances produces an order one change in the density matrix element. Therefore we must keep all terms in

$$\frac{i\vec{E}}{\hbar} \cdot \sum_{l=1}^{N} \vec{p}jl\rho lk - \frac{i\vec{E}}{\hbar} \cdot \sum_{l=1}^{N} \vec{p}lk\rho jl \qquad (4.17)$$

that have frequencies equal or close to ω_{jk}. All others, and this includes most of them, we can discard. In optics the neglect of all nonresonant terms in the "raw" Bloch equations is known as the rotating wave approximation, because all nonresonant terms have frequencies of the order of ω_{jk} so that the response of ρ_{jk} to these perturbations is of the order $Ep/\hbar\omega_{jk}$, which is small; the effect of the perturbation averages out over many oscillations. The rotating wave approximation can be formally justified by using the multiple time scale methods introduced in Chapter 6, but we will avoid all this elaborate machinery and simply discard all nonresonant terms.

Step 6. Introduction of homogeneous broadening: The reader could object that while the influence of any individual nonresonant term is small, there are so many of them (of order N^2) that they still could have some cumulative effect on the small number (order n, say) of levels taking part in any particular resonant subset. The reader would be quite correct in making this objection! Moreover, there is an additional nontrivial effect due to the cumulative effect of the weak interaction of the small subset n of resonantly coupled levels with the quantum states of the electromagnetic field considered as fluctuations about the classical field. The effect of the combination of these interactions is considerable, and although the total system is conservative and the dynamics almost periodic in the Poincaré recurrence sense,[†] the net effect is an irreversible relaxation, a gradual loss of energy from the small subset of modes of interest to the heat bath consisting of all the other material states and the quantum states of the electromagnetic field. This loss is captured phenomenologically by adding the decay terms $-\gamma_{jk}\rho_{jk}$ to the right-hand side of (4.13) for those n levels j, k that are in resonance with the electric field. In this way we have reduced the set of $(N(N-1)/2) + N - 1$ "raw" Bloch equations (4.13) to a much smaller set of $(n(n-1)/2) + n - 1$ equations with loss terms added. The diagonal elements γ_{jj} are the decay

---

[†]A conservative system of interacting oscillators will return arbitrarily close to any initial starting point on the energy surface in phase space if one waits long enough. Long enough, however, for large systems, could mean the life time of the universe.

rates due to irreversible losses to the heat bath of quantum fluctuations about the classical field and to lower-energy dipole-coupled bound states. The off-diagonal decay rates γ_{jk} have an additional contribution due to the elastic collisions between the atoms. As a result γ_{jk}, $j \neq k$, is greater than γ_{jj} and in some cases significantly so. These damping effects are referred to as *homogeneous broadening*, and the coefficients γ_{jk}^{-1} are called homogeneous broadening times.

Step 7. Identification of possible resonances: Finally we list the possible resonances.

We begin with *direct resonance*. The frequency ω of the electric field $\vec{E}$ is close to ω_{jk}. This occurs in those products, like $\vec{E} \cdot \sum \vec{p}_{jk} \rho_{kk}$, containing the diagonal terms of ρ that have no linear time dependence. This is the strong interaction which occurs in a laser, for example.

D.C. rectification occurs from terms like $\vec{E} \cdot \sum \vec{p}_{jl} \rho_{lj}$ in which the frequency $-\omega$ in $\vec{E}$ cancels the frequency $\omega_{lj} \sim \omega$ in ρ_{lj} and leads to a long-time cumulative contribution to ρ_{jj}.

Parametric resonance occurs when any of the binary combinations of the field frequencies $\pm \omega_r$ with the dipole allowed level difference frequencies $\pm \omega_{lk}$ are equal to other level difference frequencies $\pm \omega_{jk}$. This creates a strong interaction between different density matrix elements mediated by the wavepacket in the electric field. There is, however, no direct feedback to the electric field unless other effects (such as in Raman or Brillouin scattering) are important in computing the polarization. We call this a parametric resonance.

The Bloch equations for the n resonantly interacting levels are

$$\frac{\partial \rho_{jk}}{\partial t} = -(\gamma_{jk} + i\omega_{jk})\rho_{jk} + \frac{i\vec{E}}{\hbar} \cdot \sum_{l=1}^{n} \vec{p}_{jl}\rho_{lk} - \frac{i\vec{E}}{\hbar} \cdot \sum_{l=1}^{n} \vec{p}_{lk}\rho_{jl}. \quad (4.18)$$

Since the electric field $\vec{E}(\vec{r}, t)$ depends on position $\vec{r}$ and time t, so will the density matrix elements. The polarization field $\vec{P}$ is given by

$$\vec{P} = n_a \sum \vec{p}_{jk}\rho_{kj} = n_a Tr \vec{p}\rho. \quad (4.19)$$

Having computed $\vec{P}$, we now append the Maxwell equation (2.9) in its various envelope forms.

Remark.   In Chapter 5 on applications we meet another kind of broadening effect called inhomogeneous broadening. The origin of this effect is the random motion of the atoms in the medium, which means that a stationary observer sees an effective Doppler shifted frequency $\omega_{12} + kv$ where v, the velocity of the atoms, is a random variable with a Maxwellian distribution. Because of this, there is always a detuning $\omega - \omega_{12}$ between the frequency of the transition $E_2 \to E_1$ and the frequency of the applied field. For low-amplitude external waves, this leads to an incoherent interaction between the field and the atoms, resulting in an irreversible loss of energy from the field. For sufficiently large pulses, however, an extremely novel response occurs and the medium can become completely transparent. This is described in Section 5e.

4b.   The Maxwell Equations

The spatial structure of the fields is dictated by the geometrical nature of the environment in which they propagate. As always, the governing equation for a singly polarized electric field is

$$\nabla^2 \vec{E} - \frac{1}{c^2} \frac{\partial^2 \vec{E}}{\partial t^2} = \frac{1}{\epsilon_0 c^2} \frac{\partial^2 \vec{P}}{\partial t^2}. \tag{4.20}$$

If we are considering a particular TEM mode in a cavity, say the TEM_{00} mode, then we write the electric field $\vec{E}$ and polarization $\vec{P}$ as

$$\vec{E} = \vec{A}(t) V(x, y, z) e^{-i\omega_c t} + (*) \tag{4.21}$$

$$\vec{P} = \vec{\Lambda}(t) V(x, y, z) e^{-i\omega_c t} + (*) \tag{4.22}$$

where $V(x, y, z)$ has the shape of the TEM_{00} mode derived in Section 6f and ω_c its corresponding frequency. On the other hand, if $\vec{E}$ consists of a singly polarized, unidirectional wave traveling in an open medium, the frequency ω is arbitrary, the shape

$$V(x, y, z) = e^{i(\omega/c)z} \tag{4.23}$$

and the envelope A can depend on the transverse coordinates x, y and weakly on the propagation direction z. Making the slowly varying envelope approximation in either case leads to

$$\frac{1}{c} \frac{d\vec{A}}{dt} + \frac{\kappa}{c} \vec{A} = \frac{i\omega_c}{2\epsilon_0 c} \vec{\Lambda} \tag{4.24}$$

for the amplitude of the cavity mode (recall $\nabla^2 V + (\omega^2/c^2)V = 0$) and to

$$\frac{\partial \vec{A}}{\partial z} + \frac{1}{c}\frac{\partial \vec{A}}{\partial t} + \frac{\kappa}{c}\vec{A} - \frac{ic}{2\omega}\nabla_{\perp}^2\vec{A} = \frac{i\omega}{2\epsilon_0 c}\vec{\Lambda} \tag{4.25}$$

for the envelope of the purely propagating unidirectional wave. In both (4.24) and (4.25) we include an attenuation term $(\kappa/c)\vec{A}$ where $\kappa = (\sigma/\epsilon_0) + (c/L)\ln(1/\sqrt{R})$. In the cavity, the most important losses are due to cavity mirrors (reflectivity R, L is the distance between mirrors), but there is also the weaker loss due to a finite conductivity σ (dimension of C v^{-1}m^{-1}s^{-1}) in the medium.

4c. The Maxwell-Bloch Equations for A Gas of Two-Level Atoms

We now derive the equations that govern the interaction of a unidirectional electromagnetic wave in an open medium of a gas of two-level atoms. At this stage we will ignore the Doppler or inhomogeneous broadening effect. The polarization

$$\vec{P} = n_a(\vec{p}_{12}\rho_{12} + \vec{p}_{21}\rho_{21}). \tag{4.26}$$

We will assume for simplicity that the electric field is singly polarized in the direction $\hat{e}(\vec{A}(x, y, t) = \hat{e}A(x, y, t))$ and the direction of the dipole matrix element is parallel to that of the electric field. If not, we have to include the two transverse polarizations of the field and a birefringent cross-coupling. It is clear, however, that the field in the direction of $\vec{p}_{12}$ is the most important, so we take $\hat{e}$ to have this direction. We will also take $\vec{p}_{12}$ to be real. There is no loss of generality here because we can include the phase of a complex p_{12} in the density matrix element ρ_{12}. We write $\vec{p}_{12} = \vec{p}_{21} = p\,\hat{e}$. Then, from (4.22) and (4.26), we see we can identify

$$n_a p\,\rho_{12} = \Lambda(x, y, z, t)e^{i(\omega/c)z - i\omega t} \tag{4.27}$$

where

$$\vec{\Lambda}(x, y, z, t) = \hat{e}\Lambda(x, y, z, t). \tag{4.28}$$

Recall that by definition $\rho_{21} = \rho_{12}^*$. The Bloch equation for ρ_{12} is

$$\frac{\partial \rho_{12}}{\partial t} + (\gamma_{12} + i\omega_{12})\,\rho_{12} = \frac{iEp}{\hbar}(\rho_{22} - \rho_{11}) \tag{4.29}$$

and the equation for $\rho_{22} - \rho_{11}$ is

$$\frac{\partial(\rho_{22} - \rho_{11})}{\partial t} + \gamma_{11}(\rho_{22} - \rho_{11}) = 2\frac{iEp}{\hbar}(\rho_{12} - \rho_{12}^*), \tag{4.30}$$

and we have taken the diagonal homogeneous broadening elements γ_{11} and γ_{22} to be the same. Not every term on the right-hand side of (4.30) will be relevant because they do not all resonate with the "zero" frequency of $\rho_{22} - \rho_{11}$. The terms that do have the form

$$2i \frac{p}{\hbar} \left(\frac{A^*}{n_a p} \Lambda - \frac{A}{n_a p} \Lambda^* \right). \tag{4.31}$$

The other two, which we ignore, contain the second harmonics $e^{\pm 2\, i\omega t}$. It is convenient to define

$$N = n_a (\rho_{22} - \rho_{11}), \tag{4.32}$$

which measures the total difference in occupation probabilities between the two levels 1 and 2. Since we will take

$$\omega_{12} = \omega_1 - \omega_2 \tag{4.33}$$

to be positive, level 2 is of lower energy than level 1. Thus a negative N will measure the two-level medium inversion, and we will call N the inversion number. Because of the relaxation terms, energy must be supplied to the medium in order to keep it active. Specifically, a proportion of the atoms have to be pumped to their excited states. We simulate this constant energy supply by adding a constant pumping rate $\gamma_{11} N_0$ to the right-hand side of the equation for the medium inversion. The inversion number N has to be pumped to a significantly negative value in order that the input of energy can overcome the relaxational losses.

The Maxwell-Bloch equations for a unidirectional, singly polarized wave in a medium of two-level atoms are

$$\frac{\partial A}{\partial z} + \frac{1}{c} \frac{\partial A}{\partial t} - i \frac{c}{2\omega} \nabla_1^2 A + \frac{\kappa}{c} A = \frac{i\omega}{2\epsilon_0 c} \Lambda \tag{4.34}$$

$$\frac{\partial \Lambda}{\partial t} + (\gamma_{12} + i(\omega_{12} - \omega)) \Lambda = \frac{ip^2}{\hbar} A N \tag{4.35}$$

$$\frac{\partial N}{\partial t} + \gamma_{11}(N - N_0) = \frac{2i}{\hbar} \left(A^* \Lambda - A \Lambda^* \right). \tag{4.36}$$

4d. Steady-State Response and the Susceptibility Near Resonance

If the attenuation processes are much faster than the times over which the electric field envelope changes, then it is reasonable to ignore the time derivatives of Λ and N and solve for both variables in terms of the electric field envelope A. We find

$$N = \frac{N_0}{1 + \frac{4p^2|A|^2\gamma_{12}}{\gamma_{11}\hbar^2(\gamma_{12}^2 + (\omega_{12} - \omega)^2)}} \tag{4.37}$$

and

$$\Lambda = \frac{p^2\frac{N_0}{\hbar}}{\gamma_{12}^2 + (\omega_{12} - \omega)^2 + \frac{4p^2}{\hbar^2}\frac{\gamma_{12}}{\gamma_{11}}|A|^2}(\omega_{12} - \omega + i\gamma_{12})A. \tag{4.38}$$

By analogy with the usual definition of susceptibility $1/\epsilon_0\vec{P} = \chi\vec{E}$, we can define $\hat{\chi}(\omega)$ to be given by

$$\frac{1}{\epsilon_0}\Lambda = \hat{\chi}(\omega)A, \tag{4.39}$$

which gives us that

$$\hat{\chi}(\omega, |A|^2) = \hat{\chi}'(\omega, |A|^2) + i\hat{\chi}''(\omega, |A|^2)$$

$$= \frac{p^2\frac{N_0}{\hbar\epsilon_0}}{\gamma_{12}^2 + (\omega_{12} - \omega)^2 + \frac{4p^2}{\hbar^2}\frac{\gamma_{12}}{\gamma_{11}}|A|^2}((\omega_{12} - \omega) + i\gamma_{12}). \tag{4.40}$$

The envelope equation (4.34) then becomes

$$\frac{\partial A}{\partial z} + \frac{1}{c}\left(\frac{\partial A}{\partial t} + \kappa A\right) - i\frac{c}{2\omega}\nabla_{\perp}^2 A = \frac{i\omega}{2c}\left(\chi'(\omega, |A|^2) + i\chi''(\omega, |A|^2)\right)A. \tag{4.41}$$

The medium acts as an *amplifier* if N_0, and therefore $\chi''(\omega, |A|^2)$, is sufficiently negative to overcome finite conductivity losses. Otherwise the medium acts as an *attenuator* and absorbs energy from the field. We note in particular the nonlinear nature of the susceptibility, which is inversely proportional to the field intensity and for large intensities causes $|\chi|$ to tend to zero. This is called the saturable nonlinearity. For small intensities

$$\frac{4p^2}{\hbar^2}\frac{\gamma_{12}}{\gamma_{11}}|A|^2 \ll \gamma_{12}^2 + (\omega_{12} - \omega)^2, \tag{4.42}$$

one can expand the saturable nonlinear susceptibility in a Taylor series about $|A|^2 = 0$, and then (4.41) takes on the form of the nonlinear Schrödinger equation with additional linear excitation and cubic saturation terms. We shall see in Section 6j that it is also possible in certain circumstances to obtain a diffusion term on the left-hand side, in which case (4.41) becomes the celebrated complex Ginzburg–Landau (CGL) equation [56].

In the limit of zero intensity, we graph $\hat{\chi}'(\omega)$ and $\hat{\chi}''(\omega)$ as a function of ω in Figure 4.3 for an absorbing ($N_0 > 0$) medium.

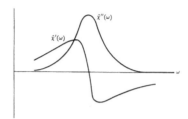

FIGURE 4.3

4e. Counter-Propagating Waves

The derivation parallels that in 4c except that now we assume an optical field consisting of two identically polarized counter-propagating waves of the same frequency, $\vec{E}(\vec{r}, t) = \hat{e}(A_1(\vec{r}, t)e^{i(kz-\omega t)} + A_2(\vec{r}, t)e^{-i(kz+\omega t)} + (*))$. This is the situation commonly encountered in a Fabry–Perot cavity or in an experiment where a beam splitter separates a single laser beam into two beams and simple optics devices are used to focus each into the opposite ends of a cell. The dipole interaction energy now contains both counter-propagating waves so that the material variables will couple to both. If we substitute the expression for the total field in the wave equation (2.9), we obtain

$$\left[2ik \left(\frac{\partial}{\partial z} + \frac{1}{c} \frac{\partial}{\partial t} \right) + \nabla_1^2 \right] A_1 \, e^{i(kz-\omega t)}$$

$$+ \left[-2ik \left(\frac{\partial}{\partial z} - \frac{1}{c} \frac{\partial}{\partial t} \right) + \nabla_1^2 \right] A_2 \, e^{-i(kz+\omega t)} + (*)$$

$$= -\frac{\omega^2}{\epsilon_0 c^2} \tilde{\Lambda}(\vec{r}, t) e^{-i\omega t} + (*).$$

The explicit "fast" time dependence is removed from the total polarization, but we must be careful when treating the z dependence. The polarization may exhibit a complicated dependence on z, but we anticipate that terms showing an $e^{ikz}(e^{-ikz})$-dependence will couple most strongly to the $A_1(A_2)$ wave when we average over many wavelengths. We will denote this average by an angle bracket and write the resulting coupled wave equations as

$$\left[2ik\left(\frac{\partial}{\partial z} + \frac{1}{c}\frac{\partial}{\partial t}\right) + \nabla_1^2\right]A_1 = -\frac{\omega^2}{\epsilon_0 c^2}\langle\tilde{\Lambda}e^{-ikz}\rangle \qquad (4.43)$$

$$\left[-2ik\left(\frac{\partial}{\partial z} - \frac{1}{c}\frac{\partial}{\partial t}\right) + \nabla_1^2\right]A_2 = -\frac{\omega^2}{\epsilon_0 c^2}\langle\tilde{\Lambda}e^{ikz}\rangle, \qquad (4.44)$$

where $\tilde{\Lambda}(\vec{r}, t)$ does not yet have the fast z-dependence removed. Similarly, the optical Bloch equations must be derived using the expression for the total field $\vec{E}(\vec{r}, t)$ given above. With this substitution the material equations become

$$\frac{\partial N}{\partial t} + \gamma_{11}[N - N_0] =$$

$$= \frac{2i}{\hbar}[\tilde{\Lambda}(A_1^*e^{-ikz} + A_2^*e^{ikz}) - \tilde{\Lambda}^*(A_1e^{ikz} + A_2e^{-ikz})] \qquad (4.45)$$

$$\frac{\partial}{\partial t}\tilde{\Lambda} + (\gamma_{12} + i(\omega_{12} - \omega))\tilde{\Lambda} = \frac{ip^2}{\hbar}(A_1e^{ikz} + A_2e^{-ikz})N. \quad (4.46)$$

Recalling that the slow envelopes A_j, $j = 1, 2$, already have the high spatial frequencies removed, the angle brackets on the right-hand side of the field equations (4.43) and (4.44) signify that we must extract those parts of the high spatial frequency from the total polarization $\tilde{\Lambda}(\vec{r}, t)$ for which the product Pe^{-ikz} changes slowly. This is achieved below by expanding the material variables in a Fourier series and comparing like powers of $e^{\pm inkz}$. This procedure yields an infinite hierarchy of coupled ordinary differential equations (o.d.e.'s) for higher-order spatial harmonics of the material variables. Let

$$\tilde{\Lambda} = e^{ikz}\sum_{p=0}^{\infty}\Lambda_1^{(p)}e^{2ipkz} + e^{-ikz}\sum_{p=0}^{\infty}\Lambda_2^{(p)}e^{-2ipkz} \qquad (4.47)$$

and

$$N = N^{(0)} + \sum_{p=1}^{\infty}\left(N^{(p)}e^{2ipkz} + (*)\right). \qquad (4.48)$$

The reader can readily show that the resulting coupled-field two-level medium equations follow:

$$\left(\frac{\partial}{\partial z} + \frac{1}{c}\frac{\partial}{\partial t}\right) A_1 - \frac{i}{2k}\nabla_1^2 A_1 = \frac{i\omega}{2\epsilon_0 c}\Lambda_1^{(0)} \tag{4.49}$$

$$\left(-\frac{\partial}{\partial z} + \frac{1}{c}\frac{\partial}{\partial t}\right) A_2 - \frac{i}{2k}\nabla_1^2 A_2 = \frac{i\omega}{2\epsilon_0 c}\Lambda_2^{(0)} \tag{4.50}$$

$$\frac{\partial N^{(0)}}{\partial t} + \gamma_{11}\left[N^{(0)} - N_0\right] = \frac{2i}{\hbar}\left[A_1^*\Lambda_1^{(0)} + A_2^*\Lambda_2^{(0)} - (*)\right] \tag{4.51}$$

$$\left(\frac{\partial}{\partial t} + \gamma_{11}\right) N^{(p)} = \frac{2i}{\hbar}\left[A_1^*\Lambda_1^{(p)} + A_2^*\Lambda_1^{(p-1)} - A_1\Lambda_2^{(p-1)*} - A_2\Lambda_2^{(p)*}\right] \tag{4.52}$$

$$\left[\frac{\partial}{\partial t} + (\gamma_{12} + i(\omega_{12} - \omega))\right]\Lambda_1^{(p)} = \frac{ip^2}{\hbar}\left[A_1 N^{(p)} + A_2 N^{(p+1)}\right] \tag{4.53}$$

$$\left[\frac{\partial}{\partial t} + (\gamma_{12} + i(\omega_{12} - \omega))\right]\Lambda_2^{(p)} = \frac{ip^2}{\hbar}\left[A_1 N^{(p+1)*} + A_2 N^{(p)*}\right]. \tag{4.54}$$

Note that if we take the single beam limit (i.e., $A_2 = 0$, $\Lambda_2^{(p)} = 0$, $\Lambda_1^{(p)} = 0(p \geq 1)$, $N^{(p)} = 0(p \geq 1)$), we recover (4.34–4.36). Equations (4.49) to (4.54) are typically solved as an initial-boundary value problem in which the $A_1(A_2)$ field is given as a function of time and the transverse coordinates at $z = 0(z = L)$ and the initial values of $N^{(p)}$, $P^{(p)}$ are given as functions of x and y by taking the Fourier series of $P(x, y, z, 0)$ and $N(x, y, z, 0)$. One must retain a sufficient number of Fourier modes in the calculation to ensure convergence of the field-matter variables over physical times of interest. By retaining terms up to $p = 1$ in the system of equations above, one accounts at lowest order for the presence of an induced spatial grating with a half wavelength period due to the mutual interference of the counter-propagating waves. These equations provide the mathematical description of a variety of interesting nonlinear phenomena both in finite length nonlinear media and in Fabry–Perot cavities. Some of the phenomena are highlighted under applications in Chapter 5, although the reader should be aware that counter-propagating beam induced instabilities is a very active area of current research and promises many exciting discoveries in the near future.

4f.　The Maxwell–Bloch Equations for a Three-level Atom

Extension of the previous analysis to the case of two optical waves with different carrier frequencies ω_1 and ω_2, each of which is near a resonance with one of two quantum transitions sharing a common intermediate level, is straightforward. The energy level scheme depicting this stepwise resonant excitation is given in Figure 4.4 with dipole coupled transitions induced by the optical wave at frequency ω_1, causing excitation of population from level 1 to level $m(p_{1m} \neq 0)$ followed by excitation from intermediate level m to level $2(p_{m2} \neq 0)$ by the optical wave at frequency ω_2.

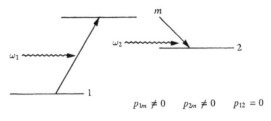

FIGURE 4.4

Physically, this interaction can be realized in a number of ways. Let two laser beams be incident on the material and suppose that one beam with frequency ω_1 is more intense than the other with frequency ω_2. Then by holding the intense field frequency ω_1 fixed and resonant with the $1 \rightarrow m$ quantum transition, we can study the nature of this strong interaction by tuning the weaker field frequency ω_2 about its resonance $m \rightarrow 2$. This experimental technique, called pump-probe laser spectroscopy, enables one to measure directly the Rabi frequency $(A_1 p_{1m}/\hbar)$ of the $1 \rightarrow m$ quantum transition. What happens is that the pump wave at the first frequency ω_1 splits the gain curve $\hat{\chi}''(\omega_2)$ for the second transition into a double peaked curve. We guide the reader through this calculation in Exercise 4.1.

A second scenario occurs when only a single intense pump wave at frequency ω_1 is incident on the sample and the second wave at ω_2 is allowed to grow from background fluctuations. This situation requires an additional ingredient in order that the $m \rightarrow 2$ transition be selected from a possibly large subset of lower dipole coupled states. If, for example, the sample is contained between two flat mirrors located at $z = 0$ and $z = L$, as in a Fabry–Perot cavity, light spontaneously emitted along the z-axis can build up as a standing wave with z-dependence $\sin kz$ if the wavelength of the emitted light (determined by the transition frequency $\omega_{m2} = (E_m - E_2)/\hbar$) satisfies $k = 2\pi n/\lambda = n\pi/L$. This match is achieved by adjusting the mirror separation L to accommodate an integer number of wavelengths (typically $n \approx 10^6$). This frequency selectability and optical feedback afforded by a Fabry–Perot or ring cavity can lead to a strong amplification of light at frequency ω_2. This is the basis for optically pumped lasers. The principle of laser operation is discussed in Chapter 5.

We next turn to the derivation of the equations for the three-level atom model. We assume for simplicity that both waves are singly polarized in the direction $\hat{e}$ and write the total optical field in (4.20) as

$$\vec{E} = \hat{e}(A_1 e^{-i\omega_1 t} + A_2 e^{-i\omega_2 t} + (*)). \tag{4.55}$$

The density matrix equations for the three-level atom configuration depicted above follow immediately from (4.18) by truncating the summation to just three terms (namely, those with no fast oscillations) and retaining those near-resonant contributions involving the linear combinations $(\omega_1 + \omega_{1m})$, $(\omega_2 + \omega_{2m})$, and $(\omega_1 - \omega_2 + \omega_{12})$. It is left as an exercise for the reader to show, after transformation to a rotating frame by defining $\rho_{1m} = \sigma_{1m} e^{i\omega_1 t}$, $\rho_{2m} = \sigma_{2m} e^{i\omega_2 t}$, $\rho_{12} = \sigma_{12} e^{i(\omega_1 - \omega_2)t}$, $\rho_{ii} = \sigma_{ii}(i = 1, 2, 3)$, and retaining only resonant contributions, that the following set of coupled o.d.e.'s with constant coefficients obtains:

$$\frac{d}{dt}\sigma_{11} = \frac{i}{\hbar} A_1^* p_{1m}\sigma_{m1} - \frac{i}{\hbar} A_1 p_{m1}\sigma_{1m} \tag{4.56}$$

$$\frac{d}{dt}\sigma_{22} = \frac{i}{\hbar} A_2^* p_{2m}\sigma_{1m} - \frac{i}{\hbar} A_2 p_{m2}\sigma_{2m} \tag{4.57}$$

$$\frac{d}{dt}\sigma_{mm} =$$

$$\frac{i}{\hbar}(A_1^* p_{1m}\sigma_{1m} + A_2 p_{2m}\sigma_{2m}) - \frac{i}{\hbar}(A_1^* p_{1m}\sigma_{m1} + A_2^* p_{2m}\sigma_{m2}) \tag{4.58}$$

$$\frac{d}{dt}\sigma_{1m} + i(\omega_1 + \omega_{1m})\sigma_{1m} =$$

$$\frac{i}{\hbar} p_{1m} A_1^* (\sigma_{mm} - \sigma_{11}) - \frac{i}{\hbar} p_{2m} A_2^* \sigma_{12} \tag{4.59}$$

$$\frac{d}{dt}\sigma_{2m} + i(\omega_2 + \omega_{2m})\sigma_{2m} =$$

$$\frac{i}{\hbar} p_{2m} A_2^* (\sigma_{mm} - \sigma_{22}) - \frac{i}{\hbar} p_{2m} A_1^* \sigma_{21} \tag{4.60}$$

$$\frac{d}{dt}\sigma_{12} + i(\omega_1 - \omega_2 + \omega_{12})\sigma_{12} = \frac{i}{\hbar} p_{1m} A_1^* \sigma_{m2} - \frac{i}{\hbar} p_{m2} A_2 \sigma_{1m} \tag{4.61}$$

with $\sigma_{im} = \sigma_{mi}^*$ and $\rho_{11} + \rho_{22} + \rho_{mm} = \sigma_{11} + \sigma_{22} + \sigma_{mm} = 1$, a statement of the conservation of total probability. Conservation of total probability $\int \psi^* \psi d\vec{x}$ means that $\rho_{11} + \rho_{22} + \rho_{mm} = 1$. Spontaneous losses to all other states are accounted for through the introduction of homogeneous broadening terms. Conservation of total probability generally no longer holds when we add decay terms, although there are physical situations where probability is conserved as the system remains closed. The detuning terms exhibit the near-resonant nature of the light-matter interaction (i.e., $\Delta = \omega_1 + \omega_{1m} = \omega_2 + \omega_{2m} = 0$ and $\omega_1 - \omega_2 + \omega_{12} = 0$ at resonance). The reader should be cautioned that the nice symmetries of the two-level atom model, which allowed us to absorb the dipole matrix element p into the definition of the polarization envelope, no longer exist. Here the distinct nonzero dipole matrix elements (p_{1m}, p_{2m}) are explicitly retained throughout (recall that $p_{12} = 0$ here). Moreover, the energy level scheme depicted in Figure 4.4 has level 1 as the lowest energy state in contrast to the two-level atom picture. This affects the signs appearing in the exponential terms above defining the transformation to a rotating frame.

The total polarization of this three-level medium follows from (4.19):

$$\vec{P} = \hat{e}n_a(p_{1m}\rho_{m1} + p_{m1}\rho_{1m} + p_{2m}\rho_{m2} + p_{m2}\rho_{2m})$$

$$= \hat{e}n_a(p_{1m}\sigma_{m1}e^{-i\omega_1 t} + p_{m1}\sigma_{1m}e^{i\omega_1 t}) \tag{4.62}$$

$$+ \hat{e}n_a(p_{2m}\sigma_{m2}e^{-i\omega_2 t} + p_{m2}\sigma_{2m}e^{i\omega_2 t}).$$

The second line shows explicitly that the polarization now drives two waves at frequency ω_1 and ω_2 and allows us to immediately identify the "slow" material variables acting to drive the individual field envelopes A_j, $j = 1, 2$. Introducing phenomenological damping terms as before to account for irreversible losses to a heat bath plus linear absorption of the individual light fields and using the expression for $\vec{E}$ above in the wave equation (4.20), we end up with the Maxwell-Bloch equations for a three-level atom:

$$\left(\frac{\partial}{\partial z} + \frac{1}{c}\frac{\partial}{\partial t}\right)A_1 + \frac{i}{2k}\nabla_\perp^2 A_1 + \alpha_1 A_1 = i\frac{\omega_1}{2\epsilon_0 c}n_a p_{1m}\sigma_{m1} \tag{4.63}$$

$$\left(\pm\frac{\partial}{\partial z} + \frac{1}{c}\frac{\partial}{\partial t}\right)A_2 + \frac{i}{2k}\nabla_\perp^2 A_2 + \alpha_2 A_2 = i\frac{\omega_2}{2\epsilon_0 c}n_a\, p_{2m}\sigma_{m2} \tag{4.64}$$

$$\frac{\partial}{\partial t}\sigma_{11} + \gamma_\parallel^1(\sigma_{11} - \sigma_{11}^0) = \frac{i}{\hbar}A_1^* p_{1m}\sigma_{m1} - \frac{i}{\hbar}A_1 p_{m1}\sigma_{1m} \tag{4.65}$$

$$\frac{\partial}{\partial t}\sigma_{22} + \gamma_\parallel^2(\sigma_{22} - \sigma_{22}^0) = \frac{i}{\hbar}A_2^* p_{2m}\sigma_{m2} - \frac{i}{\hbar}A_2 p_{m2}\sigma_{2m} \tag{4.66}$$

$$\frac{\partial}{\partial t}\sigma_{mm} + \gamma_\parallel^3(\sigma_{mm} - \sigma_{mm}^0)$$

$$= \frac{i}{\hbar}(A_1 p_{1m}\sigma_{1m} + A_2 p_{m2}\sigma_{2m}) - \frac{i}{\hbar}(A_1^* p_{1m}\sigma_{m1} + A_2^* p_{2m}\sigma_{m2}) \qquad (4.67)$$

$$\frac{\partial}{\partial t}\sigma_{1m} + (\gamma_{1m} + i(\omega_1 + \omega_{1m}))\sigma_{1m}$$

$$= \frac{i}{\hbar} p_{1m} A_1^*(\sigma_{mm} - \sigma_{11}) - \frac{i}{\hbar} p_{2m} A_2^* \sigma_{12} \qquad (4.68)$$

$$\frac{\partial}{\partial t}\sigma_{2m} + (\gamma_{2m} + i(\omega_2 + \omega_{2m}))\sigma_{2m}$$

$$= \frac{i}{\hbar} p_{2m} A_2^*(\sigma_{mm} - \sigma_{22}) - \frac{i}{\hbar} p_{2m} A_1^* \sigma_{21} \qquad (4.69)$$

$$\frac{\partial}{\partial t}\sigma_{12} + (\gamma_{12} + i(\omega_1 - \omega_2 + \omega_{12}))\sigma_{12}$$

$$= \frac{i}{\hbar} p_{1m} A_1^* \sigma_{m2} - \frac{i}{\hbar} p_{m2} A_2 \sigma_{1m} \qquad (4.70)$$

The diagonal and off-diagonal relaxation terms have the same physical interpretation as before and carry a superscript where necessary to emphasize that they may differ considerably in magnitude. This system of equations describes two co-propagating or counter-propagating ($\pm$ sign on $\partial/\partial z$) optical waves of different carrier frequency and includes finite pulse and diffraction effects.

If we proceed as before to compute the steady-state response from (4.63–4.70), it is clear that we can now define resonant susceptibilities through the definition $1/\epsilon_0 \vec{P}(\omega_i) = \hat{\chi}(\omega_i)\vec{E}(\omega_i)$, at two distinct optical frequencies ω_1 and ω_2. The real and imaginary parts of the weak probe dielectric susceptibility $\hat{\chi}(\omega_2) = \hat{\chi}'(\omega_2) + i\hat{\chi}''(\omega_2)$ are sketched in Figure 4.5 for two extremely important physical situations: a strong fixed-frequency resonant pump ($\omega_1 = \omega_{1m}$) and a strongly detuned fixed-frequency pump ($\Delta = \omega_1 + \omega_{1m} \gg \gamma_{1m}^{-1}, \gamma_{2m}^{-1}$). In the first case we observe that the absorption (amplification) profile can be split into two symmetric peaks whose separation is the Rabi frequency of the pump transition ($\omega_{\text{Rabi}} = p_{1m}A_1/\hbar$). This splitting has been measured experimentally. If the pump wave frequency is now tuned away from resonance, the two peaks become asymmetric with one shifting further away from the transition frequency ω_{1m}, while the second remains in its vicinity.* The relatively recent rediscovery that one laser beam could be used to modify the optical properties of a three-level medium for

---

*The peak that remains close to the transition frequency is called the Rayleigh scattering contribution. This is a useful technique to measure dipole matrix elements.

a second propagating optical wave has led to the emergence of a new field of research called electromagnetic induced transparency (EIT). The interested reader should consult review articles on EIT in optics and Photonics News (September 2002, p45) and Physics Today (Volume 50, Number 7, p36, 1997). The extreme case shown on the right of the figure corresponds to either two-photon absorption (if two waves at ω_1 and ω_2 are incident on the three-level medium) or stimulated Raman scattering, to be discussed later (if a single intense pump wave at ω_1 is incident on the three-level medium). This important class of nonlinear optical processes comes under the heading of three-wave interactions and involves the presence of terms quadratic in the interacting fields, as we shall now see.

EXERCISE 4.1

Solve the algebraic system obtained by setting time derivatives to zero in the system of o.d.e.'s (4.65–4.70). Use the final expressions for σ_{m1} and σ_{m2} to construct the nonlinear susceptibilities $\chi(\omega_1)$ and $\chi(\omega_2)$ by using the definition $1/\epsilon_0 \vec{P}(\omega_i) = \chi(\omega_i)\vec{E}(\omega_i)$. The algebra is rather messy. The reader is encouraged to explore the behavior of the $\chi'(\omega_i)$ and $\chi''(\omega_i)$ curves $(\chi(\omega_i) = \chi'(\omega_i) + i\chi''(\omega_i))$ by adjusting the ratio of the pump $(|A_1|)$ to probe $(|A_2|)$ amplitudes. □

4g. Indirect Excitation: Two-Photon Absorption and Stimulated Raman Scattering (SRS)

Two three-level schemes appropriate to the investigation of indirect excitations involving two optical waves are sketched in Figure 4.6. The energy level scheme on the left is the detuned version of Figure 4.5, except that now label m may refer physically to a manifold of excited intermediate levels that are dipole-coupled to the initial (level 1) and final (level 2) states. This interpretation of m is important especially when the detuning $\Delta = \omega_1 + \omega_{1m}$ is large as then no single level can be singled out in its contribution to the interaction. The level scheme on the right in Figure 4.6 is the classic picture for degenerate two-photon absorption $(\omega_1 = \omega_2)$. The important physical consideration here is that the detuning $\Delta = (|\omega_{1m}| - \omega_1, \omega_{1m} < 0)$ is so large that no population is created in any one of the manifold of the intermediate level(s) m. Notice that while the detuning Δ may be arbitrarily large, a resonance, albeit a weak one, can still occur as $\omega_1 - \omega_2 + \omega_{12}$ can be approximately zero. This weaker interaction leads to a cumulative resonant growth of the appropriate density matrix elements on a slower time scale than the direct interactions discussed earlier in this chapter. The latter are often referred to as single-photon

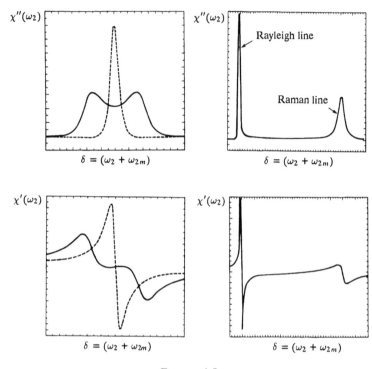

FIGURE 4.5

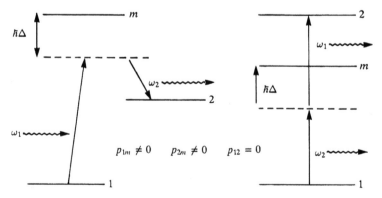

FIGURE 4.6   (a) Inverted *V* energy level configuration for nondegenerate ($\omega_1 \neq \omega_2$) two-photon absorption or stimulated Raman scattering. (b) Energy level scheme for degenerate ($\omega_1 = \omega_2$) two-photon absorption.

processes in order to distinguish them from the weaker two-photon processes of interest in this section.*

As in the previous section, we need to distinguish between two physical scenarios, namely, two optical waves incident on the material (two-photon absorption) and a single intense pump wave at frequency ω_1, scattering a downshifted Stokes wave at frequency $\omega_2(<\omega_1)$ from material fluctuations at frequency ω_{12} (stimulated Raman scattering). Mathematically, both physical situations are indistinguishable until we consider the initial conditions for each problem. As the intermediate level m is not occupied ($\sigma_{mm} = 0$) we can decouple the σ_{mm} equation from the rest in (4.63–4.70), and moreover, we can formally integrate the σ_{1m} and σ_{2m} equations (4.68–4.69), factoring out the slow variables σ_{12} and $\sigma_{ii}(i = 1, 2)$ from the resulting integrals relative to the rapidly varying exponentials $e^{\pm i\Delta t}$ to obtain

$$\sigma_{1m}(t) \simeq -\frac{p_{1m}A_1^*\sigma_{11}}{\hbar\Delta} - \frac{p_{2m}A_2^*\sigma_{12}}{\hbar\Delta} \tag{4.71}$$

$$\sigma_{2m}(t) \simeq -\frac{p_{2m}A_2^*\sigma_{22}}{\hbar\Delta} - \frac{p_{1m}A_1^*\sigma_{21}}{\hbar\Delta}. \tag{4.72}$$

In deriving these approximate expressions, we have assumed that the detuning from the intermediate levels is so large that $|\Delta| \gg \gamma_{1m}^{-1}, \gamma_{2m}^{-1}$. The homogeneous broadening terms proportional to $\gamma_{1m}^{-1}, \gamma_{2m}^{-1}$ have been dropped in the denominator. Substituting these expressions into the remaining equations for σ_{11}, σ_{22}, and σ_{12} leads to a much simpler set of "effective" two-level atom equations describing two-photon absorption and stimulated Raman scattering:

$$\frac{d}{dt}\sigma_{11} = -i\hbar q_{12}\left(A_1^*A_2\sigma_{21} - A_1A_2^*\sigma_{12}\right) \tag{4.73}$$

$$\frac{d}{dt}\sigma_{22} = -i\hbar q_{21}\left(A_1A_2^*\sigma_{12} - A_1^*A_2\sigma_{21}\right) \tag{4.74}$$

$$\frac{d}{dt}\sigma_{12} + i\{(\omega_1 - \omega_2 + \omega_{12}) + q_{11}|A_1|^2 - q_{22}|A_2|^2\}\sigma_{12}$$

$$= -i\,q_{12}A_1^*A_2(\sigma_{22} - \sigma_{11}) \tag{4.75}$$

with $\sigma_{21} = \sigma_{12}^*$, $\sigma_{11} + \sigma_{22} = 1$, and the effective coupling parameter $q_{ij} = p_{im}p_{mj}/\hbar^2(\omega_i + \omega_{im})$. If we assume that m represents a manifold of dipole-coupled discrete and continuum energy levels, then the repeated index m

---

*Multi-photon processes ($N > 2$) are commonly observed when an intense laser beam ionizes an atom. These indirect excitations show an intensity dependent ionization probability of $I^N = |A|^{2N}$ where N can be large.

implies summation and integration assuming none of the states in the manifold m interact with each other.

While these "effective" two-level atom equations bear some resemblance to those derived in Section 4b, we observe some differences. Firstly, the interaction terms are quadratic in the optical fields, and the strength of the coupling parameters depend inversely on the detuning Δ. Therefore strong two-photon interactions are expected in materials with level configurations such that at least one intermediate level is not too far away. The energy level distributions in atoms or molecules are dictated by electromagnetic forces binding electrons to nuclei, nuclei to nuclei, and so on. These energy levels are fixed for a particular species (they are determined as eigenvalues and eigenfunctions of the unperturbed problem defined by the time-dependent Schrödinger equation (4.1)) and are only weakly perturbed by external influences. It is fortuitous that the alkali vapors, and rubidium in particular, have an energy level configuration that closely matches that depicted in Figure 4.6b and hence display a strong two-photon absorption. The second notable difference is that the two-photon frequency detuning term $(\omega_1 - \omega_2 + \omega_{12})$ appearing in the σ_{12} equation is augmented by an intensity dependent $(q_{11}|A_1|^2 - q_{22}|A_2|^2)$ term in equation (4.75). This latter contribution is called the two-photon A.C. Stark shift because the effective resonance is now $\omega_1 - \omega_2 + \omega_{12} + q_{11}|A_2|^2 - q_{22}|A_2|^2 = 0$. This intensity dependent detuning means that different parts of a propagating optical pulse envelope experience a varying degree of resonant interaction with the material resulting in strong distortion and possible self-steepening on its leading or trailing edge, leading to optical shock formation.

The Maxwell-Bloch equations describing two-photon absorption or stimulated Raman scattering follow from substituting the above expressions for σ_{1m} and σ_{2m} in the polarization expansion (4.62). We again allow for either co- or counter-propagating optical pulses.

$$\left(\frac{\partial}{\partial z} + \frac{1}{c}\frac{\partial}{\partial t} \right) A_1 + \frac{i}{2k}\nabla_T^2 A_1 + \alpha_1 A_1$$

$$= \frac{i\omega_1}{2\epsilon_0 c} n_0 \hbar q_{11}\sigma_{11}A_1 - \frac{i\omega_1}{2\epsilon_0 c} n_0 \hbar q_{12}A_2\sigma_{21} \quad (4.76)$$

$$\left(\pm\frac{\partial}{\partial z} + \frac{1}{c}\frac{\partial}{\partial t} \right) A_2 + \frac{i}{2k}\nabla_T^2 A_2 + \alpha_2 A_2$$

$$= \frac{i\omega_2}{2\epsilon_0 c} n_0 \hbar q_{22}\sigma_{22}A_2 - \frac{i\omega_2}{2\epsilon_0 c} n_0 \hbar q_{12}\sigma_{12}A_1 \quad (4.77)$$

$$\frac{\partial}{\partial t}\sigma_{12} + \left[\gamma_{12} + i\left((\omega_1 - \omega_2 + \omega_{12}) + q_{11}|A_1|^2 - q_{22}|A|^2\right)\right]\sigma_{12}$$

$$= i\, q_{12} A_1^* A_2 (\sigma_{22} - \sigma_{11}) \qquad (4.78)$$

$$\frac{\partial}{\partial t}\sigma_{11} + \gamma_{11}\left(\sigma_{11} - \sigma_{11}^0\right) = -i\, q_{12}\left(A_1^* A_2 \sigma_{21} - A_1 A_2^* \sigma_{12}\right) \qquad (4.79)$$

$$\frac{\partial}{\partial t}\sigma_{22} + \gamma_{11}\left(\sigma_{22} - \sigma_{22}^0\right) = -i\, q_{12}\left[A_1 A_2^* \sigma_{12} - A_1^* A_2 \sigma_{21}\right]. \qquad (4.80)$$

These equations are rather complicated. How can we reduce them to the three-wave interaction form for stimulated Raman scattering discussed later in Chapter 5 and commonly encountered in the literature? First of all, let us ignore linear absorption ($\alpha_i = 0$) and assume infinite plane waves by dropping the diffraction terms in the field envelope equations (4.76) and (4.77). Next, we assume exact resonance ($\omega_1 - \omega_2 + \omega_{12} = 0$) and assume that the A.C. Stark shift terms ($q_{11}|A_1|^2$ and $q_{22}|A_2|^2$) make no significant contribution for very large detunings. The final approximation is to assume that the material excitation is so weak that the initial state occupation probability σ_{11} remains close to unity so that we can assume that it is a constant, replace ($\sigma_{22} - \sigma_{11}$) in (4.78) by ($2\sigma_{22} - 1$) using the approximate conservation of total probability, and drop σ_{22} relative to unity. The resulting stimulated Raman three-wave interaction equations then follow:

$$\left(\frac{\partial}{\partial z} + \frac{n(\omega_1)}{c}\frac{\partial}{\partial t}\right) A_1 = i\frac{\hbar\omega_1}{2\epsilon_0 c} n_a q_{12}\sigma_{21} A_2 \qquad (4.81)$$

$$\left(\pm\frac{\partial}{\partial z} + \frac{n(\omega_2)}{c}\frac{\partial}{\partial t}\right) A_2 = i\frac{\hbar\omega_2}{2\epsilon_0 c} n_a q_{12}\sigma_{21}^* A_1 \qquad (4.82)$$

$$\frac{\partial}{\partial t}\sigma_{21} + \gamma_{12}\sigma_{21} = iq_{12}A_1 A_2^*. \qquad (4.83)$$

These equations apply equally to the study of two-photon absorption/amplification and, with a change in physical interpretation of the coupling parameters, to stimulated Brillouin scattering. They describe the interaction of wavepackets or optical pulses and should be compared to the three-wave interaction equations for wavetrains presented in Section 6g. In their present form they describe both co-propagating and counter-propagating optical pulses (an initial-boundary value problem with $A_1(A_2)$ specified at $z = 0(z = L)$ and σ_{21} given on $0 \leq z \leq L$ at $t = 0$). They also illustrate a number of important physical phenomena observed in pulse propagation studies. The alert reader will have noticed that we have replaced $1/c$ by $n(\omega_i)/c$ in equations (4.81) and (4.82). This accounts for the fact that we cannot simply ignore dispersion from the much larger set of nonresonant states when we deal with very short optical pulses of duration $\sim 10^{-12}$ s. This background contribution is captured by a weak frequency dependence of the linear refractive index $n(\omega)$. The weak dispersion of the refractive index means that if the two optical pulses are sufficiently short and sufficiently well separated in frequency, then both pulses gradually separate in time due to their different group velocities $v_g = c/n(\omega)$, and as a consequence their mutual interaction weakens. This phenomenon is called "walk-off." On the other hand, for long optical pulses of duration $T_p \gg \gamma_{12}^{-1}$ we can set $\partial \sigma_{21}/\partial t = 0$ in the material equation, substitute the resulting expression for $\sigma_{21}(t)$ into the right-hand side of both field equations, and recover the classical $\chi^{(3)}$ interaction terms associated with the general polarization expansion for instantaneously responding media presented in Section 2d. We shall see equations of this form in Chapter 5 when we discuss stimulated Raman and Brillouin scattering. An additional reference is [57].

4h.   Other Nonlinear Mechanisms: The Condensed Phase

The physical origins of optical nonlinearity are many and varied. The quantum mechanical treatment presented above allows one to determine the nonlinear coupling parameters for the electronic contribution to the optical nonlinearity. This nonresonant electronic contribution to the n_2 coefficient for glass in an optical fiber is very small and very fast. The relaxation to the steady-state response is of the order of 10^{-14} s. However, there are other physical mechanisms that give coupling parameters many orders of magnitude greater than the electronic one when light couples to matter in the condensed phase (liquid or solid). In this section we will discuss two, *molecular reorientation* and *electrostriction*. Both provide a significantly enhanced contribution to an effective n_2 coefficient and, moreover, are responsible for extremely important stimulated scattering effects. The characteristic material response times for molecular reorientation are $\sim 10^{-12}$ s and for electrostriction $\sim 10^{-9}$ s. Other physical mechanisms, not discussed here, are thermal nonlinearities, which are usually very slow, $\sim 10^{-6}$ to 1 s, derived from the incoherent absorption of light by the material. In contrast to the electronic case, these latter phenomena allow a completely classical description.

Molecular Reorientation (Debye Equation)

The general problem of the reorientation of *polar molecules* (with a permanent rather than induced dipole moment) in a static electric field was considered by Debye in 1912. The average orientation of these molecules takes on a Maxwell–Boltzmann energy distribution

$$f(\delta V) = \frac{\exp{-\delta V/k_B T}}{\int \exp{-\delta V/k_B T}} d\Omega \qquad (4.84)$$

where the integral is taken over the entire solid angle and δV is the energy of the molecule in the presence of the applied field; $k_B = 1.3806 \times 10^{-23}$ JK^{-1} is the Boltzmann constant, and T refers to absolute temperature in kelvins. The potential energy for a dipole (permanent or induced) has been given in (4.5) as $\delta V = -\vec{p} \cdot \vec{E}(\vec{p} = e\,\vec{R})$, and if we assume that the E field, whether static or time varying, is polarized in the z-direction, then from Figure 4.7 we can write $\delta V = -pE \cos\theta$.

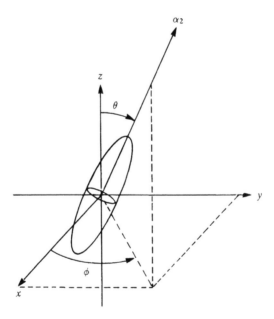

FIGURE 4.7   Reorientation of the CS$_2$ molecule.

When the orienting field is varying in time, the equilibrium distribution function f is no longer applicable, and, as shown by Debye, the distribution function $f(\theta, t)$ obeys the rotational diffusion equation

$$\zeta \frac{\partial f}{\partial t} = \frac{1}{\sin\theta} \frac{\partial}{\partial \theta} \left[\sin\theta \left(kT \frac{\partial f}{\partial \theta} - Mf \right) \right] \qquad (4.85)$$

where $M = -(\partial \delta V / \partial \theta) E \sin \theta$ is the torque on the molecule due to the applied field and ζ is the damping constant of inner friction. This coefficient has been estimated by Stokes to be given by $\zeta = 8\pi \eta a^2$ for a liquid of viscosity η consisting of spherical molecules of radius a. Since the torque on a permanent dipole is seen to reverse its sign as the direction of the applied field is reversed, it is clear that such a permanent dipole plays no part in contributing to the refractive index change induced by optical fields as it simply cannot follow the rapid oscillations of the field.

The reorientation of anisotropic polarizable molecules, such as the cigar-shaped molecule CS_2 depicted in Figure 4.7, in a time-varying field, results in an induced birefringence due to the inherent anisotropy in the polarizability of the individual molecules. The polarizability tensor in its principal coordinate system for such a molecule is

$$\underset{=}{\overset{\alpha}{=}} = \begin{pmatrix} \alpha_1 & 0 & 0 \\ 0 & \alpha_1 & 0 \\ 0 & 0 & \alpha_2 \end{pmatrix} \tag{4.86}$$

which, for a general orientation of the molecule, takes the form

$$\underset{=}{\overset{\alpha}{=}}(\theta, \phi) = \underset{=}{\overset{A}{=}}^T \cdot \underset{=}{\overset{\alpha}{=}} \cdot \underset{=}{\overset{A}{=}}$$

$$= \begin{bmatrix} \alpha_1 + (\alpha_2 - \alpha_1) \sin^2 \theta \cos^2 \phi & (\alpha_2 - \alpha_1) \sin^2 \theta \sin \phi \cos \phi & (\alpha_2 - \alpha_1) \cos \theta \sin \theta \cos \phi \\ (\alpha_2 - \alpha_1) \sin^2 \theta \sin \phi \cos \phi & \alpha_1 + (\alpha_2 - \alpha_1) \sin^2 \theta \sin^2 \phi & (\alpha_2 - \alpha_1) \cos \theta \sin \theta \sin \phi \\ (\alpha_2 - \alpha_1) \sin \theta \cos \theta \cos \phi & (\alpha_2 - \alpha_1) \sin \theta \cos \theta \sin \phi & \alpha_1 + (\alpha_2 - \alpha_1) \cos^2 \theta \end{bmatrix} \tag{4.87}$$

where the rotation matrix $\underset{=}{\overset{A}{=}}$ is expressed in terms of the Euler angles shown in the figure. The induced dipole moment $\vec{P}$ may be written in the form

$$P_i(t) = \alpha_{ij}(\theta, \phi) E_j(t) \tag{4.88}$$

and the energy of the molecule in the field $E(t)$ as

$$\delta V_i(t) = -\tfrac{1}{2} \alpha_{ijk}(\theta, \phi) E_j(t) E_k(t). \tag{4.89}$$

Assuming for simplicity that the electric field is polarized along the z-axis in the figure, the torque M experienced by the molecule is given by

$$M = -\frac{\partial \delta V}{\partial \theta} = \tfrac{1}{2} \left(\tfrac{\partial}{\partial \theta} \alpha_{zz} \right) E^2(t)$$

$$= -\frac{(\alpha_2 - \alpha_1)}{2} E^2(t) \sin 2\theta \tag{4.90}$$

Substituting this expression into the rotational diffusion equation, we find

$$\frac{\zeta}{k_B T}\frac{\partial f}{\partial t} = \frac{1}{\sin\theta}\frac{\partial}{\partial\theta}\left(\sin\theta\frac{\partial f}{\partial\theta} + \frac{(\alpha_2 - \alpha_1)E^2(t)\sin 2\theta}{2k_B T}f\right) \qquad (4.91)$$

for the field along the z-axis. As a first step in simplifying this equation, let us derive an approximate expression for the distribution function f in a constant electric field E_0 by using a Taylor series expansion

$$
\begin{aligned}
f &= \frac{\exp(-\delta V/k_B T)}{\int \exp(-\delta V/k_B T)d\Omega} \\[2mm]
&= \frac{\exp(1/2k_B T(\alpha_1 + (\alpha_2 - \alpha_1)\cos^2\theta)E_0^2)}{\int \exp(1/2k_B T(\alpha_1 + (\alpha_2 - \alpha_1)\cos^2\theta)E_0^2)d\Omega} \\[2mm]
&\simeq \frac{1 + 1/2k_B T(\alpha_1 + (\alpha_2 - \alpha_1)\cos^2\theta)E_0^2 + \cdots}{4\pi(1 + 1/6k_B T(\alpha_2 - \alpha_1)E_0^2 + \cdots)} \\[2mm]
&\simeq \frac{1}{4\pi} + \frac{(\alpha_2 - \alpha_1)E_0^2(\cos^2\theta - 1/3)}{8\pi k_B T}
\end{aligned}
\qquad (4.92)
$$

where we have used the relation $\delta V = -1/2\alpha_{zz}E_z^2$. The dropping of higher-order terms is justified since α is typically of the order of 10^{-29} m^{-3}, so that even for a power density $(n\epsilon_0 c/2|E_0|^2)$ of 10^6 W/m^2 we have $\alpha|E|^2/k_B T \simeq 10^{-2}$ at $T = 300$ K. The physical interpretation of this expression is now straightforward. The static field E_0 preferentially orients the molecules along its direction, and when it is switched off the molecules relax to a random isotropic distribution $f = 1/4\pi$.

EXERCISE 4.2

By considering a field of the form

$$E(t) = \begin{cases} E_0, & t < 0 \\ 0, & t \geq 0 \end{cases}$$

with E_0 oriented along the z-axis and a form for the solution $f(t) = A(t) + B(t)\cos^2\theta$, as suggested by the expression for f above, show that

$$f(\theta, t) = \frac{1}{4\pi} + \frac{(\alpha_2 - \alpha_1)E_0^2}{8\pi k_B T}(\cos^2\theta - 1/3)e^{-t/\tau_R} \qquad (4.93)$$

with $\tau_R = \zeta/6k_B T$. □

EXERCISE 4.3

By adopting a trial solution of the form $f(\theta, t) = A(1 + \beta(\cos^2\theta - 1/3)\Phi(t))$, $\beta = 1/k_B T$ for the more general case of an arbitrary but linearly polarized field $E(t)$, show that $\Phi(t)$ obeys the following o.d.e.:

$$\tau_R \frac{d}{dt}\Phi + \Phi = \alpha E^2(t) \tag{4.94}$$

with $\alpha = (\alpha_2 - \alpha_1)/2$ by keeping terms up to first order in β.   □

Notice that the function $\Phi(t)$ replaces $(\alpha_2 - \alpha_1)E_0^2/2$ for the static field case in Exercise 4.2. Equation (4.94), called the Debye equation, shows clearly that the mechanism for the optical nonlinearity is the intrinsic anisotropy of the polarizability $(\alpha_2 \neq \alpha_1)$, which causes the applied optical field to orient preferentially an initially random distribution of molecules along its direction. The characteristic response time τ_R is typically $\sim 10^{-12}$ s for this reorientational nonlinearity, and the nonlinear coefficient is typically orders of magnitude greater than the electronic one. The Debye equation is the analog for liquids of the population inversion equation in the Bloch equations when the detuning from resonance is very large relative to the homogeneous line width γ_{12}^{-1} so that we can solve (4.35) algebraically for Λ in terms of A and N.

For references, the reader might begin with [92].

EXERCISE 4.4

Show that an equation of the form (4.94) can be derived from the two-level atom optical Bloch equations if one assumes a large enough detuning such that the initial level population remains unchanged. (*Hint*: Go to (4.35) and eliminate Λ adiabatically and assume that $N \simeq N_0$.)   □

Electrostriction (Acoustic Wave Equation)

In the derivation of the material dispersion relation using the simple model of a diatomic crystal lattice in Section 6g, we observe that the acoustic branch of the dispersion relation is related to a collective displacement of the atoms as a whole along a line. In the presence of an inhomogeneous electric field, dielectrics are subjected to a volume force, called electrostriction, that induces macroscopic density changes in the material. A spectacular manifestation of the electrostriction force are the damage tracks formed in many solid materials when an intense laser beam collapses into multiple filaments. The loud bang associated with the onset of damage is clear evidence for coupling of the electromagnetic wave to acoustic waves in the material.

The coupling parameters for electrostrictive self-focusing and stimulated Brillouin scattering can be calculated classically by considering the change in refractive index caused by the density variation in the material. This index change modifies the polarization of the material through the relation

$$\vec{P} = \epsilon_0 n^2 \vec{E} = \epsilon_0 (n_0 + \delta n)^2 \vec{E} \simeq \epsilon_0 n_0^2 \vec{E} + 2\epsilon_0 n_0 \delta n \vec{E}, \qquad (4.95)$$

where $\delta n \ll n_0$ is the refractive index change due to density variations. It remains now to derive the driving term for the material waves in the acoustic wave equation (4.96). The reader will have to accept this equation as given; a derivation from the linearized Navier–Stokes equation is given in reference [49].

$$\nabla^2 \Delta_\rho - \frac{2\Gamma_B}{v^2} \frac{\partial}{\partial t} \Delta_\rho - \frac{1}{v^2} \frac{\partial^2}{\partial t^2} \Delta_\rho = \frac{1}{v^2} \nabla \cdot \vec{F}_v \qquad (4.96)$$

where Δ_ρ is defined as the increase in density, $\Gamma_B = \eta k_a^2 / \rho$ (η = viscosity, ρ = density, v = acoustic wave velocity, and k_a = magnitude of acoustic wavevector) is the acoustic damping coefficient, and $\vec{F}_v$ is the force per unit volume. In obtaining an expression for $\vec{F}_v$ we need to relate the change δn in refractive index to the variation $\delta \rho$ in density. We proceed as follows.

If $\vec{u}$ is an arbitrary velocity field within a dielectric, then the rate at which energy U is lost by the field is given by

$$\frac{dU}{dt} = -\int \vec{F}_v \cdot \vec{u} \, dv \qquad (4.97)$$

where we integrate over the volume v. Since

$$U = \frac{1}{2} \int \vec{E} \cdot \vec{D} \, dv = \frac{1}{2} \int \frac{1}{\epsilon} \vec{D} \cdot \vec{D} \, dv \qquad (4.98)$$

where $\epsilon = \epsilon_0 n^2$ is the dielectric permittivity, we obtain

$$\delta U = \frac{1}{2} \int D^2 \delta \left(\frac{1}{\epsilon} \right) dv + \int \vec{E} \cdot \delta \vec{D} \, dv$$

$$= -\frac{1}{2} \int E^2 \delta \epsilon \, dv + \int \vec{E} \cdot \delta \vec{D} \, dv. \qquad (4.99)$$

Here the first term represents the energy change caused by the change in dielectric permittivity, and the second term corresponds to the energy change caused by the displacement of sources.*

Hence, in the absence of free changes,

$$\frac{dU}{dt} = -\frac{1}{2} \int E^2 \left(\frac{\partial \epsilon}{\partial t} \right) dv. \qquad (4.100)$$

To arrive at an expression for the volume force we must get this expression into a form involving the velocity field $\vec{u}$. This can be done by means of the hydrodynamic equation of continuity

$$\nabla \cdot (\rho \vec{u}) + \frac{\delta \rho}{\partial t} = 0 \qquad (4.101)$$

representing the conservation of mass. The total derivative of ρ, when evaluated such that the observation point moves with a chosen volume element in a velocity field, can be written as

$$\frac{D\rho}{Dt} = (\nabla \rho) \cdot \vec{u} + \frac{\partial \rho}{\partial t}. \qquad (4.102)$$

We now must use a dielectric equation of state, such as the Clausius–Mossoti equation below, which gives the dependence of dielectric constant on density, then

$$\frac{D\epsilon}{Dt} = \frac{d\epsilon}{d\rho} \frac{D\rho}{Dt} \qquad (4.103)$$

where $d\epsilon/d\rho$ is derived from the equation of state. Now $D\epsilon/Dt = \partial \epsilon/\partial t$ as we are considering a homogeneous dielectric medium (i.e., $\nabla \epsilon = 0$) and so,

---

*This can be seen as follows:

$$\int \vec{E} \cdot \delta \vec{D} \, dv = -\int \nabla \phi \cdot \delta \vec{D} \, dv \qquad \text{(for an electrostatic field)}$$

$$= \int \phi \nabla \cdot (\delta \vec{D}) \, dv - \int \nabla \cdot (\phi \delta \vec{D}) \, dv$$
$$\text{(second integral is zero as it can be transformed to a surface integral)}$$

$$= \int \phi \, \delta (\nabla \cdot \vec{D}) \, dv$$

$$= 4\pi \int \phi \, \rho \, dv = 0) \qquad \text{(because } \nabla \cdot \vec{D} = \rho \text{ and we assume no charges)}$$

on using the equation of continuity above, we obtain

$$\frac{\partial \epsilon}{\partial t} = -\frac{d\epsilon}{d\rho}\rho(\nabla \cdot \vec{u}). \qquad (4.104)$$

Therefore we can write

$$\frac{dU}{dt} = \int E^2 \rho \frac{d\epsilon}{d\rho}(\nabla \cdot \vec{u}) \, dv$$

$$= \int \nabla \cdot \left(E^2 \rho \frac{d\epsilon}{d\rho}\vec{u}\right) dv - \int \nabla \left(E^2 \rho \frac{d\epsilon}{d\rho}\right) \cdot \vec{u} \, dv. \qquad (4.105)$$

The first integral can be converted to a surface integral that vanishes in the limit, and so we can immediately identify the volume force $\vec{F}_v$ as

$$\vec{F}_v = \nabla \left(E^2 \frac{d\epsilon}{d\rho}\rho\right). \qquad (4.106)$$

Substituting this expression back into the acoustic wave equation, we get

$$\nabla^2 \Delta\rho - \frac{2\Gamma_B}{v^2}\frac{\partial}{\partial t}\Delta\rho - \frac{1}{v^2}\frac{\partial^2}{\partial t^2}\Delta\rho = \frac{2\epsilon_0 n_0}{v^2}\rho \, dn \, d\rho \nabla^2(E^2). \qquad (4.107)$$

Note that this equation is strictly valid for a static inhomogeneous electric field, but it can also be employed for high-frequency optical fields, provided we average the square of the total field over many wavelengths. This electrostrictive forcing term involving the gradient of the electromagnetic field intensity can contribute significantly to the self-focusing effect (n_2 coefficient) that dominates over the usual electronic contribution to n_2 for long optical pulses. For the purposes of the present discussion, we observe that it can also resonantly drive a much lower frequency acoustic wave if we consider the total field $\vec{E}$ to consist of a superposition of two optical waves whose beat frequency is close to the acoustic wave frequency. This is the physical mechanism for *stimulated Brillouin scattering*. Because of the smallness of the acoustic wave frequency, the Stokes shift of the scattered wave is much smaller than that for stimulated Raman scattering.

Finally, the Clausius–Mosotti state equation relating refractive index changes to density changes is given by (the reader will just have to accept this formula)

$$\frac{n^2 - 1}{n^2 + 2} = \frac{1}{3}\rho\alpha \qquad (4.108)$$

and therefore

$$(\Delta n)_\rho = \frac{(n_0^2 - 1)(n_0^2 + 2)}{6n_0} \frac{\Delta\rho}{\rho_0}. \qquad (4.109)$$

This allows us to determine the coupling coefficients for electrostriction from first principles.

This ends the derivation of the material equations, which will be used in the remainder of this chapter and in Chapter 5 to discuss stimulated light scattering, laser operation, laser instabilities, optical bistability, counter-propagating beam instabilities, and self-induced transparency.

4i.   Maxwell–Debye Equations

The Maxwell–Debye equations represent the starting point for the study of the interaction of one or more electromagnetic waves with nonresonant media where the finite material response time τ is relevant. In many instances the importance of this response time is not readily apparent, especially when dealing with cw (continuous wave) beam interactions. These interactions involve changes induced in the material refractive index due to the presence of an applied electromagnetic wave and are examples of the polarization expansion including retardation effects discussed in Chapter 2. Writing the electric displacement vector $\vec{D} = \epsilon_0 n^2(\omega)\vec{E}$ and $n(\omega) = n_0(\omega) + \delta n(E)$, where n is the field-dependent induced change in refractive index, we obtain using the definition $P = P_L + P_{NL}$,

$$P_{NL} = 2\epsilon_0 n_0 \delta n(E) \; E. \qquad (4.110)$$

The change δn, which is small relative to n_0, is governed by the Debye equation derived in Section 4h.

Unidirectional Wave   The coupled field–matter equation follows directly from the above definition of the induced polarization and the wave equation (2.9) where $\vec{E} = \vec{e} A(\vec{r}, t) e^{i(kz-\omega t)} + (*)$,

$$\left(\frac{\partial}{\partial z} + \frac{n_0}{c} \frac{\partial}{\partial t} \right) A - \frac{i}{2k} \nabla_\perp^2 A + i \frac{\omega_0}{c} \delta n \; A = 0,$$

$$\tau \frac{\partial \delta n}{\partial t} + \delta n = n_2 |A|^2. \qquad (4.111)$$

It is easy to see that the NLS equation follows directly from this equation by setting $\tau = 0$. This equation has been used in self-focusing theory to study the transient behavior of short intense optical pulses. For pulses of duration $\tau_p \geq \tau$, the critical self-focusing (blow-up singularity of the two-dimensional NLS equation) is delayed but not prevented.

Counter-Propagating Waves Substituting $\vec{E} = \hat{e}(A_1 e^{i(kz-\omega t)} + A_2 e^{-i(kz+\omega t)} + (*))$ into the wave equation (2.9) and the Debye equation (4.94) and expanding the induced refractive index change in an even Fourier series ($\delta n = \delta n_0 + \sum_{p=1}^{\infty}(\delta n_1 e^{2\,ipkz} + (*))$), one obtains

$$\left(\frac{\partial}{\partial z} + \frac{n_0}{c}\frac{\partial}{\partial t}\right)A_1 - \frac{i}{2k_0}\nabla_1^2 A_1 = -i\frac{\omega_0}{c}(\delta n_0 A_1 + \delta n_1 A_2) \qquad (4.112)$$

$$\left(-\frac{\partial}{\partial z} + \frac{n_0}{c}\frac{\partial}{\partial t}\right)A_2 - \frac{i}{2k_0}\nabla_1^2 A_2 = +i\frac{\omega_0}{c}(\delta n_0 A_2 + \delta n_1^* A_1) \qquad (4.113)$$

$$\tau\frac{\partial}{\partial t}\delta n_0 + \delta n_0 = n_2(A_1 A_1^* + A_2 A_2^*) \qquad (4.114)$$

$$\tau\frac{\partial}{\partial t}\delta n_1 + \delta n_1 = n_2 A_1 A_2^*. \qquad (4.115)$$

In (4.112–4.115), the Fourier series expansion has been truncated at the second term so as to include the spatial grating effect. Diffusion effects have been ignored.

4j. The Born–Oppenheimer (Adiabatic) Approximation

The study of the dynamical motion of a large collection of electrons and nuclei, bound by electrostatic forces, presents a formidable problem in both classical and quantum mechanics. Nevertheless, spectacular success has been made in elucidating the internal structure of complex molecules, leading to a detailed understanding of chemical and biological systems. Much of this progress can be traced to the work of Born and Oppenheimer in 1927 [58] who, by employing perturbation theory, showed that the Schrödinger equation describing these complex interactions is approximately separable. The small parameter in their analysis is $\eta \propto [\mu/M]^{1/4} \ll 1$, the ratio of the electron to nuclear mass. Since atomic nuclei are ten thousand to a few hundred thousand times heavier than electrons, this parameter will be small in general. It is reasonable therefore to assume as a zeroth approximation that these massive nuclei are at rest relative to the much more energetic electrons (recall that their kinetic energies are $\hbar^2/2M\nabla^2$ and $\hbar^2/2\mu\nabla^2$ respectively) and to take their motion into account in a higher-order approximation. Born and Oppenheimer showed that the electronic energy is of order η, the vibrational energy of the nuclei of order η^2, and the rotational energy of the molecule as a whole of order η^4. Our picture of a complex molecule is one of a large sluggish object with electrons buzzing around the nuclei (binding them together in chemical bonds) with the bound nuclei vibrating near their equilibrium separation values on a much slower time scale and the molecule as a whole undergoing an even slower rotation. The above observations suggest that one can make a separation of variables approximation so that, at lowest order, we have separate subproblems for each of the electronic, vibrational, and rotational motions. The essential ideas behind the Born-Oppenheimer approximation can be sketched by appealing to the methods already used earlier in this chapter. The full technical details can be found in references [58, 59].

The total Hamiltonian determining the internal state of an arbitrary molecular system is of the form

$$H = T_R + T_r + V(r, R) \tag{4.116}$$

where $T_r = -\hbar^2/2\mu \sum_i \partial^2/\partial r_i^2$ is the kinetic energy operator for the electrons and $T_R = -\hbar/2M \sum_j \partial^2/\partial R_j^2$ is the kinetic energy operator for the nuclei. The potential energy function $V(r, R)$ involves summation over all two-body Coulombic interactions between positively charged nuclei and electrons (attractive forces due to opposite charges) and mutual interactions between the positively charged nuclei and negatively charged electrons themselves (repulsive forces between like charges). The details of these interactions are not relevant to the following discussion. The adiabatic approximation (Born-Oppenheimer) is based on the assumption that the kinetic energy operator for the heavy particles acts as a small perturbation, and so we write

$$H = H_0 + T_R \tag{4.117}$$

with the unperturbed Hamiltonian $H_0 = T_r + V(r, R)$. In zeroth order, when the mass of the heavy particle is assumed infinitely large, the problem of finding stationary states of the Schrödinger equation reduces to the following eigenvalue problem:

$$(H_0 - \epsilon_n(R))\phi_n(R, r) = 0, \tag{4.118}$$

in which the heavy particle coordinate R appears as a parameter. This eigenvalue problem yields the electronic energy levels $\epsilon_n(R)$ of the system and their corresponding eigenfunctions $\phi_n(R, r)$, where the label n distinguishes different bound states. These electronic energies are a function of the nuclear coordinate R, and this equation is simply (4.1). Assuming that we have solved this eigenvalue problem, we now seek a solution of

$$(H - E)\Psi(R, r) = 0 \qquad (4.119)$$

in the form

$$x\Psi(R, r) = \sum_n \Phi_n(R)\phi_n(R, r)$$

by direct analogy with (4.6) where $\phi_n(R, r)$ are the eigenfunctions of H_0 with R appearing as a parameter. Substituting for $\Psi(R, r)$ in (4.119), multiplying by $\phi_m^*(R, r)$ and integrating over the light particle coordinates r, we obtain the set of equations

$$(T_R + \epsilon_m(R) - E)\Phi_m(R) = \sum_n \Lambda_{mn}\Phi_n(R) \qquad (4.120)$$

where the operator Λ_{mn} is given by

$$\Lambda_{mn} = \frac{\hbar^2}{M} \sum_j \int \phi_m^*(R, r)\frac{\partial}{\partial R_j}\phi_n(R, r)dr\frac{\partial}{\partial R_j}$$

$$- \int \phi_m^*(R, r)T_R\phi_n(R, r)dr. \qquad (4.121)$$

This set of equations is exact with the right-hand side reflecting the coupling between electronic, vibrational, and rotational degrees of freedom. There now remains the technical question of showing that the right-hand side of (4.120) is small enough that it can be set to zero as a first approximation. We avoid this question here and simply set it to zero, to obtain

$$[T_R + \epsilon_m(R)]\Phi_{mv}^{(0)}(R) = E_{mv}^{(0)}\Phi_{mv}^{(0)}, \qquad (4.122)$$

an eigenvalue problem for each state of motion of the heavy particles (nuclei). Notice how the eigenvalue of the electronic problem (4.118) now acts as an effective potential for the nuclear problem.

The kinetic energy operator T_R for the heavy nuclei describes both vibrational and rotational motion (translational motion introduces no quantum features), and the eigenfunction $\Phi_{mv}^{(0)}(R)$ can be further factored depending on the symmetry of the molecule in question. The label "v" denotes the set

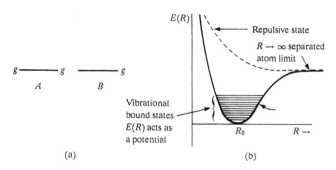

FIGURE 4.8 (a) Separated atoms in their ground electronic states E_g. (b) Diatomic molecule electronic states.

of vibrational and rotational quantum numbers. In summary, then, the total wavefunction for a molecule can be written approximately in separable form

$$\Psi = \Psi_{el}(r, R_0)\Psi_{vib}(R)\Psi_{rot}(\theta_i) \tag{4.123}$$

where r is the electron coordinate, R is the displacement of the nucleus from its equilibrium position R_0, and θ_i are the Euler angles determining the orientation of the molecule in space. Figure 4.8 shows pictorially the effect of bringing two well-separated atoms A and B in their electronic ground states together to form a diatomic molecule *A-B*.

Challenge: The nature of the Born-Oppenheimer approximation suggests another intriguing possibility which deserves to be explored. If the external field $\vec{E}(\vec{r}, t)$ is a wavepacket rather than a wavetrain, then it may very well be the case that the ratio of the wavelength (frequency) to the distance (time) over which the envelope of the field $\vec{E}$ varies is comparable to the ratios of atomic to molecular scales and the frequencies of electronic to vibrational or rotational transitions. In this case, one should in fact allow the probability amplitude $a_j(t)$ in (4.6) be a slowly varying function of position to reflect the fact that atoms in different parts of the molecule, while experiencing the same local Coulomb potential, would see different large-scale influences. This would lead to the addition of an $i\nabla^2 a_j$ term in (4.12) and a corresponding flux divergence $i\nabla \cdot (a_n^*\nabla a_j - a_j\nabla a_n^*)$ in (4.13), the equation for the density matrix element ρ_{jk}. There is an obvious closure difficulty in handling these extra terms because the flux vector $\vec{F}_{kj} = i(a_k^*\nabla a_j - a_j\nabla a_k^*)$ cannot be directly expressed as a function of ρ_{jk}. However, one might make an approximation that $\vec{F}_{kj}$ is proportional to the gradient of ρ_{jk} so that one would thus obtain the additional terms $-i\nabla \cdot f_{jk}\nabla\rho_{jk}$ on the left-hand side of (4.13). Alternatively, one could simply express the polarization and population inversion as a quadratic product of the probability amplitudes and work directly with them. It would be interesting to explore the effects of the presence of such terms in the two-level laser.

Chapter 5

APPLICATIONS: LASERS, OPTICAL BISTABILITY, DISTRIBUTED FEEDBACK, SELF-INDUCED TRANSPARENCY, AND STIMULATED SCATTERING

Overview

THIS CHAPTER IS ABOUT APPLICATIONS, and we will use the mathematical apparatus developed in the previous chapters to discuss lasers, the propagation of optical beams in feedback cavities and through refractive index gratings. We will also see that solitons and other special solutions (e.g., self-similar solutions) of soliton equations play important roles in helping us understand the transverse structure of optical beams in optical cavities, the longitudinal structure of pulses in resonant media, superfluorescence, and the production of the Stokes wave in Raman scattering processes.

For the most part, the field equations we employ are either the Maxwell–Bloch or Maxwell–Debye equations, which describe the propagation of optical beams in resonant and nonresonant media respectively. In optical feedback geometries such as ring or Fabry–Perot cavities, these equations are augmented by boundary conditions that describe how the field at the beginning of the $(n + 1)$th passage of the beam through the cavity depends on conditions generated at the end of the nth pass. The simplest such boundary condition is the reduction of the intensity of the beam at a reflecting mirror. The advantage of feedback geometries is that one can effectively achieve extremely long interaction distances, thereby allowing for more efficient field–matter coupling.

At the outset we want to distinguish between what we call passive and active media. In the passive medium, the atoms making up the medium are initially in their ground states and the source of energy to the system is external, usually a laser beam or mixture of laser beams containing only one or a small number of discrete frequencies. In the active medium, the atoms are initially prepared, by some means we shall shortly discuss, in their excited states, and the light that is initially spontaneously emitted by the atoms is amplified preferentially along the cavity axis. A fundamental difference between passive and active systems is that the former can operate and give interesting behavior even when the frequency of the external laser source is tuned away from the natural frequencies of the medium whereas the latter must operate fairly close to resonant conditions. In this chapter, the medium is considered active when we discuss lasers and the phenomenon of superfluorescence. The discussion concerning optically bistable cavities and self-induced transparency considers passive media.

The outline of the chapter is as follows.

5a. Here we discuss lasers. We begin by stressing the fundamental ingredients required for laser operation. We then introduce the simplest model of a laser involving a medium of two-level atoms. At first we consider a single-mode laser in which the spatial structure of the lasing mode is determined a priori to be the TEM$_{00}$ mode, derived in Section 6f. In this case, the active order parameters of the system are the time-dependent amplitudes of the electromagnetic, polarization, and population inversion fields. They satisfy a set of three coupled ordinary differential equations (6.196–6.198) with $a = 0$ that are closely related to the famous Lorenz equations. Next we relax the spatial structure restriction and allow several transverse TEM$_{rs}$ modes with the same longitudinal wavenumber and natural frequencies close to that of the basic TEM$_{00}$ mode to compete. Close to the lasing threshold we find that the system can be described by one order parameter, which depends on the transverse coordinates and time and which obeys the celebrated complex Ginzburg–Landau (CGL) equation introduced in Section 6i. We consider the three-level model and show how this system reveals a wealth of rich dynamical behavior as the population inversion is increased. In particular, we see how the laser can suddenly shut off as the gain bandwidth of the laser system undergoes a Rabi splitting in which two lobes separate away from the "on resonance" frequency and move further and further away from each other as the external pump is increased.

5b. In this section, we discuss the manifestation of optical bistability in ring and Fabry–Perot cavities. The basic idea is fairly simple. One can think of a tuned cavity as a forced and damped oscillator. As is well known, nonlinearity distorts the output versus input curve so that there can be several equilibrium output states for a single set of input conditions. These equilibrium states are alternately stable and unstable. We examine the nature of the stable solutions of such arrangements and their possible roles as the 0, 1 states of a logic element in an all-optical processor. We also discuss the stability of the "stable" solutions and find that they can undergo a series of period-doubling bifurcations leading to a temporally chaotic output. These transitions are dramatically affected by time delay, and so we have then to reexamine the validity of adiabatic elimination of the matter variables. When the input beam has a Gaussian rather than plane wave shape, further interesting events occur. If part of the beam is above the critical threshold at which the output switches from one state to another, then different parts of the beam attempt to access different intensity levels, and this leads to large electric field gradients and therefore large diffraction. A rich set of transverse spatial structures arises as a result of a balance of diffraction and nonlinear feedback. In these situations, the optical beam not only displays chaotic temporal behavior but also can lose spatial coherence. In short, it can be considered to be fully turbulent.

5c. Of related interest to the material covered in 5b is the nature of the response of an optical beam in a medium with periodic refractive index and the interaction of two counter-propagating light beams.

5d. To this point, much of the analysis has involved field equations of nonlinear Schrödinger type coupled to either the Bloch or Debye equations describing the medium response. Losses are accounted for by homogeneous broadening and/or mirror effects. In this section, we look at the remarkable behavior of a pulse propagating through an almost perfectly tuned resonant and initially passive medium. The lack of perfect tuning is due to the random motion of the atoms in the medium, and the resulting Doppler shifts cause the frequency mismatch $\omega_{12} - \omega = kv$ to be a random variable with a Maxwellian distribution. This random frequency mismatch, called inhomogeneous broadening, produces an effect, analogous to Anderson localization in solid-state physics, in which all wave functions are localized in a (one-dimensional) medium with random potential. The analogous behavior for waves is that they cannot propagate through a one-dimensional random medium. A small pulse interacts incoherently with the medium, and the electromagnetic field decays in a localization distance roughly proportional to the inverse of the width of the distribution of the frequency mismatch. The medium itself is left in a ringing state. But if the pulse is sufficiently nonlinear, a miracle occurs. We will see that there is a special set of nonlinear wavepackets that phaselock the material response. The leading edge of one of these pulses excites the medium, and its tail regains all the energy given up to the medium coherently. The atoms are all returned to their ground states. The pulse propagates coherently in an otherwise opaque medium. It has created its own (self-induced) transparency. We discuss the ramifications of this result on the much broader topic of Anderson localization.

We discuss this phenomenon and also mention what one can expect in the superfluorescence case, where the medium starts off in the excited state.

It turns out that this case is precisely equivalent to the Raman scattering of a pump wave by a material vibration when attenuation effects are ignored.

5e-f. The final sections deal with Raman and Brillouin scattering in two limits. First, we consider the case for which the material response is very fast and so the matter variables can be adiabatically eliminated to give a pair of coupled-mode equations for the pump and Stokes waves. In the case of Brillouin scattering, the Stokes wave travels backwards, and this case is briefly analyzed. We then look at the other limit in the context of Raman scattering where the relaxation time of the material is infinite, and discuss the nature of the solutions of the resulting three-wave interaction equations. Mathematically, this problem is the same as the superfluorescence problem.

5a. Lasers

The invention of the laser in the early 1960s led to an explosive growth in the field of nonlinear optics. Here at last was a source of highly intense, coherent, and monochromatic light that could be used to probe matter like a fine tuning fork. Already the use of lasers in modern technology is commonplace, ranging in application from high-density data storage on optical disks to greatly improved surgical techniques in ophthalmology, neurosurgery, and dermatology. Before beginning a quantitative treatment of the various laser models, it is worthwhile to consider the basic ingredients and fundamental principles that underlie this wonderful invention [2–5, 60].There are three.

First is an external pumping mechanism to feed energy (usually incoherently) into the lasing medium. The latter must be capable of storing this energy in a long-lived excited (metastable) state before returning back to the original equilibrium state through the release of energy as light. This procedure is referred to as causing an _inversion_ in the atoms/molecules/semiconductor bands making up the laser medium. Usually, the pump process is highly unselective, depositing energy over a broad band of states, most of which do not participate in the lasing process. For example, in gas lasers, an electrical discharge ionizes the molecules in the gas, and the ions in turn recombine with the electrons to emit light of different frequency in a cascade back to their original undisturbed equilibrium state. In a solid-state laser a mercury flashlamp deposits energy within a broad band of energy states, some of which couple directly to those states involved in the laser transition. In a semiconductor laser, an electrical current excites electrons from the valence to conduction band, leaving positive holes in the former. This sea of electrons (called carriers) eventually recombines with the positive holes after some characteristic time called the carrier recombination time, emitting light in the process of frequency $\hbar\omega = (E_c - E_v)$ corresponding to the energy difference between electrons in the conduction and valence bands.

Second is a large gain per unit length to amplify the emitted light energy (spontaneous emission). This requires that the transition selected for lasing have a large electric dipole matrix element (the p_{jk}'s of Chapter 4) coupling the upper and lower laser levels.

Third is an optical feedback mechanism, usually consisting of nearly flat mirrors of high reflectivity. The resonator is tuned to support frequencies close to that of the laser transition. Initially light is emitted spontaneously in all directions. However, light emitted along the laser axis is fed back into the lasing medium by the mirrors to stimulate even more light emission from the continuously pumped medium. Moreover, this highly directional light is further amplified on each pass back and forth in the resonator as long as the losses due to finite output mirror transmission, nonperfect reflectivity, and light diffraction of off-axis rays over the edges of the mirror can be overcome by a sufficiently large population inversion.

The Two-Level Laser [6f, 6i]

The open cavity resonator shown in Figure 5.1 was the brainchild of Schawlow and Townes and marked the critical breakthrough in the design of the working laser because its geometry gives a preferential treatment to the fundamental TEM$_{00}$ mode. We will see in Section 6f how to use the paraxial approximation to construct solutions U_{rs} ($k = n\omega/c$) to the Helmholtz equation

$$\nabla^2 U + \frac{n^2\omega^2}{c^2} U = 0. \tag{5.1}$$

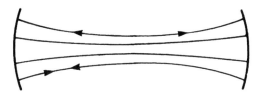

FIGURE 5.1   Schematic laser resonator.

The set of allowed wavevectors k and frequencies ω is determined by insisting that the total phase change $kd - (r + s + 1)\tan^{-1} d/2\, z_0$ is $q\pi$, q an integer.

This gives the mode resonance frequencies

$$\omega_{qrs} = \frac{c}{nd}\left(q\pi + 2(r+s+1)\tan^{-1}\frac{d}{2\,z_0}\right)$$ (5.2)

for the special case of a confocal cavity, where d is both the distance between, and the radii of curvature of, two spherical mirrors. The open cavity spectrum is shown in Figure 6.13, and the standing-wave combinations of $U_{rs}(q)$ and $U_{rs}(-q)$ we call V_{rs}. A number of important physical conclusions may be drawn from an inspection of the mode frequency expression (5.2). Firstly, the longitudinal mode spectrum indexed by the integer q has a frequency separation between nearest neighbors of magnitude $c\pi/nd$. Control over the longitudinal mode spacing can be simply achieved by increasing (compress the mode spectrum) or decreasing (expand the mode spectrum) the mirror separation d. Secondly, the transverse mode frequency spectrum indexed by the integer pair (r, s) is clustered near each longitudinal mode extending to the high-frequency side. The appearance of the sum $r+s$ in the frequency spectrum means that there is an increasingly high degeneracy in the transverse mode spectrum for higher-order modes as indicated in Figure 6.13. The transverse mode frequency spacing depends on the mirror radius R, and in the special case of perfectly flat mirrors the modes are completely degenerate. The usual choice for mirror curvature places the center of the radius of curvature of each mirror at the location of the other; this cavity is called confocal, and the transverse mode separation is exactly one-half of the longitudinal mode spacing.

In order to derive the simplest model, we will assume that a single TEM$_{00}$ mode with a specific longitudinal wavenumber q is preferentially excited. Of course, in reality all the modes V_{qrs} can be excited from the background noise, but what we are saying is that the one we choose will be the first to be excited. Moreover, as the population inversion is increased, this preferred mode $V(x, y, z)$ will suppress all others. We will return to this assumption later. We will also assume that the electric field is singly polarized, a situation that can be experimentally achieved by placing a polarizer in the laser cavity. We write

$$\begin{Bmatrix}\vec{E}\\\vec{P}\end{Bmatrix} = \hat{e}\begin{Bmatrix}F(t)\\\Lambda(t)\end{Bmatrix} V(x, y, z)e^{-i\omega_c t} + (*).$$ (5.3)

The equation for $F(t)$ follows from substituting (5.3) into (2.9), ignoring d^2F/dt^2, and approximating $\partial^2 P/\partial t^2$ by $-\omega_c^2\Lambda\, V\, e^{-i\omega_c t} + (*)$. We obtain

$$\frac{dF}{dt} = \frac{i\omega_c}{2\epsilon_0}\Lambda.$$ (5.4)

We include losses that are principally due to imperfect mirror reflection by adding the term κF to the left-hand side of (5.4),

$$\frac{dF}{dt} + \kappa F = \frac{i\omega_c}{2\epsilon_0}\Lambda \tag{5.5}$$

with κ given by

$$\kappa = \frac{\sigma}{\epsilon_0} + \frac{c}{L}\ln\frac{1}{\sqrt{R}}. \tag{5.6}$$

The first term is due to finite conductivity effects, which are very weak. The rationale for the major component of the loss is that the amplitude of each component of the standing wave $FV\,e^{-i\omega_c t} + F^*V^*e^{i\omega_c t}$ undergoes a loss by the factor $\sqrt{R}$ every L/c units of time, which corresponds to an average exponential loss rate of $(c/2L)\ln(1/R)$.

EXERCISE 5.1

Suppose the loss in amplitude is exponential, i.e., $A(t) = A_0\,e^{-\kappa t}$. Calculate κ as follows. Because $A(t + L/c) = \sqrt{R}A(t)$, we have $e^{-\kappa(L/c)} = \sqrt{R}$ from which we see $\kappa = (-c/L)\ln\sqrt{R}$ which is positive because $R < 1$. $\square$

The material equations are derived from (4.29) and (4.30), the equations for the density matrix elements ρ_{12} and $\rho_{22} - \rho_{11}$. Recall that $\vec{P} = \hat{e}n_a p(\rho_{12} + \rho_{12}^*)$, $p = p_{12} = p_{21}$, and that therefore we can identify

$$n_a p\,\rho_{12} = \Lambda\,V\,e^{-i\omega_c t}. \tag{5.7}$$

Substitute (5.7) in (4.29), multiply across the equation by $V^*e^{i\omega_c t}$ and integrate over the cavity volume to obtain

$$\frac{d\Lambda}{dt} + (\gamma_{12} + i(\omega_{12} - \omega_c))\Lambda = \frac{ip^2}{\hbar}FN \tag{5.8}$$

where

$$N = \frac{n_a \int (\rho_{22} - \rho_{11})VV^*d\vec{x}}{\int VV^*d\vec{x}} \tag{5.9}$$

measures the excess number of atoms in the lower energy state, and $\hbar\omega_{12} = E_1 - E_2$ is the energy level difference. Because we take ω_{12} positive, $E_1 > E_2$, and therefore we have taken 2 to be the lower level and 1 the upper level.

The equation for N is achieved by multiplying the equation for $\rho_{22} - \rho_{11}$ by VV^* and integrating over the cavity volume. We obtain

$$\frac{dN}{dt} + \gamma_{11}(N - N_0) = \frac{2\,i\,A_0}{\hbar}(F^*\Lambda - F\Lambda^*) \tag{5.10}$$

where A_0 is the normalization constant

$$A_0 = \frac{\int(VV^*)^2 d\vec{x}}{\int VV^* d\vec{x}} \tag{5.11}$$

and N_0 is the externally provided population inversion. The medium has more atoms in the excited state if $N_0 < 0$, but N_0 must be sufficiently negative so as to overcome both mirror and homogeneous broadening losses for the system to act as a laser.

Equations (5.5), (5.8), and (5.10) are the equations for the two-level laser. The nonlasing solution

$$F = \Lambda = N - N_0 = 0 \tag{5.12}$$

becomes unstable through a forward bifurcation to the lasing solution (the reader should check that these are solutions):

$$F = F_0 e^{-i\nu t}, \quad \Lambda = \frac{ip^2}{\hbar}\frac{F_0 N e^{-i\nu t}}{\gamma^{12} + i(\omega_{12} - \omega_c)}$$

$$N = \frac{N_0}{1 + \frac{4p^2 A_0}{\hbar^2}\frac{\gamma_{12}}{\gamma_{11}}\frac{|F_0|^2}{\gamma_{12}^2 + (\omega_{12} - \omega_c)^2}}, \tag{5.13}$$

with ν and $|F_0|$ given by the expressions

$$\omega_L = \omega_c + \nu = \frac{\gamma_{12}\omega_c + \kappa\omega_{12}}{\gamma_{12} + \kappa},$$

$$\kappa = \frac{\omega_c p^2 \gamma_{12}|N_0|}{2\epsilon_0\hbar(\gamma_{12}^2 + (\omega_{12} - \omega_c)^2) + \frac{4p^2}{\hbar^2}A_0\frac{\gamma_{12}}{\gamma_{11}}|F_0|^2}. \tag{5.14}$$

This solution exists (i.e. $|F_0| > 0$) as soon as the population inversion $|N_0|$ exceeds the critical value

$$N_c = \frac{2\epsilon_0\hbar\kappa(\gamma_{12}^2 + (\omega_{12} - \omega_c)^2)}{\omega_c p^2 \gamma_{12}}. \tag{5.15}$$

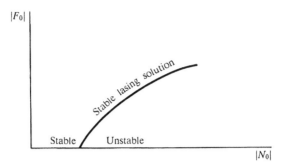

FIGURE 5.2   Transition of two-level laser.

The phase of F_0 is undetermined. This transition can be represented by the graph of Figure 5.2.

In analyzing the equations for a two-level atom, it is convenient to make the change of variables

$$t = \frac{1}{\gamma_{12}}t', \quad F = \frac{i\hbar\gamma_{12}}{2\sqrt{A_0 p}}A, \quad \Lambda = \frac{\hbar\gamma_{12}\epsilon_0\kappa}{\omega_c\sqrt{A_0 p}}P, \quad N - N_0 = \frac{2\epsilon_0\kappa\hbar\gamma_{12}}{\omega_c p^2}n,$$

$$(5.16)$$

whence (5.5), (5.8), and (5.10) become

$$\frac{dA}{dt'} = -\sigma A + \sigma P$$

$$\frac{dP}{dt'} = rA - P(1 + i\Omega) - An \qquad (5.17)$$

$$\frac{dn}{dt'} = -bn + \tfrac{1}{2}(AP^* + A^*P)$$

where

$$\sigma = \frac{\kappa}{\gamma_{12}}, \quad \Omega = \frac{\omega_{12} - \omega_c}{\gamma_{12}}, \quad b = \frac{\gamma_{11}}{\gamma_{12}}, \quad r = \frac{\omega_c p^2 |N_0|}{2\epsilon_0\hbar\kappa\gamma_{12}}. \qquad (5.17')$$

In the perfectly tuned laser, $\Omega = 0$, and thus (5.17) are the Lorenz equations; when $\Omega \neq 0$ they are called the complex Lorenz equations. The behavior of the solutions for values of r (the nondimensional pumping $|N_0|$) near r_c, the value at which the system becomes a laser, is analyzed in detail in Section 6i. Let us now preview these results and make some additional remarks.

- If the beam is wide ($z_0 = \tfrac{1}{2}w_0^2 k \gg d$), then from (5.2) we see that many transverse modes have almost the same frequency and can be excited when $|N_0|$ exceeds its critical value. In that case, we can take $V(x, y, z) = \exp ikz$, $k = \frac{1}{c}\omega_q = qx/d$, d the intermirror distance, and let $F(t) \to F(x, y, t)$. Scaling $(x, y) \to w_0(x', y')$, w_0 being the beam width, the first equation in (5.17) becomes

$$\frac{\partial A}{\partial t} - ia\nabla_1^2 A = -\sigma A + \sigma P \qquad (5.18)$$

$a = c^2/2\omega\gamma_{12}w_0^2$. We analyze the behavior of this "sideband" laser in Section 6i when a is both zero and nonzero.

- when $a = 0$, the nonlasing solution $A = P = n = 0$ goes unstable at $r = r_c = 1 + \Omega^2/(\sigma + 1)^2$ or from (5.17'),

$$\frac{\omega_c p^2 |N_0|}{2\epsilon_0 \hbar \kappa \gamma_{12}} = 1 + \left(\frac{\omega_{12} - \omega_c}{(\sigma + 1)\gamma_{12}}\right)^2$$

$$= 1 + \frac{(\omega_{12} - \omega_L)^2}{\gamma_{12}^2}$$

because $\omega_{12} - \omega_L = (\omega_{12} - \omega_c)/(\sigma + 1)$, $\sigma = \kappa/\gamma_{12}$. This gives us that the critical $|N_0|$, called N_c, is

$$N_c = \frac{2\epsilon_0 \hbar \kappa}{\omega_c p^2 \gamma_{12}} \left(\gamma_{12}^2 + (\omega_{12} - \omega_L)^2\right),$$

which is (5.15).

- The lasing solution is

$$\begin{pmatrix} A \\ P \\ n \end{pmatrix} = \begin{pmatrix} \sqrt{b(r - r_c)}\ e^{-i(\sigma\Omega)/(\sigma+1)t} \\ (1 - i\frac{\Omega}{\sigma+1}\sqrt{b(r - r_c)}\ e^{-i(\sigma\Omega)/(\sigma+1)t} \\ r - r_c \end{pmatrix} \qquad (5.19)$$

which is exactly (5.13).

- We now discuss some limits. Suppose the homogeneous broadening γ_{12} associated with the polarization is much larger than either cavity damping κ or the damping coefficient γ_{11} associated with the population inversion N. In (5.17), this means that σ and b, the inverse time scales governing the behavior of the electric field A and the population inversion n, respectively, are small. In that case we may consider the approximation of adiabatically eliminating Λ from (5.8) or P from the second equation in (5.17) by ignoring $d\Lambda/dt$ and dP/dt because they are small when compared to $(\gamma_{12} + i(\omega_{12} - \omega))$ Λ or $(1 + i\Omega)$ P respectively. We then get for $1 \gg \sigma, b$ (drop prime from t)

$$P = \frac{1}{1 + i\Omega}(r - n)A$$

$$\frac{dA}{dt} = -\sigma A \left(1 - \frac{1}{1 + i\Omega}(r - n)\right) \tag{5.20}$$

$$\frac{dn}{dt} = \frac{AA^*r}{1 + \Omega^2} - n\left(b + \frac{AA^*}{1 + \Omega^2}\right).$$

Equations (5.20) are called the laser rate equations. A further reduction is possible if $1 \gg b \gg \sigma$, whence

$$n = \frac{AA^*r}{(1 + \Omega^2)b + AA^*}, \tag{5.21}$$

and then A satisfies

$$\frac{dA}{dt} = -\sigma A \left(1 - \frac{(1 - i\Omega)br}{(1 + \Omega^2)b + AA^*}\right)$$

$$= b\sigma A \frac{(r - (1 + \Omega^2) - i\Omega r - \frac{1}{b}AA^*)}{(1 + \Omega^2)b + AA^*}. \tag{5.22}$$

From (5.22) it is easy to see that

$$A \to \pm\sqrt{b(r - \bar{r}_c)}\, e^{-i\sigma\Omega t}, \quad \bar{r}_c = 1 + \Omega^2, \tag{5.23}$$

and from (5.20), (5.21) we see

$$n = r - \bar{r}_c, \quad P = \frac{\bar{r}_c}{1 + i\Omega}A = (1 - i\Omega)A. \tag{5.24}$$

The reader should compare these approximations with the exact solutions:

$$A = \pm\sqrt{b(r - r_c)}\exp{-i\frac{\sigma\Omega}{\sigma + 1}t},$$

$$P = \pm\left(1 - i\frac{\Omega}{\sigma + 1}\right)\sqrt{b(r - r_c)}\exp{-i\frac{\sigma\Omega}{\sigma + 1}t}, \tag{5.25}$$

$$n = r - r_c, \quad r_c = 1 + \frac{\Omega^2}{(\sigma + 1)^2}.$$

Note that the neglect of σ with respect to 1 in (5.25) gives (5.23), (5.24). The reader should also carry out the following exercise.

EXERCISE 5.2

Let $\bar{r}_c = 1+\Omega^2$, $\bar{x}_0 = 1-i\Omega$. Show that if we set $A = \sqrt{(r-\bar{r}_c)}B_1 e^{-i\sigma\Omega t}$ in (5.22) and take r close to $\bar{r}_c$, we get

$$\frac{dB_1}{dt} = (r-\bar{r}_c)\frac{\sigma\bar{x}_0}{\bar{r}_c}\left(B_1 - \frac{1}{b}B_1^2 B_1^*\right),\qquad(5.26)$$

which is exactly (6.227) with $\sigma \ll 1$.   ☐

- We want to stress that as σ and r become larger, the behavior of the solutions (5.19) and (5.20) and (5.22) is completely different from that of (5.17). The latter always relaxes to the laser solution. In the former, on the other hand, and in particular in the zero detuning limit $\Omega = 0$ (whence A, P, n are real), the lasing solution eventually becomes unstable, and one can realize an asymptotic state in which the phase point $A(t)$, $P(t)$, $n(t)$ never settles down but continually makes excursions about one of the two laser solutions ($\pm$ in (5.25)) with what appear to be random jumps from circling one fixed point to circling the other. The strange set is known as the Lorenz attractor and is shown in Figure 5.3. The values of σ for which this can occur must be greater than $b+1$ so that we are in what is called the bad cavity (i.e., $\kappa > \gamma_{11} + \gamma_{22}$) limit. Despite the fact that at these parameter values the laser is inefficient, one can nevertheless carry out an experiment to verify this behavior. This has been done by Weiss and Brock [61], using the optically pumped three-level laser configuration discussed in this section, and the output of his experiment is shown in Figure 5.4. As the detuning Ω is increased away from zero, it is found that the chaotic behavior is inhibited, much as the presence of strong rotation inhibits turbulence in fluids by making it difficult for the field variables to change along the directions of rotation (see [62]).

EXERCISE 5.3

Carry out a stability analysis of the solution (5.25) of the equations in (5.17). For a reference, see [62].   ☐

- There is a danger in adiabatic elimination that we want to stress. Observe that if we include the effects of diffraction by the addition of the terms

$$-i\,a\nabla_1^2 A$$

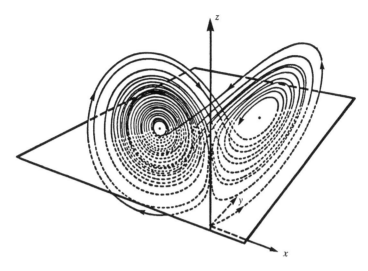

FIGURE 5.3   The Lorenz attractor.

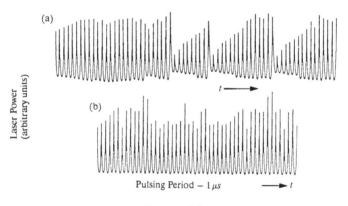

FIGURE 5.4

to the first equation in (5.17) where

$$a = \frac{c^2}{2\omega\gamma_{12}\omega_0^2} \tag{5.27}$$

and w_0 is the beam waist size, the adiabatic elimination of P and n means that the equation for B_1 with $A = \sqrt{r - \bar{r}_c} B_1 e^{-i\sigma\Omega t}$ is

$$\frac{\partial B_1}{\partial T} - ia\nabla_1^2 B_1 = (r - \bar{r}_c)\frac{\sigma\bar{x}_0}{\bar{r}_c}\left(B_1 - \frac{1}{b}B_1^2 B_1^*\right), \tag{5.28}$$

which is (6.227) with $\sigma + 1$ and $1 + \sigma x_0^2/r_c$ each replaced by 1. But this presents a problem, because there is now no diffusion term in (5.28).

We should have had as coefficient of $\nabla_1^2 B_1$

$$\frac{-ia}{1 + \frac{\sigma x_0^2}{r_c}} = \frac{-ia}{1 + \frac{\sigma}{r_c}\left(1 - \frac{\Omega^2}{(\sigma+1)^2} - \frac{2i\Omega}{\sigma+1}\right)}$$

which one would think might be approximated by

$$-ia\left(1 + \frac{2\,i\sigma\Omega}{\overline{r}_c}\right)$$

so that (5.28) is (remember $\Omega < 0$)

$$\frac{\partial B_1}{\partial t} - ia\nabla_1^2 B_1 + \frac{2a\sigma\Omega}{\overline{r}_c}\nabla_1^2 B_1 = (r - \overline{r}_c)\frac{\sigma \overline{x}_0}{\overline{r}_c}\left(B_1 - \frac{1}{b}B_1^2 B_1^*\right),$$
(5.29)

so that the effects of diffusion are included. Its importance is in distinguishing between the growth rate of small-amplitude disturbances with different wavenumber k. (Clearly for stability we must have $\Omega < 0$. The case when $\Omega > 0$ is covered in the last remark and in more detail in Section 6i.) For example, if we set

$$B_1 = B\,e^{i\vec{k}\cdot\vec{x}},$$

it is clear from (5.29) that the mode with wavenumber $k = 0$ grows the fastest.

One might therefore argue that because the $k = 0$ wavenumber grows the fastest, the space-independent solution

$$B_1 \to \sqrt{b}\,e^{i\phi}$$

is achieved. But let us test the stability of (there is no loss of generality in taking $\phi = 0$)

$$B_1 = \sqrt{b}$$
(5.30)

in (we now use the equation (6.240), which is absolutely correct for r close to r_c)

$$\frac{\partial B_1}{\partial t} - \frac{ia}{1 + \frac{\sigma x_0^2}{r_c}}\nabla_1^2 B_1 = \frac{(r - r_c)\frac{\sigma x_0}{r_c}}{1 + \frac{\sigma x_0^2}{r_c}}\left(B_1 - \frac{1}{b}B_1^2 B_1^*\right).$$
(5.31)

In Exercise 6.15 we ask you to show that the space-independent solution to (6.241) and (5.31) is unstable if

$$\beta_r \gamma_r + \beta_i \gamma_i = \text{Real}(\beta \gamma^*) < 0$$

In this case

$$\gamma = \frac{ia}{1 + \dfrac{\sigma x_0^2}{r_c}}, \quad \beta = \frac{(r - r_c)\dfrac{\sigma x_0}{r_c}}{b(1 + \dfrac{\sigma x_0^2}{r_c})}.$$

But

$$\beta \gamma^* = \frac{(r - r_c)\dfrac{\sigma}{b r_c} a}{\left| 1 + \dfrac{\sigma x_0^2}{r_c} \right|^2} \left(-i - \frac{\Omega}{\sigma + 1} \right),$$

so that for $\Omega < 0$, $\text{Real}\beta\gamma^* > 0$. Because there is no modulational instability, it is reasonable to expect that the space-independent solution (often called a condensate) is realized.

- Because the CGL equation (5.31) has the modulationally stable form, the vortices we refer to in Section 6i are not initiated spontaneously. Nevertheless, as we discuss in Section 6i, they are stable solutions for a certain range of parameters.

- Further, when $\Omega > 0$, the first modes to go unstable are $k = \pm\sqrt{\Omega/a}$ and not $k = 0$ because these sidebands have a natural frequency Ω that when coupled to ω_c allows the lasing field to oscillate at ω_{12}, the natural frequency of the two-level medium. See Section 6i.

The Optically Pumped Three-Level Laser (OPL) [6l, 6m]

Despite the fact that the two-level, single-mode laser model does not exhibit rich dynamical behavior for any practical ranges of parameter values, other laser systems, in particular inhomogeneously broadened two-level lasers and optically pumped three-level lasers, have been shown both experimentally and theoretically to exhibit complex chaotic dynamics at easily accessible physical parameter values. The resonantly pumped and resonant lasing three-level laser model is briefly discussed here as it naturally extends the two-level Haken–Lorenz model to a higher dimensional phase space, and its bifurcation study is well within the capabilities of the AUTO package designed to follow solution branches of systems of o.d.e.'s and discussed in Section 6m. We will see that there is a pronounced change, relative to the Haken–Lorenz model, in the behavior of the equilibrium solutions of this laser system with increasing pump amplitude. Assuming, as before, a single-mode homogeneously broadened system and a plane polarized field, we expand the latter as

$$E = A_1 e^{i(k_1 z - \omega_1 t)} + A_2 e^{i(k_2 z - \omega_2 t)} + (*). \tag{5.32}$$

This expression contains the pump wave A_1 of frequency ω_1, which is a well-defined field external to the laser cavity shown in Figure 5.1. The second field A_2 at frequency ω_2 represents the generated laser field in the ring cavity. A fundamental difference between this laser system and the two-level laser just discussed is that there is a direct coherence between the pump and lasing fields in the present system. This coherent interaction is reflected in the "two-photon" off-diagonal density matrix element ρ_{12} appearing in the three-level density matrix equations derived in Section 4f.

The basic ingredients for building an optically-pumped laser are the same as before. The novel feature of this system is that a laser, external to a ring or Fabry–Perot cavity coherently and selectively pumps atoms or molecules from a filled ground level into an initially empty excited level (labeled "m" in Figure 4.4), thereby causing an inversion between levels m and 2 of the three-level medium. We now design our cavity to amplify selectively and feed back light initially spontaneously emitted from the upper laser level m to the final lasing level 2. The laser light is emitted from the cavity at a frequency close to ω_{m2}. Notice that the pump wave at frequency ω_1 plays a passive role in the laser dynamics, merely pumping the atoms or molecules into an excited state while suffering no reflections (feedback) itself. The pump amplitude A_1 therefore appears as a parameter in the mathematical description of the optically pumped laser. Let us now derive the simplest mathematical description of this laser system by making a number of physical assumptions.

In the material equations describing a three-level medium derived in Section 4f, we assume that the external pump laser is exactly resonant with the pump transition ($\omega_1 + \omega_{1m} = 0$) and moreover that the laser emission is also exactly on resonance (ω_2 is now identified with ω_c). Furthermore, we assume that all off-diagonal density matrix relaxation rates are identical by setting $\gamma_\perp = \gamma_{im}, i = 1, 2, 3$ in equations (4.68–4.70) and also that all diagonal density matrix elements are equal by setting $\gamma_{ii} = \gamma_{11}, i = 1, 2, 3$. This assumption is believed to be valid for the special case of a far-infrared optically pumped laser, which has shown a wide variety of unstable chaotic oscillations. The final assumption is that the laser is arranged to emit in a single

TEM$_{00}$ mode so that we can carry over the analysis starting at equation (5.3). We now make the identification

$$n_a p_{2m} \rho_{m2} = n_a p_{m2} \sigma_{m2} e^{-i\omega_c t}$$
$$= \Lambda V e^{-i\omega_c t} \tag{5.33}$$

which replaces equation (5.7) for the two-level laser case. The field equation for the laser emission is now simply

$$\frac{dA_2}{dt} + \kappa A_2 = \frac{i\omega_2}{2\epsilon_0} n_a p_{2m} \sigma_{m2}, \tag{5.34}$$

which exhibits explicitly the coupling to the material oscillation. As in the Lorenz case, the full system of field–matter equations reduce in number as we go from the detuned to the exact resonance case. By scaling time to the common polarization decay rate $\gamma_\perp (t' = \gamma_\perp t)$ and defining the following dimensionless quantities

$$\alpha = -\frac{A_1 p_{1m}}{\hbar\gamma_\perp}, \quad \beta = -\frac{A_2 p_{2m}}{\hbar\gamma_\perp}, \quad b = \frac{\gamma_{11}}{\gamma_\perp}, \quad \sigma = \frac{K}{\gamma_\perp}, \quad g = \frac{n_a \omega_c p_{2m}^2}{2\epsilon_0 \hbar\gamma_\perp^2},$$

we obtain the following system of six coupled ordinary differentiable equations describing resonant emission of the optically pumped laser:

$$\dot{x}_1 = -\sigma x_1 + g x_3$$
$$\dot{x}_2 = -x_2 - x_1 x_4 + \alpha x_5$$
$$\dot{x}_3 = -x_3 + x_1 x_6 - \alpha x_4$$
$$\dot{x}_4 = -x_4 + x_1 x_2 + \alpha x_3 \tag{5.35}$$
$$\dot{x}_5 = -b(1 + x_5) - 4\alpha x_2 - 2x_1 x_3$$
$$\dot{x}_6 = -b x_6 - 4x_1 x_3 - 2\alpha x_2$$

where

$$x_1 = \beta \text{(real)}, \quad x_2 = \text{Im}\sigma_{m1}, \quad x_3 = \text{Im}\sigma_{m2},$$
$$x_4 = \text{Re}\sigma_{21}, \quad x_5 = (\sigma_{mm} - \sigma_{11}),$$
$$x_6 = (\sigma_{mm} - \sigma_{22}).$$

We have assumed that α and β are real without loss of generality and that the system is initially in its ground state ($\sigma_{11}^{(0)} = 1, \sigma_{mm}^{(0)} = \sigma_{22}^{(0)} = 0$). Stationary solutions to the above system of coupled o.d.e.'s are straightforward to compute by setting $\dot{x}_i = 0 + (i = 1, 6)$ and solving the resulting algebraic system of equations. As in the Lorenz case, there exists a nonlasing solution given by

$$x_1 = 0, \quad x_2 = -\frac{\alpha b}{4\alpha^2 + b}, \quad x_3 = x_4 = 0,$$

$$x_5 = -\frac{b}{4\alpha^2 + b}, \quad x_6 = \frac{2\alpha^2}{4\alpha^2 + b} \tag{5.36}$$

and a pair of lasing solutions ($\pm x_1$) given by

$$x_2 = -\frac{\sigma}{g}\left(\frac{4\beta^2 + (\alpha^2 + 1)b}{\alpha(2 + b)}\right)$$

$$x_3 = \frac{\sigma}{g}x_1$$

$$x_4 = -\frac{\sigma}{g}\left(\frac{4\beta^2 + (b - 2\alpha^2)}{\alpha(2 + b)}\right)x_1 \tag{5.37}$$

$$x_5 = -1 + \frac{\sigma}{g}\left(\frac{2(-b)x_1^2 + 4(\alpha^2 + 1)b}{b(2 + b)}\right)$$

$$x_6 = -\frac{\sigma}{g}\left(\frac{4\beta^2 - 2(\alpha^2 + 1)}{2 + b}\right)$$

which exist only for x_1 a real root of the biquadratic equation

$$\left(1 + \alpha^2 + \frac{4x_1^2}{b}\right)\left(1 + \frac{4\alpha^2}{b} + x_1^2\right) = \frac{\alpha^2 g}{\sigma b}(2 + b)\left(1 + \frac{\sigma}{g}(2 + b)x_1^2\right). \tag{5.38}$$

This latter restriction limits the lasing solution to a finite window in the pump parameter space (α), and a typical equilibrium bifurcation diagram is sketched in Figure 5.5.

We notice an immediate difference between the equilibrium bifurcation behavior of this optically pumped system and the Lorenz bifurcation diagram sketched in Figure 5.2. While the laser turns on in a similar fashion in both cases, in the optically pumped system it turns off again. The explanation for

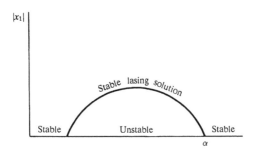

FIGURE 5.5 Bifurcation diagram for an optically pumped three-level laser.

this behavior is simple and is contained in the plot of $\hat{\chi}''(\omega_2)$ shown on the left in Figure 4.5 for the three-level atom. Recalling that $\hat{\chi}''(\omega_2)$ is proportional to the gain of the lasing emission, we see that the gain profile can be split into two peaks, which move apart as the pump amplitude is gradually increased. Eventually the gain near line center, where the laser wants to emit, drops below the cavity losses and the laser turns off.

This is only a part of the story, and a detailed analysis employing the bifurcation package AUTO and a variety of numerical methods shows that this laser system exhibits a wide spectrum of local and global bifurcation behavior. A brief glimpse of some of this fascinating behavior is provided by the bifurcation diagram in Figure 5.6, which was generated using the AUTO package starting at a fixed value of α on the nonlasing branch. This skeletal bifurcation diagram highlights just a small part of the complex bifurcation behavior of the optically pumped laser model under one-parameter variation. Shown for positive x_1 are the equilibrium branches as before, plus a blowup of the principal branches of periodic solutions emanating from two Hopf bifurcation points labeled HB1 and HB2, with one occurring on the lasing (HB1) and the other on the nonlasing branch (HB2). Bifurcation of stationary solutions from the nonlasing branch occur at the points BP1 and BP2. Further bifurcations along the periodic branches of solutions involve period-doubling cascades and symmetry-breaking bifurcations. A proper dynamical systems interpretation of this and other such figures requires a careful two-parameter unfolding of both local and global co-dimension 2 bifurcations of this system. The reader can gain an appreciation of the power of the computer in unraveling much of the complex dynamical behavior of the Lorenz equations by reading the book by Sparrow [63].

5b. Optical Bistability

Optical bistability is a term used to describe the nonunique response of an optical feedback system for which there is more than one stable output state [64].

kindly prov
caption.

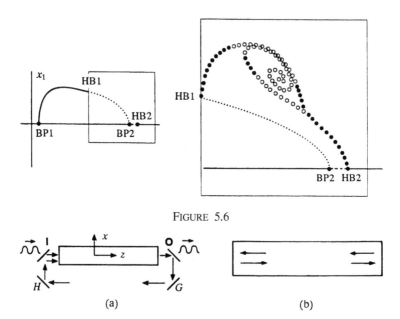

FIGURE 5.6

FIGURE 5.7 (a) Ring cavity geometry showing externally injected laser field propagating through the nonlinear medium with part of it being output to a detector and the remainder recirculated about the cavity. The mirrors at I and O are partially reflecting and partially transmitting ($R = 1 - T < 1$). The mirrors at G and H are 100% reflecting. (b) Fabry–Perot cavity showing counter-propagating waves between two almost flat mirrors. Standing wave effects are important in this configuration.

The phenomenon can be realized in both the ring and Fabry–Perot cavity geometries shown in Figure 5.7. The medium is passive, and the atoms are excited from their ground states by the passage of an electromagnetic wave.

    The beam propagation problem in optical cavities requires that we augment the appropriate field–matter equations (Maxwell–Bloch equations for a two-level medium or Maxwell–Debye equations for a nonresonant medium) with a set of boundary conditions appropriate to the type of cavity of interest.

The Ring Cavity

For the ring cavity of Figure 5.7(a) with an injected external field, the appropriate boundary condition on the field is best expressed in coordinates z and $\tau = t - (z/c)$ (see Figure 5.8):

$$\vec{E}(0, x, y, \tau = t) = \sqrt{T}\,\vec{E}_{\mathrm{in}}(0, x, y, \tau = t) + R\vec{E}\left(L + l, x, y, \tau - \frac{L + l}{c}\right).$$

$$(5.39)$$

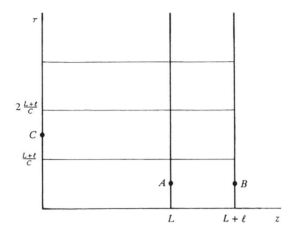

FIGURE 5.8   Configuration of cavity in z, τ space.

In (5.39), $L_T = L + l$ is the total length of the cavity; L is the length of the nonlinear medium and l is the length of the return path. We assume that in the return portion the beam does not diffract and the electromagnetic field obeys the linear wave equation so that

$$\vec{E}\left(L + l, x, y, \tau - \frac{L+l}{c}\right) = \vec{E}\left(L, x, y, \tau - \frac{L+l}{c}\right)$$

or $\vec{E}(B) = \vec{E}(A)$. The boundary condition (5.39) says that $\vec{E}(C) = \sqrt{T}\vec{E}_{in}(C) + R\vec{E}(A)$.

Let us now assume that the field is well approximated by a slowly varying wavepacket linearly polarized in the x–y plane and propagating along the cavity axis, which we take to be the z-direction. Therefore we set

$$\vec{E}(z, x, y, \tau) = \hat{e}F(z, x, y, \tau)e^{-i\omega\tau} + (*).$$

The boundary condition on the envelope $F(z, x, y, \tau)$ is ($k = \omega/c$)

$$F(0, x, y, \tau) = \sqrt{T}\,F_{in}(x, y, \tau) + Re^{ikL_T} F\left(L, x, y, \tau - \frac{L_T}{c}\right) \qquad (5.40)$$

where F_{in} refers to the injected laser field amplitude, which we take to be real without loss of generality. This boundary condition relates the field at the start of the nonlinear medium on each pass of the cavity to the sum of the input field multiplied by the mirror amplitude transmission coefficient ($\sqrt{T}$) and the time-retarded field from the previous pass. The phase term e^{ikL_T} is the effective detuning of the running wave frequency from the cavity resonance;

we shall see that if the argument of the exponential is an integer multiple of π it corresponds to the maximum transmission of the ring cavity. In fact, the empty cavity transmission function can be simply derived from this boundary condition by setting $F(0, x, y, t) = F(L, x, y, t-l/c) = F$ and solving for the ratio of the modulus squared of this latter quantity to the input field intensity. A little calculation shows that the result for the empty cavity transmission is

$$T = \frac{I_T}{I_{\text{in}}} = \frac{|F|^2}{|F_{\text{in}}|^2} = \frac{1}{1 + R^2 - 2R\cos\beta_0}, \quad \beta_0 = kL_T. \tag{5.41}$$

This transmission function is plotted against the cavity detuning ($\beta_0 = kL_T$) in Figure 5.9. We already know that a refractive index change induced through off-resonant coupling of light to matter can modify the propagation constant of the light field, so we anticipate that this intensity-dependent phase change can lead to a nonlinear transmission function in steady state leading to possible hysteretic behavior. This is the basic principle underlying dispersive optical bistability. In practice, when we consider realistic incident laser beams such as the TEM$_{00}$ mode of a laser, we need to account for diffraction of the collimated light beam as it propagates both in the nonlinear medium and in free space. To simplify the analysis, we will always neglect the latter. This is reasonable because the diffraction and/or the return path l can be small. We shall see, however, that it is not possible under any circumstances to neglect the former. The basic equations for the envelope fields in the nonlinear cavity are either the Maxwell-Bloch equations (4.34–4.36) (see (4c) for derivation)

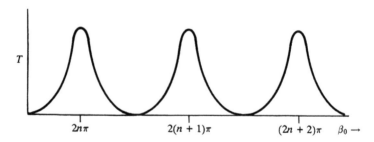

FIGURE 5.9 Empty cavity transmission function plotted against cavity detuning.

$$\frac{\partial F}{\partial z} + \frac{1}{c}\frac{\partial F}{\partial t} - \frac{ic}{2\omega}\nabla_{\perp}^2 F = \frac{i\omega}{2\epsilon_0 c}\Lambda \tag{5.42}$$

$$\frac{\partial \Lambda}{\partial t} + (\gamma_{12} + i(\omega_{12} - \omega))\Lambda = \frac{ip^2}{\hbar}FN \tag{5.43}$$

$$\frac{\partial N}{\partial t} + \gamma_{11}(N - N_0) = \frac{2i}{\hbar}(F^*\Lambda - F\Lambda^*) \tag{5.44}$$

or the Maxwell-Debye equations (4.111)

$$\frac{\partial F}{\partial z} + \frac{n_0}{c}\frac{\partial F}{\partial t} - \frac{ic}{2n_0\omega}\nabla_\perp^2 F = -i\frac{\omega}{c}\delta n F \tag{5.45}$$

$$\tau\frac{\partial \delta n}{\partial t} + \delta n = n_2 F F^*. \tag{5.46}$$

Because the atoms in the medium are all initially in their ground states, we have that $N_0 = n_a$. The initial-boundary value problem is most conveniently visualized by appealing to Figure 5.8 showing the $(z, \tau = t - z/c)$ plane.

The equations (5.42–5.44) or (5.45–5.46) describe the evolution of the fields in the nonlinear medium $0 < z < L$. The initial condition given as a function of z on $0 < z < L + l$ at $\tau = 0$ is that either

$$\Lambda = N - n_a = 0 \tag{5.47}$$

$$\text{or,} \quad \delta n = 0. \tag{5.48}$$

It remains to specify the electromagnetic field envelope $F(0, x, y, t)$ at the beginning of the medium $z = 0$. For $0 < z < (L + l)/c$,

$$F(0, x, y, z) = \sqrt{T} F_{\text{in}}(x, y, z). \tag{5.49}$$

For $nL + l/c < t < (n + 1)L + l/c, n \geq 1$,

$$F(0, x, y, t) = \sqrt{T} F_{\text{in}}(x, y, t) + Re^{ikL_T} F\left(L, x, y, z - \frac{L_T}{c}\right), \tag{5.50}$$

that is, the new starting envelope is given by the envelope on the previous pass at the end of the nonlinear medium. The field $F(z = L, x, y, z - L_T/c)$ is found as a functional of the initial field $F(z = 0, x, y, \tau - L_T/c)$ by solving the partial differential equations (p.d.e.'s) (5.42–5.44) or (5.45–5.46). The goal is to find the asymptotic form of the output field

$$F(L, x, y, \tau) \tag{5.51}$$

as $\tau \to \infty$. The initial-boundary value problem defined by (5.42–5.44), or (5.45–5.46) and (5.47–5.50) is now well posed. We return to analyze the solution after formulating the analogous problem for the Fabry–Perot cavity.

The Fabry–Perot Cavity

The flow of both counter-propagating field envelopes and material variables in the (z, t) plane along their respective characteristics $\tau_F = t - (z/c) =$ constant, $\tau_B = t + (z/c) =$ constant, and $z =$ constant, is depicted in Figure 5.10. It has been assumed, for simplicity in interpretation, that the Fabry–Perot cavity is filled with a two-level atom medium. As in Figure 5.8 for the ring cavity, all material variables evolve vertically along their characteristics and these in turn couple to the forward and backward electromagnetic wave envelopes A_F and A_B on the characteristics $t - (z/c) = \tau_F, t + (z/c) = \tau_B$ respectively. In the case of a passive cavity, the injected field flows in from the left along the A_F characteristic, as in the ring cavity case. The material variables are again specified on the interval $0 < z < L$ at $t = 0$. The initial medium inversion $N = N_0 > 0$, signifying an absorber and the intra-cavity field, is zero on this characteristic as before. Once the field A_F reaches the right-hand mirror $(T_2 = 1 - R_2)$, part of it is transmitted as

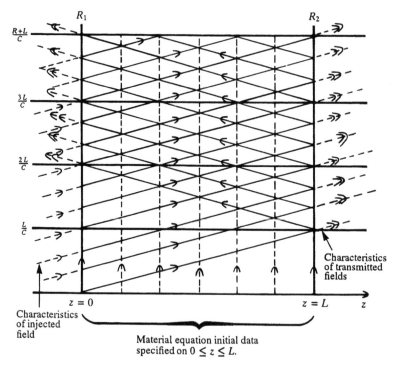

FIGURE 5.10 Flow of counter-propagating fields and material variables in the $z - t$ plane for an injected field incident from the left in a Fabry–Perot cavity. After a complete roundtrip $(t_r = 2L/c)$ both fields are coupled to the material variable, moving along vertical characteristics. Partially transmitted fields exit from the left and right as $R_1, R_2 < 1$.

$\sqrt{T_2} A_F(L, t)$ along an extension of the A_F characteristic to the right, and the remainder $A_B(L, t) = -\sqrt{R_2} A_F(L, t)$ now progresses to the left along the A_B characteristic. Both forward and backward waves are now coupled to the material variables as they evolve in time. When A_B reaches the left mirror at $z = 0$ it is partially transmitted as $\sqrt{T_1} A_B(0, t)$, and the reflected part $A_F(0, t) = -\sqrt{R} A_B(0, t)$ is added onto the continually flowing injected field $\sqrt{T_1} A_{\text{in}}(0, t)$ entering from the left. At this stage the electromagnetic signal has completed one roundtrip of duration $t_r = 2L/c$. The procedure now repeats and the time evolution is tracked until the system reaches an asymptotic state, which may be stationary, oscillatory, or chaotic. In the case of an active medium, such as a laser, a similar situation holds except that the injected field is absent and a background noise due to spontaneous emission is specified on the terminal characteristic, which is amplified by the inverted medium $N = N_0 < 0$.

The importance of pictures such as Figures 5.8 and 5.10 in understanding laser beam propagation in nonlinear optical media cannot be overstated. For example, if we remove the mirrors we obtain the appropriate initial-boundary value problem for unidirectional or counter-propagating beams in extended media. A host of nonlinear optical phenomena associated with these problems already exist, and new discoveries are being made. In particular, the recent discovery that the counter-propagating beam problem does not require mirror feedback in order to generate complex dynamical behavior in space and time opens up new challenges in nonlinear mathematics. The role of the linear diffraction term in the time-space evolution is to cause the beam to spread out in a direction orthogonal to its propagation direction. This outward diffusion of rays causes the amplitude of the electromagnetic waves on each z, t slice of the full three-dimensional space to gradually reduce unless the ray diffusion can be countered by total internal reflection, such as occurs from a discontinuity in refractive index in a waveguiding medium, or, more interestingly, by nonlinear self-focusing in one or two transverse space dimensions. We will rely heavily on these pictures when discussing optical bistable and lasing systems in the following sections.

Analysis of Ring Cavity

Although in principle both the problems for the ring and Fabry–Perot cavities are equivalent and in practice the Fabry–Perot cavity is an easier arrangement to implement (because of the lack of a need for a return path), the analysis for the ring cavity problem is much more tractable. We will carry out the calculations for the Maxwell–Bloch equations. A completely parallel set of calculations can be carried out for the Maxwell–Debye equations and we leave this to the reader. It is convenient to make several further simplifying assumptions.

Our first assumption is that the material relaxation times γ_{11}^{-1} and γ_{12}^{-1} are much faster than the times L_T/c over which the field envelope changes. This, as we shall see, can be a very risky assumption indeed, but let us return to the reason later. Second, we will make the more innocuous assumption that the input field is continuous and so $F_{in}(x, y, t)$ is independent of time t. The first assumption allows us to eliminate the polarization and population inversion fields

$$\Lambda = \frac{1}{\gamma_{12} + i(\omega_{12} - \omega)} \frac{ip^2}{\hbar} F N \tag{5.52}$$

$$N = \frac{N_0}{1 + \dfrac{4p^2}{\hbar^2} \dfrac{\gamma_{12}}{\gamma_{11}} F F^*(\gamma_{12}^2 + (\omega_{12} - \omega)^2)^{-1}} \tag{5.53}$$

whereupon the evolution equation for the electric field envelope $F(z, x, y, \tau)$ is given by

$$\frac{\partial F}{\partial z} - i \frac{c}{2\omega} \nabla_1^2 F = \frac{i\omega}{2c} \hat{\chi}(\omega, |F|^2)F, \tag{5.54}$$

where the complex susceptibility

$$\hat{\chi}(\omega) = \hat{\chi}'(\omega) + i\hat{\chi}''(\omega) = \frac{\omega_{12} - \omega + i\gamma_{12}}{\gamma_{12}^2 + (\omega_{12} - \omega)^2 + \dfrac{4p^2}{\hbar^2} \dfrac{\gamma_{12}}{\gamma_{11}}|F|^2} \frac{p^2 n_a}{\epsilon_0 \hbar}. \tag{5.55}$$

Because $F_{in}(x, y)$ is independent of t, the solution $F(z, x, y, \tau)$ also is independent of τ in each of the parallelograms

$$(n - 1)\frac{L + l}{c} < \tau < n\frac{L + l}{c}, \quad 0 < z < L + l$$

drawn in Figure 5.8. We designate the solution $F(z, x, y, \tau)$ in the nth parallelogram (which is now a rectangle)

$$0 < z < L + l, \quad (n - 1)\frac{L + l}{c} < \tau < n\frac{L + l}{c}, \quad n > 1, \tag{5.56}$$

by $F_n(z, x, y)$ which is now independent of τ and satisfies the equation

$$\frac{\partial F_n}{\partial z} - i \frac{c}{2\omega} \nabla_1^2 F_n = \frac{i\omega}{2c} \hat{\chi}(\omega, |F_n|^2) F_n. \qquad (5.57)$$

The boundary condition (5.50) now becomes the map

$$F_{n+1}(0, x, y) = \sqrt{T} \, F_{in}(x, y) + R e^{ikL_T} F_n(L, x, y), \quad n \geq 0 \qquad (5.58)$$

with $F_0 = 0$. $F_n(L, x, y)$ is found as a functional of $F_n(0, x, y)$ by solving (5.57) with initial condition $F_n(0, x, y)$. One can think of (5.58) as an infinite-dimensional map for the complex field $F_n(0, x, y)$. Our goal is to find the asymptotic behavior of the output

$$F_n(L, x, y) \qquad (5.59)$$

as $n \to \infty$.

To get some feel of what the solutions might be like, we begin by taking the input field to be a plane wave

$$\sqrt{T} \, F_{in}(x, y) = a \qquad (5.60)$$

where a is a real constant. It follows that $F_n(z, x, y)$ is independent of the transverse coordinates x and y, and then it is easy to solve (5.57). We make this task even easier by assuming that the detuning $\omega_{12} - \omega$ is much larger than the homogeneous broadening γ_{12} so that the loss over the distance L due to a finite $\hat{\chi}''(\omega)$ is negligible when compared with the mirror loss. The case in which we neglect absorption is called the dispersive bistability case. The opposite case, in which we include the effects of absorption but neglect dispersion, is called the absorptive bistability case. In either case, the output versus input curve is multivalued.

Let $F_n(0) = g_n$ and then from (5.57)

$$F_n(z) = g_n \exp(i\omega/2c)\hat{\chi}'(\omega, |g_n|^2)L \qquad (5.61)$$

and the map (5.58) becomes

$$g_{n+1} = a + g_n R \, \exp i(kL_T + (\omega L/2c)\hat{\chi}'(\omega, |g_n|^2)), \quad n \geq 0 \qquad (5.62)$$

with $g_0 = 0$, a (nonanalytic) map of one complex number g_n to another g_{n+1} known as the Ikeda map [65]. The map consists of applying to the complex number g_n three simple operations (Figure 5.11): rotation through an angle that depends on the magnitude of $g_n(g_n^{(1)})$, a contraction by $R(g_n^{(2)})$, and a

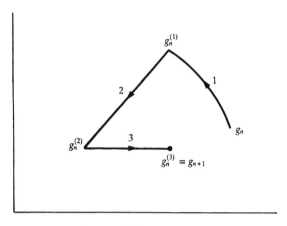

FIGURE 5.11   Ikeda map.

translation by a ($g_n^{(3)} = g_{n+1}$). We can look for fixed point solutions of the map by assuming

$$\lim_{n \to \infty} g_n = g \qquad (5.63)$$

and solving the nonlinear algebraic equation

$$g = a + Rg\, e^{i\phi}, \quad \phi = kL_T + \frac{\omega L}{2c}\hat{\chi}'(\omega, |g|^2). \qquad (5.64)$$

For $kL_T = 0.4$, $R = 0.9$, we plot in Figure 5.12(a) the output $|g|$ versus the input a. It has a multivalued response characteristic of a forced, damped nonlinear oscillator. A slight adjustment in the physical parameter setting ($kL_T = 0.2$, $R = 0.9$) leads to the single-valued response curve with a very steep rise as shown in Figure 5.12(b). This is the optical analog of the electrical transistor! Following electronics, we can cascade a number of these optical transistor response functions and build all of the basic logic elements ("OR," "AND," "NOR," etc.) existing in modern electronic computers. The multivalued response function sketched in Figure 5.12(a) can, on the other hand, act as an all-optical memory for long-term storage of information. Let us consider that the external field amplitude (a) is initially held low so that the transmitted light from the cavity corresponds to the point E on the curve. As we gradually increase a, the output of the cavity proportional to $|g|$ gradually increases along the branch EB until we reach the point $B(a = b)$ on the curve. Beyond this point, the cavity transmission suddenly jumps onto the upper branch ("suddenly" may mean in 10^{-9} s), and the transmitted light intensity remains strong, following the upper branch in the direction of the point labeled D. If we now gradually decrease a, the transmitted light remains

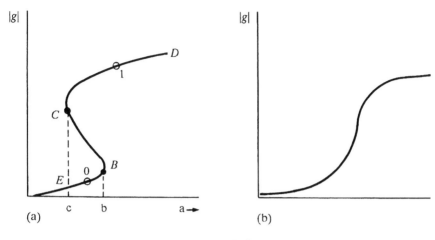

FIGURE 5.12

strong, gradually decreasing until we reach the point C on the upper branch. Below this point the intensity drops suddenly (at $a = c$) and returns to the initial value E on the lower branch. This type of behavior is called optical bistability because the portions of both branches EB and CD in the multivalued region of the curve are stable and can be accessed by different routes. How might we use this bistable response to store a bit of information represented, say, as a short optical pulse incident on the cavity? First of all we introduce a "holding" beam ($a = $ constant) to maintain the light transmission halfway between C and B on the lower branch. We call this operating point logic level "0" representing a zero bit (absence of the pulse). If a short optical pulse is now incident on the cavity such that the sum of its peak amplitude and that of the holding beam exceeds the value B the transmission of the cavity jumps to the upper branch and after the pulse has passed, it sits there at the point labeled logic level "1" as long as the holding beam is maintained. Logical level "0" can be restored by reducing the holding beam intensity below $a = c$ and then incrementing it back again.* Note that we have avoided referring to the middle branch BC so far in the discussion. This is because this branch is always unstable, as can be shown by linearizing about the fixed point g for any location along the branch.

In fact, a closer inspection of the properties of the Ikeda map shows that it displays many of the dynamical characteristics of the better known Henon map, a two-dimensional invertible mapping. In particular, the Ikeda map is a two-dimensional invertible, area contracting mapping of the complex plane.

---

*The above picture represents just one way of achieving a switch between logic levels "0" and "1." Recall that g is a complex variable so that, in principle, one can switch between states using a phase rather than intensity adjustment.

By defining the Jacobian of the map as

$$J = \begin{pmatrix} \partial F/\partial g_n & \partial F/\partial g_n^* \\ \partial F^*/\partial g_n & \partial F^*/\partial g_n^* \end{pmatrix},$$

with F the right-hand side of equation (5.62), it is easy to show that the determinant of J,

$$\det(J) = R^2 < 1. \tag{5.65}$$

EXERCISE 5.4

Obtain the result (5.65).   □

This means that areas contract by R^2 on each application of the map (5.62), and since $\det J = \lambda_1 \lambda_2$, where λ_1 and λ_2 are the roots obtained by linearizing the map about the fixed point (g), the only types of bifurcation allowed for (5.62) are saddle-node and period-doubling bifurcations. This latter result can be most easily established by considering the "Kerr" limit of the susceptibility expression following equation (5.54). Assuming $|\omega_{12} - \omega| \gg \gamma_{12}$ in this expression, we can drop the absorption terms ($\chi''(\omega)$) and obtain an expression for $\chi'(\omega)$ in the following form:

$$\chi'(\omega) \simeq \frac{p^2 n_a}{\epsilon_0 \hbar(\omega_{12} - \omega)} \cdot \frac{1}{1 + \alpha|g|^2} \simeq \frac{p^2 n_a}{\epsilon_0 \hbar(\omega_{12} - \omega)}(1 - \alpha|g|^2) \tag{5.66}$$

where $\alpha = 4p^2 \gamma_{12}/\hbar^2 \gamma_{11}(\omega_{12} - \omega)^2$ and the last line follows from assuming $\alpha|g|^2 \ll 1$. We can now write the map (5.62) in explicit form:

$$g_{n+1} = a + g_n R e^{i(\phi - \alpha_0|g_n|^2)} \tag{5.67}$$

where

$$\phi = kL_T + \frac{\omega L}{2c} \frac{p^2 n_a}{\epsilon_0 \hbar(\omega_{12} - \omega)} \quad \text{and} \quad \alpha_0 = \frac{p^2 n_a}{\epsilon_0 \hbar(\omega_{12} - \omega)} \alpha.$$

EXERCISE 5.5

Show that the eigenvalues of the linearization of the plane wave Kerr map (5.67) are given by

$$\lambda_{\frac{1}{2}} = R(D - \sqrt{D^2 - 1}) \tag{5.68}$$

where $D = \cos(\phi + \alpha_0|g|^2) - \alpha_0|g|^2 \sin(\phi + \alpha_0|g|^2)$, by solving $\det|J - \lambda I| = 0$. □

As the roots λ_1, λ_2 are either both real or complex conjugate pairs and because of the condition $\det J = \lambda_1 \lambda_2 = R^2 < 1$, the only ways in which the plane wave fixed point g, obtained as a solution to equation (5.64), can go unstable is by one of the roots crossing the unit circle along the positive or negative real axis. The eigenvalue behavior as a function of parameter variation is sketched in Figure 5.13. The situation in Figure 5.13(a) depicts one eigenvalue, say λ_1, passing through $+1$ while maintaining $\lambda_1 \lambda_2 = R^2$ (i.e., $\lambda_2 = R^2$ at this point) and corresponds to a saddle-node bifurcation (turning points on the bistable curve). In Figure 5.13(b) the eigenvalue, say λ_1, is passing through -1 while keeping $\lambda_1 \lambda_2 = R^2$ (i.e., $\lambda_2 = -R^2$ at this point) and corresponds to a flip or period-doubling bifurcation. In fact, it is easy to show that this can represent the first in a period-doubling cascade to a chaotic attractor.

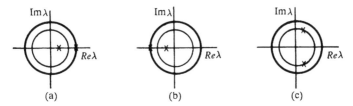

FIGURE 5.13 (a) Saddle-node bifurcation. (b) Period doubling bifurcation. (c) Complex conjugate roots confined to circle of radius $R < 1$.

EXERCISE 5.6

Show that the inverse of the plane wave Kerr map (5.67) is given by

$$g_n = \frac{(g_{n+1} - a)}{R} e^{-i\phi} e^{i\alpha_0|g_{n+1} - a|^2/R^2}. \tag{5.69}$$

□

The local analysis presented here can be extended to provide a global picture of the dynamics by using the fact that the only unstable fixed points allowed are saddle points and that the stable and unstable manifolds (W^s and W^u) of these saddle points are invariant under the mapping

$$F(W^u) = W^u = F^{-1}(W^u)$$

$$F(W^s) = W^s = F^{-1}(W^s).$$

The manifolds W^s and W^u are the sets of points that approach the saddle point under forward or backward iteration; $F(F^{-1})$ denotes the forward mapping as defined by equation (5.67) (inverse mapping by (5.69)) for the Kerr map applied to this set. Although the local bifurcations are restricted to saddle-node or period-doubling type, the global dynamics is extremely rich. Figure 5.14(a) illustrates one important global feature of the mapping showing the formation of a homoclinic tangency in the complex g-plane, where the stable and unstable manifolds just touch. Increasing the parameter "a" further causes the manifolds to intersect (it can be proved that there must be infinitely many such intersections), signifying the birth of infinitely many new stable periodic orbits.[†] These new orbits are then forced to undergo period-doubling cascades. Figure 5.14(b) shows a picture of the chaotic attractor for the Ikeda map at the end of a period-doubling cascade. The interested reader can find a detailed exposition on the Ikeda map in reference. [66]

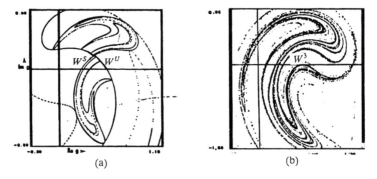

(a) (b)

FIGURE 5.14

The danger of adiabatic elimination can be best appreciated by taking the Debye equation (5.45–5.46) for the material response together with the ring cavity boundary conditions (5.40). Recall that the Debye equation describes the off-resonant limit of the Maxwell-Bloch equations and hence is the appropriate description of dispersive optical bistability with finite material response included. By assuming an infinite plane wave, we can integrate the electric field envelope equation (5.45) along its characteristics and substitute the result into the ring cavity boundary condition to obtain the delay differential equation

---

[†]The reader might worry that we are violating uniqueness by allowing W^u and W^s to intersect. Note that these manifolds are not solution curves of a flow and so uniqueness is not violated. It is best to think of these manifolds as the crossing points on a Poincaré surface of section of a continuous time flow embedded in one higher dimension.

$$F(t) = a + RF(t - t_R)e^{-i(\beta(t) - \beta_0)} \tag{5.70}$$

$$\gamma^{-1}\frac{d\beta(t)}{dt} = -\beta(t) + |F(t - t_R)|^2, \tag{5.71}$$

where $\beta_0 = -kL_T$, $\beta(t) = -(\omega L/c)\delta n(t)$, $t_R = L/c$, the intensity $|F(t)|^2$ has been scaled to the dimensionless intensity $|F_s|^2 = c/n_2\omega L$, and $\tau = \gamma^{-1}$. In the limit of an infinitely fast material response ($\gamma \to \infty$) we recover the plane wave Kerr map limit (5.67). Fixed points of the delay differential equations (5.70–5.71) are computed by setting $F(t) = F(t - t_R)$ and $\frac{d\beta}{dt} = 0$, and these are obviously identical to fixed points of the original map.

Differences between the map and delay differential equation limit arise when we linearize about these fixed point solutions. Setting $F(t) = F_s + \delta F(t)$, $\beta(t) = \beta_s + \delta\beta(t)$ in equations (5.70–5.71), where ($F_s = a + RF_s e^{i(|F_s|^2 - \beta_0)}$, $\beta_s = |F_s|^2$) and ($\delta F(t), \delta\beta(t)$) are small perturbations, we obtain a linearized set of equations for the perturbations:

$$\delta F(t) = iRF_s e^{i(|F_s|^2 - \beta_0)}\delta\beta(t) + Re^{i(|F_s|^2 - \beta_0)}\delta F(t - t_R) \tag{5.72}$$

$$\gamma^{-1}\frac{d\delta\beta}{dt} = -\delta\beta + F_s\delta F^*(t - t_R) + F_s^*\delta F(t - t_R). \tag{5.73}$$

Let $\delta F(t) = ue^{st}$, $\delta\beta = ve^{st}$, and $\delta F^* = we^{st}$ and substitute in (5.72–5.73) to obtain the quasi-polynomial

$$(1 + R^2 e^{-2st_R})(\gamma^{-1}s + 1) - 2Re^{-st_R}[(\gamma^{-1}s + 1)\cos(|F_s|^2 - \beta_0)$$
$$- |F_s|^2\sin(|F_s|^2 - \beta_0)] = 0, \tag{5.74}$$

which determines the stability of the fixed point (F_s, β_s) with respect to growth or decay of plane wave perturbations. It is convenient to set $\lambda = e^{st_R}$ and rewrite equation (5.74) in the form

$$(\lambda^2 + R^2)\left(\frac{\ln\lambda}{\gamma t_R} + 1\right)$$
$$- 2R\lambda\left[\left(\frac{\ln\lambda}{\gamma t_R} + 1\right)\cos(|F_s|^2 - \beta_0) - |F_s|^2\sin(|F_s|^2 - \beta_0)\right] = 0. \tag{5.75}$$

We can see the influence of finite material response directly in equation (5.75). Notice that the relevant parameter is γt_R, the ratio of the cavity roundtrip time t_R to the material response time $\tau = \gamma^{-1}$. In the limit $\gamma t_R \to \infty$, we recover the quadratic equation (5.68) determining stability of the Kerr map plane wave fixed points. One must be very careful in taking this limit, however, in the parameter range where the Kerr map displays period-doubling cascades. The basic reason is that the finite jump discontinuity in a period-doubled waveform of the map is in reality a sharp boundary layer of width γ^{-1}, in which region the nature of the solution can significantly change. Numerical solution of equations (5.70–5.71) shows high-frequency ringing oscillations near the sharp edge of the jump. These high frequencies are a consequence of the singular nature of the problem, which is reflected in the fact that the quasi-polynomial (5.75) has infinitely many roots and not just two, as in the map case. Figure 5.15 contrasts a period-2 waveform of the map with the corresponding waveform for the differential delay equation obtained directly from an optics experiment [67]

It is also very dangerous to overlook the effects of diffraction and transverse dependence. Suppose that the input wave has a Gaussian shape in the

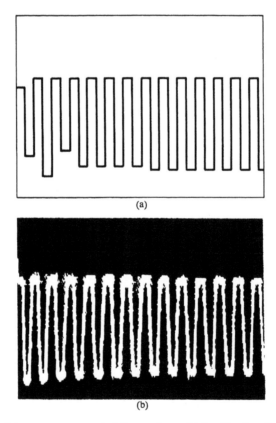

(a)

(b)

FIGURE 5.15   (a) A square wave period-2 waveform obtained by iterating the map (5.67). (b) An experimental waveform from reference [67] showing high-frequency ringing oscillations on the overall period-2 waveform.

transverse direction but with small enough gradients that diffraction can be initially ignored. Initially, the output signal follows the graph in Figure 5.16(a) with the output value at any (x, y) determined by the corresponding value of $I_{in} = (1 - R)F_{in}^2$ at the same (x, y). Observe now that if the central portion of the Gaussian shape $F(x, y) > F_B$, that value corresponding to point B in Figure 5.12(a), the center of the beam switches up to the upper branch whereas the outer portions remain on the lower branch (Fig. 5.16b). Therefore, the gradient of F becomes large on the contour in x, y where $F(x, y) = F_B$ and diffraction becomes very important. The output signal has a shape that in

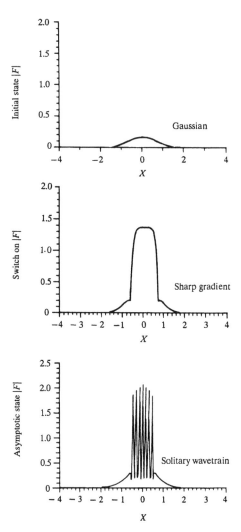

FIGURE 5.16 I_{out} as a function of x after $n = 0$, 10 and 100 passes.

one transverse dimension consists of soliton-like structures [68] (Fig. 5.16c), which may or may not settle down to a time-independent asymptotic state. In two transverse dimensions, the output states are much richer, because the focusing singularity breaks the cylindrical ring solitons that form initially into filaments, which interact and recombine into continually rearranging patterns.

5c. Instability and Hysteresis in Distributed Feedback Structures

Coupled mode/wave analysis underpins many theoretical treatments of linear and nonlinear interactions between individual propagating waves. We have already seen this analysis applied to the coupling between two polarization states of a light beam in Chapter 2 and to counter-propagating beams at the beginning of the present chapter. Let us take this analysis further now and examine a number of nonlinear optical effects whose manifestation depends on some form of distributed feedback effect. Mirrors provided a simple local feedback in our discussion of lasers and optical bistability in the earlier sections. A simple periodic modulation of the refractive index in the direction of propagation of an electromagnetic wave can, under appropriate conditions to be discussed shortly, provide a distributed feedback of light that is very sensitive to wavelength. Thus a simple adjustment in wavelength of the incident light can switch the distributed feedback device from strong reflection to strong transmission. The alert reader will have realized by now that such a switching capability can be achieved through an optical nonlinearity, thereby avoiding the necessity to tune the wavelength of light. A large variety of passive and active (laser) devices have been designed on this distributed feedback principle. It was also briefly mentioned in the introduction to this chapter and in Chapter 4 that two counter-propagating light waves of identical frequency can induce a periodic modulation in a perfectly uniform nonlinear material due to their mutual interference. This induced nonlinear spatial grating will be shown to cause a variety of instabilities in both bulk and waveguiding (fiber) materials. Our analysis is necessarily restricted to some of the simpler cases as this area is currently an active and relatively unexplored research field.

As a specific application we will consider the case of two counter-propagating beams in a Kerr nonlinear medium allowing for a finite medium response. This will give us the opportunity to introduce the Maxwell–Debye coupled equations in a number of important physical contexts.

Consider two counter-propagating infinite plane waves in a Kerr nonlinear medium where the linear background refractive index has a weak periodic modulation in the direction of propagation. A schematic of the physical set-up is shown in Figure 5.17. This experimental configuration could also refer to a waveguide or fiber, where instead of infinite plane waves we would need to consider transversely confined modes (in x and y) of the structure. Whether we are dealing with infinite plane waves or guided modes does not affect our conclusions in the least, beyond redefining the constant coefficients appearing in the equations.

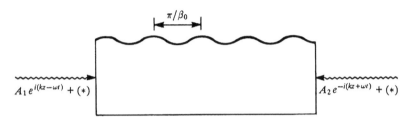

FIGURE 5.17

Allowing for a finite medium response, the propagation of the waves in this distributed feedback structure is described by the Maxwell–Debye equations

$$\frac{\partial^2 E}{\partial t} - \frac{1}{c^2}\frac{\partial^2 E}{\partial t^2} = \mu_0 \frac{\partial^2 P}{\partial t} \tag{5.76}$$

$$\tau \frac{\partial \delta n}{\partial t} + \delta n = \frac{1}{2}n_2|E|^2, \tag{5.77}$$

where the induced polarization $P = \epsilon_0 n^2 E$.

The refractive index of the medium $n = n_0 + n_1 \cos 2\beta_0 z + \delta n(E)E$, where the first term is the background linear index, the second is a weak ($n_1 \ll n_0$) linear periodic modulation of wavelength π/β_0, and the final nonlinear term accounts for the finite material response of the Kerr medium. In steady state, we recover the usual $\chi^{(3)}$ instantaneous response discussed in Section 4d. The field within the medium is taken as the sum of forward and backward propagating plane waves whose complex envelopes are slowly varying functions of space and time,

$$E = A_1(z, t)e^{i(kz - \omega t)} + A_2(z, t)e^{-i(kz + \omega t)} + (*). \tag{5.78}$$

Following the general procedure outlined earlier for counter-propagating beams, we expand the nonlinear refractive index δn in a Fourier series,

$$\delta n = \delta n_0 + \delta n_1 e^{2ikz} + \delta n_1^* e^{-2ikz} + \cdots \tag{5.79}$$

retaining just the terms shown.* The coupled wave equations follow from equations (5.76–5.77) by retaining on the right-hand side those terms that are synchronous with the appropriate driving field. Consider the linear periodic modulation appearing in the polarization $P (= \epsilon_0(n_0^2 + 2n_0 n_1 \cos 2\beta_0 z + \delta n(E))E)$ where $\delta n, n_1 \ll n_0$, and small second-order terms have been dropped, that is, $\epsilon_0 n_0 n_1 (e^{2i\beta_0 z} + e^{-2i\beta_0 z})(A_1 e^{ikz} + A_2 e^{-ikz} + (*))$. This gives a contribution $\epsilon_0 n_0 n_1 e^{2i\Delta\beta z} A_2$ as a driving term in the A_1 equation where $\Delta\beta = (\beta_0 - k)$ is the detuning of the incident field wavenumber from the so-called Bragg resonance condition. It is a straightforward exercise now to derive the following coupled equations that describe the dynamics of counter-propagating waves:

$$\frac{\partial A_1}{\partial z} + \frac{n_0}{c}\frac{\partial A_1}{\partial t} = i\kappa A_2 e^{2i\Delta\beta z} + i\gamma^{(\delta n_0 A_1 + \delta n_1 A_2)} \tag{5.80}$$

$$\frac{\partial A_2}{\partial z} - \frac{n_0}{c}\frac{\partial A_2}{\partial t} = -i\kappa A_1 e^{-2i\Delta\beta z} - i\gamma(\delta n_0 A_2 + \delta n_1^* A_1) \tag{5.81}$$

$$\tau\frac{\partial \delta n_0}{\partial t} + \delta n_0 = |A_1|^2 + |A_2|^2 \tag{5.82}$$

$$\tau\frac{\partial \delta n}{\partial t} + \delta n_1 = A_1 A_2^* \tag{5.83}$$

This set of equations describe a variety of linear and nonlinear wave coupling phenomena associated with counter-propagating beams. They differ from the Maxwell–Debye set derived in Chapter 4 due to the presence of the linear cross-coupling term arising from the imposed linear refractive index grating. We therefore encounter two spatial grating terms, a linear imposed one and a nonlinear induced one due to the interference $(A_1 A_2^*)$ between the counter-propagating waves. We start by analyzing the simple linear distributed feedback structure by dropping all nonlinear terms from the equations.

Linear Distributed Feedback Structure

In steady state, the forward and backward coupled waves obey the coupled linear set of o.d.e.'s

$$\frac{dA_1}{dz} = i\kappa A_2 e^{2i\Delta\beta z} \tag{5.84}$$

$$\frac{dA_2}{dz} = -i\kappa A_1 e^{-2i\Delta\beta z}. \tag{5.85}$$

*The fast time dependence $e^{\pm i\omega t}$ factors out, as discussed in Section 4e.

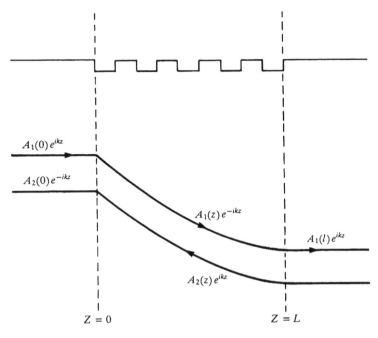

FIGURE 5.18

Consider the structure depicted in Figure 5.18 where a weak periodic modulation of the linear refractive index extends over a finite region of length L. We want to solve for the waves in this structure subject to the following boundary conditions $A_1(0) = A$, $A_2(L) = 0$. The solutions that satisfy these boundary conditions are

$$A_1(z)e^{-ikz} = Ae^{i\beta_0 z} \frac{[\Delta\beta \sinh(S(z-L)) + iS \cosh(S(z-L))]}{[-\Delta\beta \sinh SL + iS \cosh SL]} \qquad (5.86)$$

$$A_2(z)e^{ikz} = Ae^{-i\beta_0 z} \frac{i\kappa \sinh S(z-L)}{[-\Delta\beta \sinh SL + i \cosh SL]} \qquad (5.87)$$

where $S = \sqrt{|\kappa|^2 - \Delta\beta^2}$.

These solutions show that a finite corrugated section can act as a high reflectivity mirror for frequencies (wavelengths) close to the Bragg value. The fields across the structure are graphed in Figure 5.18 for the special case $\Delta\beta = 0$. For a range of frequencies such that $\Delta\beta(\omega) < \kappa$, the solutions have an exponentially decaying (growing) part. In this forbidden region (frequency stop band), the incident field amplitude $A_1(z)$ drops off exponentially at the expense of an exponentially growing backward reflected wave $A_2(z)$. Note that the quantity $d/dz(|A_2|^2 - |A_1|)^2 = 0$, indicating that the total power carried by both waves is conserved. These distributed feedback structures form the basis for many miniature semiconductor laser devices whose mathematical description follows by allowing the refractive index to become complex. The imaginary part of the refractive index being proportional to $\chi''(\omega)$ allows for amplification of light along the structure, and the refractive index as a whole depends in a complicated fashion on the density of electron-hole pairs generated in the semiconductor.

Nonlinear Induced Feedback in a Uniform Medium

If we drop the linear coupling term from the original coupled field equations we arrive at the case of two identical infinite plane waves (or waveguide modes) counter-propagating in a uniform (in z) relaxing Kerr medium [69]. The effect of the nonlocal material response may be best appreciated by rewriting the equations in the following form:

$$\frac{\partial A_1}{\partial z} + \frac{n_0}{c}\frac{\partial}{\partial t}A_1 = -i\gamma(H_{11}+H_{22})A_1 - i\gamma H_{12}A_2 \qquad (5.88)$$

$$-\frac{\partial A_2}{\partial z} + \frac{n_0}{c}\frac{\partial}{\partial t}A_2 = -i\gamma(H_{11}+H_{22})A_2 - i\gamma H_{21}A_1 \qquad (5.89)$$

where

$$H_{ij} = \frac{1}{\tau}\int_{-\infty}^{t} A_i(z,t')A_j^*(z,t')e^{-(t-t')/\tau}dt'. \qquad (5.90)$$

The diagonal terms of H_{ij} contribute to an overall phase change for each counter-propagating wave while the off-diagonal terms provide the coupling. The latter are called spatial grating terms as they exist due to an interference between counter-propagating waves. In steady state, $H_{ij} = A_i(z)A_j^*(z)$ and the solutions to (5.88–5.89) reduce to

$$A_1^0(z) = \sqrt{I_1}e^{-i\gamma(I_1+2I_2)z+i\phi_1} \qquad (5.91)$$

$$A_2^0(z) = \sqrt{I_2}e^{i\gamma(I_2+2I_1)z+i\phi_2} \qquad (5.92)$$

where ϕ_1 and ϕ_2 are determined by the initial conditions. Note that the intensities $I_i = A_i A_i^*$ do not change along the medium so that the two waves do not exchange energy. The phase shift that both waves acquire may not be equal so the nonlinear interaction induces some nonreciprocity. This effect may be important in fiber-optic rotation sensors (laser gyros) whose output is sensitive to an induced phase difference.

From the above discussion it is clear that the steady-state solutions cannot exhibit any bistable behavior. However, a linear stability analysis about the steady-state solution shows that they may be unstable. Linearizing the above equations about $A_j^0(z)$ with the perturbed solution written in the form

$$A_j(z, t) = A_j^0(1 + \epsilon F_j(z)e^{\lambda t} + \epsilon G_j^*(z)e^{\lambda^* t}), \quad j = 1, 2 \qquad (5.93)$$

yields the following linear system of o.d.e.'s for the perturbations F_j and G_j:

$$
\frac{d}{dz}\begin{pmatrix} F_1 \\ F_2 \\ G_1 \\ G_2 \end{pmatrix}
$$

$$
= \begin{pmatrix}
-\lambda' + P_1 - 2P_2 & -(P_1 + P_2) & -P_2 & -2P_2 \\
P_1 + P_2 & \lambda' - P_1 + 2P_2 & 2P_2 & P_2 \\
P_2 & 2P_2 & -\lambda' - P_1 + 2P_2 & P_1 + P_2 \\
-2P_2 & -P_2 & -(P_1 + P_2) & \lambda' + P_1 - 2P_2
\end{pmatrix}
\begin{pmatrix} F_1 \\ F_2 \\ G_1 \\ G_2 \end{pmatrix}.
$$
$$(5.94)$$

Here $\lambda' = \lambda n_0/c$, $P_1 = i\gamma I$, $P_2 = P_1/(1 + \gamma\tau)$, and equal intensities $I_1 = I_2 = I$ were assumed. The perturbations satisfy the boundary conditions $F_1(0) = G_1(0)$
$= F_2(L) = G_2(L) = 0$. The important time scales appearing in this problem are the transit time in the medium $t_r = n_0 L/c$ and the medium response time τ. Setting the determinant of the above 4×4 matrix equal to zero, it is possible to show that the steady-state solutions are unstable above a certain critical intensity threshold I_c. The characteristic frequency of oscillation of the dynamic solution near threshold is approximately $(2t_r)^{-1}$, corresponding to excitation of different longitudinal modes of the distributed feedback resonator, which is generated by the standing wave pattern, through the nonlinearity. Numerical solution of the above coupled p.d.e.'s shows that the solutions undergo period-doubling transitions to chaos. Moreover, the instability vanishes in the limits $\tau = 0$ (instantaneous response) or $\tau \gg t_r$.

These results show that mirror or linear distributed feedback is unnecessary in order to initiate optical instabilities. A recent extension of the above plane wave model to include transverse beam effects in one dimension for instantaneously responding media ($\tau = 0$) shows that a wide range of dynamic instabilities is possible for both self-focusing or self-defocusing Kerr-like optical nonlinearities [70]. The full implications of these studies remain to be explored.

Mixed Linear and Nonlinear Distributed Feedback

Returning to our original model equations, with both linear and nonlinear coupling, it is straightforward to show that the system can exhibit a bistable steady-state response [71]. In steady-state the counter-propagating waves obey the following pair of coupled o.d.e.'s:

$$-i A_1' = \kappa A_2 e^{-i2\Delta\beta z} + \gamma(|A_1|^2 + 2|A_2|^2)A_1 \tag{5.95}$$

$$i A_2' = \kappa A_1 e^{i2\Delta\beta z} + \gamma(2|A_1|^2 + |A_2|^2)A_2 \tag{5.96}$$

Setting $\rho_{1,2}(z) = A_{1,2}(z)e^{i\phi_{1,2}(z)}$ one obtains

$$\rho_1' = \kappa\rho_1 \sin\Psi \tag{5.97}$$

$$\rho_2' = \kappa\rho_2 \sin\Psi \tag{5.98}$$

$$\phi_1' = \kappa\frac{\rho_2}{\rho_1}\cos\Psi + \gamma(\rho_1^2 + 2\rho_2^2) \tag{5.99}$$

$$\phi_2' = -\kappa\frac{\rho_1}{\rho_2}\cos\Psi - \gamma(2\rho_1^2 + 2\rho_2^2). \tag{5.100}$$

It is straightforward to show that this system has two constants of motion, $\rho_1^2 - \rho_2^2 =$ constant and $\Psi(z) = 2\Delta\beta z + \phi_1(z) - \phi(z) = $ const. These constants can be used to derive an equation for the forward flux within the distributed Bragg structure,

$$(\tfrac{1}{2}Ly') = (y - J)[(\kappa L)^2 y - (y - J)(\Delta\beta L + 2y)^2]$$
$$= Q(y) \tag{5.101}$$

with $y = |A_1|^2/|A_c|^2$, $J = |A_T|^2/|A_c|^2$ with $|A_T|^2 = |A_1|^2 - |A_2|^2$ being the transmitted flux and the "critical intensity" $|A_c|^2 = 4n_0/3\pi n_2 L$. At the input to the structure we set $y = I$, the normalized input intensity, and integrate the above equation. In general this yields a complicated relation between the input and transmitted intensities in terms of elliptic functions. The expression for this transmission simplifies somewhat if we assume Bragg resonance ($\Delta\beta = 0$) as the roots of the polynomial $Q(y)$ are then:

$$y_{\genfrac{}{}{0pt}{}{1}{2}} = \tfrac{1}{2}[J \pm [J^2 + (\kappa L)^2]^{1/2}], \quad y_3 = J, \quad y_4 = 0 \tag{5.102}$$

and the transmission function becomes

$$T = \frac{J}{I} = 2\left(1 + nd\left\{\frac{2[(\kappa L)^2 + J^2]^{1/2}}{1 + (J/2\kappa L)^2}\right\}\right) \tag{5.103}$$

where $nd(u|m)$ is a tabulated Jacobian elliptic function. A graph of the input (J) versus output (J) intensities show that this device can display a bistable response for certain parameter settings. A full numerical solution [10] of the original set of coupled p.d.e.'s shows that the negatively sloped and positively sloped upper branch region of these response curves are dynamically unstable. Oscillations in the output intensity tend to lie close in frequency to the inverse of the transit time ($\tau_R = n_0 L/c$) when they are periodic but tend to show chaotic oscillations at higher input intensities with much higher mean frequencies (Figure 5.19). A detailed mathematical analysis of this nonlinear structure remains to be carried out.

5d. Coherent Pulse Propagation and Self-Induced Transparency

In 1967, McCall and Hahn [72] discovered a remarkable new mode of lossless propagation in two-level systems with inhomogeneous broadening. Inhomogeneous broadening occurs because of the random motion of the atoms in the medium, which means that the effective frequency ω_{12} of the two-level atoms is Doppler shifted to an amount $\omega_{12} + kv$ where k is the wavenumber and the velocity v can be taken to be a Maxwellian distributed random variable with mean zero. The equations that describe the propagation of the electric field envelope $F(z, t)$ are exactly (5.42), (5.43), and (5.44), except that the Λ on the right-hand side of (5.42) is replaced by an average over the ensemble of random velocities $\langle\Lambda\rangle = \int g(\alpha_1)\Lambda(z, t, \alpha_1)d\alpha_1$ where the distribution $g(\alpha_1)$, $2\alpha_1 = \omega_{12} - \omega$, is Maxwellian with mean zero and variance β^2. For convenience of calculation we will often take $g(\alpha)$ to be Lorentzian, that is, $g(\alpha) = \beta/\pi(\alpha^2 + \beta^2)$. We have

$$\frac{\partial F}{\partial z} + \frac{1}{c}\frac{\partial F}{\partial t} - \frac{ic}{2\omega}\nabla_1^2 F = \frac{i\omega}{2\epsilon_0 c}\langle\Lambda\rangle \tag{5.104}$$

$$\frac{\partial\Lambda}{\partial t} + (\gamma_{12} + i(\omega_{12} - \omega))\Lambda = \frac{ip^2}{h}FN \tag{5.105}$$

$$\frac{\partial N}{\partial t} + \gamma_{11}(N - N_0) = \frac{2iA}{\hbar}(F^*\Lambda - F\Lambda^*). \tag{5.106}$$

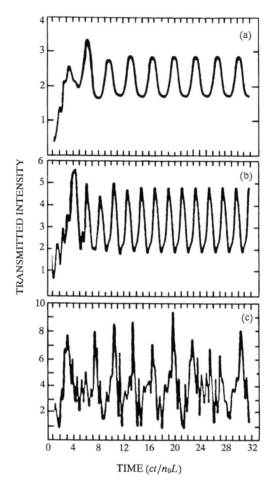

FIGURE 5.19 (a) Self-pulsing solution for a step input of intensity 4. The material relax-ation time is $\tau = 0.05$ in units of the transit time. (b) Self-pulsing solution for a step input of intensity 6. (c) Chaotic solution. Input intensity $I = 9$.

The initial-boundary value problem has the character of a Goursat problem and is well posed as follows. The electric field envelope $F(z, x, y, t)$ is given along one characteristic $z = 0$ (namely, at the beginning of the medium) as a function of time and the transverse coordinates x and y. If the transverse structure of the electric field $V(x, y, z)$ is fixed by waveguiding action, then diffraction is absent, and the reader will recall that N is the weighted average of the excess population in the ground state,

$$N = n_a \frac{\int (\rho_{22} - \rho_{11}) V V^* d\vec{x}}{\int V V^* d\vec{x}}, \tag{5.107}$$

and A is the normalization constant $\int V^2 V^{*2} d\vec{x}(\int V V^* d\vec{x})^{-1}$. If the transverse structure is not fixed, then $V(x, y, z) = e^{i(\omega/c)z}$ and the diffraction term is present. In that case $N = n_a(\rho_{22} - \rho_{11})$, the excess number of atoms in the ground state, and $A = 1$. Because $\rho_{11} + \rho_{22} \simeq 1$ from conservation of probability, the ground state values of N and Λ are n_a and 0 respectively. Since the only excitation the medium receives is the incoming field, N_0 is also n_a. The situation is pictured in Figure 5.20.

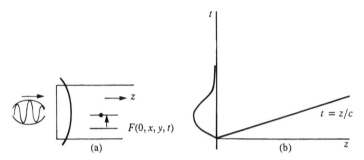

FIGURE 5.20   (a) A finite width pulse approaches the two level medium. (b) The initial-boundary value problem in (t, z) space.

The Linear Problem [6c]

Linearize the fields F, Λ, and $N - n_a$ about zero. Then we can ignore the right-hand side of (5.106) and from (5.105), with N replaced by n_a,

$$\Lambda = \frac{ip^2}{\hbar} n_a \int_\infty^t e^{-(\gamma_{12}+2i\alpha_1)(t-\tau)} F(\tau) d\tau, \quad 2\alpha_1 = \omega_{12} - \omega.$$

Then

$$F_z + \frac{1}{c} F_t - \frac{ic}{2\omega} \nabla_1^2 F = -\frac{\omega p^2 n_a}{2\epsilon_0 \hbar c} \int_{-\infty}^t d\tau F(\tau) \int g(\alpha_1) e^{-(\gamma_{12}+2i\alpha_1)(t-\tau)} d\alpha_1$$

$$(5.108)$$

Define

$$G(s) = \int g(\alpha_1) e^{-(\gamma_{12}+2i\alpha_1)s} d\alpha_1, \quad s > 0,$$

$$(5.109)$$

$$= 0, \quad s < 0,$$

its Fourier transform and the Fourier transform of $F(z, x, y, t)$

$$\begin{cases} \hat{G}(\Omega) \\ \hat{F}(\Omega) \end{cases} = \int_{-\infty}^{\infty} \begin{cases} G(t) \\ F(t) \end{cases} e^{i\Omega t} dt. \tag{5.110}$$

Then

$$F_z + \frac{1}{c} F_t - \frac{ic}{2\omega} \nabla_1^2 F = -\frac{\omega p^2 n_a}{2\epsilon_0 \hbar c} \int_{-\infty}^{\infty} G(t - \tau) F(\tau) d\tau$$

transforms to

$$\hat{F}_z - \frac{i\Omega}{c} \hat{F} - \frac{ic}{2\omega} \nabla_1^2 \hat{F} = -\frac{\omega p^2 n_a}{2\epsilon_0 \hbar c} \hat{G}(\Omega) \hat{F}$$

with solutions

$$\hat{F}(z, x, y, \Omega) = f(z, x, y, \Omega) \exp(-k_I z + i k_R z)$$

where $f(z, x, y)$ satisfies the paraxial equation $f_z - (ic/2\omega)\nabla_1^2 f = 0$ and

$$k_R = \frac{\Omega}{c} - \frac{\omega p^2 n_a}{2\epsilon_0 \hbar c} \hat{G}_I(\Omega)$$

$$\tag{5.111}$$

$$k_I = \frac{\omega p^2 n_a}{2\epsilon_0 \hbar c} \hat{G}_R(\Omega).$$

EXERCISE 5.7

Let us work out $\hat{G}(\Omega)$ for the Lorentzian line width distribution

$$g(\alpha_1) = \frac{\beta_1}{\pi(\alpha_1^2 + \beta_1^2)}.$$

The reader should show that

$$G(s) = e^{-(\gamma_{12} + 2\beta_1)s}, \quad s > 0$$

$$= 0, \quad s < 0$$

and then

$$\hat{G}(\Omega) = \frac{1}{\gamma_{12} + 2\beta_1 - i\Omega} = \frac{\gamma_{12} + 2\beta_1}{(\gamma_{12} + 2\beta_1)^2 + \Omega^2} + \frac{i\Omega}{(\gamma_{12} + 2\beta_1)^2 + \Omega^2}.$$

$\square$

The width of the loss $\hat{G}_R(\Omega)$, whose graph is shown in Figure 5.21, is proportional to the sum of the homogeneous broadening γ_{12} and twice the width of the inhomogeneous distribution β_1.

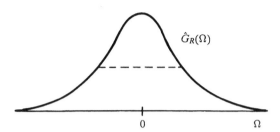

FIGURE 5.21 The graph of $\hat{G}(\Omega)$.

We now have a startling result. Even when the attenuation effects of homogeneous broadening are completely ignored, the electromagnetic wave loses energy to the medium due to the incoherent response of the medium to the incoming field.

EXERCISE 5.8

(This exercise is related to Exercise 6.5). Take $F(z, x, y, t)$ to be a plane wavepacket independent of the transverse coordinates x, y. Then $f_z = 0$ and $\hat{F}_0(z, \Omega) = \hat{F}_0(\Omega)e^{-k_I z + ik_R z}$. Ignore homogeneous broadening and use the method of steepest descent to calculate the long distance and time behavior of

$$
F(x, y, z, t) = \frac{1}{2\pi} \int e^{-i\Omega t} \, d\Omega \, \hat{F}_0(\Omega)e^{ik(\Omega)z}
$$

$$
= \frac{1}{2\pi} \int \hat{F}_0(\Omega) \, d\Omega \, e^{h(\Omega)z}
$$

(5.112)

with $h(\Omega) = ik(\Omega) - i\Omega(t/z)$. The stationary point Ω_c at which the integrand e^{hz} is maximum along an appropriate path is given by $h'(\Omega_c) = 0$. The reader will find that the field $F(z, x, y, t)$ decays in a distance of order β_1^{-1}. $\qquad\square$

Observe that in the sharp line limit $\beta_1 \to 0$ in which case $g(\alpha_1) = \delta(\alpha_1)$, the Dirac delta function, $k_I = 0$ and there is no attenuation. The field gives up photons to excite the medium, and in the absence of inhomogeneous broadening, the excited atoms eventually relax to the ground state and give back the energy to the field. The initial wavepacket is dispersed, but the energy in the field $\int FF^* dt$ integrated over all time is independent of z.

When $\beta_1 \neq 0$, the attenuation due to the random motion of the atoms is caused by the incoherent return of energy to the field. It is analogous to the phenomenon of Anderson localization. Let $\psi(x)$ be the probability amplitude of an electron with energy E in a one-dimensional random potential $V(x)$, modeling the effects of impurities. There is a theorem that the only solutions to

$$\psi_{xx} + (E - V(x))\psi = 0 \tag{5.113}$$

that are bounded at $x = \pm\infty$ are a countable set that occur at negative discrete values $E = E_j$ of the energy. Moreover, each wavefunction is localized and decays exponentially as $x \to \pm\infty$. In other words, there are no running states for positive E that look like $e^{\pm i\sqrt{E}x}$ at $x = \pm\infty$ for E positive. The electron cannot freely propagate in a one-dimensional medium of impurities. Likewise, if we ask to propagate right-going waves $e^{-iEt+i\sqrt{E}x}$, $x < 0$, of

$$i\phi_t + \phi_{xx} - V(x)\phi = 0 \tag{5.114}$$

through a medium with $V(x)$ a random function for $x > 0$ and zero for $x < 0$, then setting $\phi(x, t) = e^{-iEt}\psi(x)$ we again obtain (5.113), and $\psi(x)$ decays as $x \to \infty$. The wave is trapped in the random medium.

We now ask: what happens when we look at nonlinear wavepackets?

Nonlinear Propagation

It is convenient to make the following rescalings:

$$F = i\frac{\hbar c}{2\sqrt{Ap}}e, \quad \Lambda = \frac{n_a p}{2\sqrt{A}}\lambda, \quad N = -n_a n, \quad N_0 = n_a n_0,$$

$$t - \frac{z}{c} = \sqrt{\frac{\epsilon_0 \hbar}{\omega p \sqrt{A}}}\tau, \quad \frac{\vec{x}}{c} = \sqrt{\frac{\epsilon_0 \hbar}{\omega p \sqrt{A}}}\vec{x}', \quad (\omega_{12} - \omega) = \sqrt{\frac{\omega\sqrt{A}p}{\epsilon_0 \hbar}}2\alpha$$

$$\gamma_{12} = \sqrt{\frac{\omega\sqrt{A}p}{\epsilon_0 \hbar}}\gamma_2, \quad \gamma_{11} = \sqrt{\frac{\omega\sqrt{A}p}{\epsilon_0 \hbar}}\gamma_1 \tag{5.115}$$

and find (drop primes on x, y, z)

$$e_z - \frac{i}{F_r} \nabla_1^2 e = \langle \lambda \rangle = \int g(\alpha) \lambda(\alpha, z, \tau) d\alpha, \tag{5.116}$$

$$\lambda_\tau + (\gamma_2 + 2 i\alpha)\lambda = en, \tag{5.117}$$

$$n_\tau + \gamma_1(n - n_0) = -\tfrac{1}{2}(e\lambda^* + e^*\lambda), \tag{5.118}$$

with F_r, the Fresnel number, equal to $(4\omega\epsilon_0\hbar/p\sqrt{A})^{1/2}$ and $g(\alpha)$ is a probability distribution; that is, $g(\alpha) > 0$ and $\int g(\alpha)d\alpha = 1$. In examples for which we carry out explicit computations, it is convenient to use the Lorentzian line width (although the more realistic distribution would be Maxwellian),

$$g(\alpha) = \frac{\beta}{\pi(\alpha^2 + \beta^2)}. \tag{5.119}$$

When the medium is in the ground state $n = -1$. For the most part, we will take the Fresnel number $F_r = \infty$ and the homogeneous broadening coefficients γ_1 and γ_2 to be zero, although we will later comment on their influences.

We now look at the *sharp line limit* in which $g(\alpha) = \delta(\alpha)$, the Dirac delta function. In other words, the medium energy level difference $E_1 - E_2$ is tuned exactly to the energy $\hbar\omega$ of the incoming photons. Equations (5.116), (5.117), and (5.118) now are

$$e_z = \lambda$$

$$\lambda_\tau = en \tag{5.120}$$

$$n_\tau = -\tfrac{1}{2}(e\lambda^* + e^*\lambda)$$

with $e(0, \tau)$ given and $\lambda = n + 1 = 0$ as $\tau \to -\infty$. If $e(0, \tau)$ is real, then $e(z, \tau)$ remains real, and we may set

$$e = u_\tau, \quad \lambda = -\sin u, \quad n = -\cos u \tag{5.121}$$

and find

$$u_{z\tau} = -\sin u, \tag{5.122}$$

the sine–Gordon equation, which is an exactly integrable system [65].

EXERCISE 5.9

Show that (5.122) is the integrability condition ($V_{\tau z} = V_{z\tau}$) for the pair of equations (see [66, 67])

$$V_\tau = \begin{pmatrix} -i\zeta & \frac{1}{2}u_\tau \\ -\frac{1}{2}u_\tau & i\zeta \end{pmatrix} V, \quad V_z = -\frac{i}{4\zeta}\begin{pmatrix} \cos u & -\sin u \\ -\sin u & -\cos u \end{pmatrix} V. \tag{5.123}$$

□

The initial value problem for (5.122) is well posed as follows: Give $u(z, \tau)$ on $z = 0$ and arrange that $u(z_1, +\infty) - u(z, -\infty)$ is equal to a multiple of 2π. Equation (5.122) is solved by the inverse transform in exactly the same way as we solve the nonlinear Schrödinger equation for $q(x, t)$ in Section 6k. Here τ plays the role of x and z plays the role of time t. Let us briefly review the method (see [9–13, 73, 74]). We are given $u(0, \tau)$ and by solving the first equation in (5.123), we relate the fundamental solution matrix at $\tau = -\infty$ to that at $\tau = +\infty$. The elements of the connection matrix are what we call the scattering data $S(z = 0)$. The success of the inverse scattering transform arises because the evolution of S in z is very simple and we recover $u(z, \tau)$ from $S(z)$. In terms of the notation used in Section 6k.

$$u(0, \tau) \longrightarrow \left\{ \left(\zeta_k, \beta_k = \frac{1}{\gamma_k a_k'^2} \right)^N, \quad \frac{b^*}{a}(\xi, 0) \right\}$$

↓ evolution in z

$$\text{(5.124)}$$

$$u(z, \tau) \longleftarrow \left\{ \zeta_k = \zeta_k(0), \beta_k = \beta_k(0)e^{(i/2\zeta_k)z}, \right.$$

$$\left. \frac{b^*}{a}(\xi, z) = \frac{b^*}{a}(\xi, 0)e^{(i/2\xi)z} \right\}.$$

In this section, we are mainly interested in the soliton solutions of (5.122) The solitons have parameters given by $\zeta_k = \xi_k + i\eta_k$. If $\frac{1}{2}e(0, \tau)$ is the square well potential shown in Figure 5.22, the scattering data S is worked out in Exercise (6.21). For $e(0, \tau)$ real, the discrete eigenvalues either lie on the imaginary axis $\xi = i\eta$ or come in conjugate pairs $(-\xi + i\eta, \xi + i\eta)$. In the former case,

$$u(z, \tau) = \pm 4\tan^{-1}\exp\theta \tag{5.125}$$

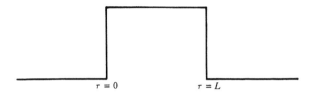

FIGURE 5.22

with

$$\theta = -2\eta\tau + \frac{z}{2\eta} + \theta_0 \tag{5.126}$$

and the electric field is

$$e(z, \tau) = \pm 4\eta \operatorname{sech}\theta. \tag{5.127}$$

The soliton (5.125) is called a 2π pulse (also called a kink (antikink) because it represents $a + 2\pi(-2\pi)$ twist in the field $u(z, \tau)$) because the area under the electric field

$$\text{Area} = \int_{-\infty}^{\infty} e(z, \tau)d\tau = \pm 2\pi. \tag{5.128}$$

In the latter case,

$$u(z, \tau) = 4\tan^{-1}\left(\frac{\eta}{\xi}\cos\phi\operatorname{sech}\theta\right) \tag{5.129}$$

where

$$\phi = -2\xi\tau - \frac{\xi}{2(\xi^2 + \eta^2)}z + \phi_0$$
$$\theta = -2\eta\tau + \frac{\eta}{2(\xi^2 + \eta^2)}z + \theta_0 \tag{5.130}$$

The conjugate pairs are called breathers. Note that in the small amplitude limit, $\eta/\xi \ll 1$, the breathers take the form

$$u(z, \tau) = \frac{4\eta}{\xi}\operatorname{sech}\theta\cos\phi \tag{5.131}$$

of a wavepacket. They have zero area

$$\text{Area} = \int_{-\infty}^{\infty} e(z, \tau)d\tau = 0$$

and are therefore sometimes called 0π pulses.

Therefore an initial pulse $e(0, \tau)$ breaks up into a series of 2π and 0π pulses (for the square well potential, there are no breathers; however, the chirped square well potential—see Exercise 6.23—can have breathers) and radiation measured by $b^*/a(\xi, z = 0)$. The velocity of the 2π pulses, namely the locus of constant θ, is given by $c/(1 + (c/4\eta^2)) < c$, and the velocity of breathers is

$$\frac{c}{1 + \frac{c}{4(\xi^2+\eta^2)}} < c. \tag{5.132}$$

What is absolutely remarkable is that even in the broadline, nontuned case, the equations

$$e_z = \langle \lambda \rangle,$$

$$\lambda_\tau + 2i\alpha\lambda = en, \tag{5.133}$$

$$n_\tau = -\tfrac{1}{2}(e^*\lambda + e\lambda^*),$$

the inhomogeneously broadened Maxwell–Bloch equations for a two-level atom, are still exactly integrable [75, 76]. Moreover, unlike the fate of the small-amplitude radiation discussed above, *the solitons and breathers do not attenuate as they propagate through the medium.* Thus even though the medium is random, the nonlinear soliton pulses find a way to synchronize the different frequencies $\omega_{12} + kv$ associated with the moving atoms. The radiation component of the initial pulse $e(0, \tau)$ is attenuated, however, in much the same way as the linear fields are.

Before we look at the mathematics describing this behavior (which the reader may wish to postpone), there are several important points we wish to make:

- There is a special class of pulses, namely the 2π and 0π solitons, to which the medium is transparent. Their formulae are given in (5.155–5.159).

- An initial pulse $e(0, \tau)$ breaks up into soliton components $e_s(0, \tau)$ and radiation $e_R(0, \tau)$. The latter is trapped in a distance proportional to β^{-1}, the inverse of the width of the distribution $g(\alpha)$. The medium is left in a permanently "ringing" state within this region.

- An initial pulse $e(0, \tau)$ quickly (in a distance, $z = 0(\beta)$, proportional to the width of the distribution) reshapes so that its area

$$\text{Area} = \int_{-\infty}^{\infty} e(z, \tau) d\tau$$

  becomes a multiple of 2π. This is called the McCall–Hahn area theorem. Note that, as a corollary, if $\beta = 0$, then the initial condition itself $e(0, \tau)$ must have an area that is a multiple of 2π.

- The soliton pulses are unstable to the effects of diffraction, much like what would happen to a soliton of

$$i q_z + q_{tt} + 2q^2 q^* = 0$$

  if we were to add diffraction terms either $\pm(q_{xx} + q_{yy})$. (Therefore, these coherent soliton wavepackets, if made to resonate with one of the energy level transitions of atmospheric CO_2, will not make reliable ray guns!)

- The effects of small homogeneous broadening terms ($\beta \gg \gamma_1, \gamma_2$) can be captured by perturbation theory. Essentially the soliton amplitudes decay slowly in a manner similar to that discussed in Section 6k.

- Perhaps the most interesting ramification of lossless propagation by certain types of nonlinear pulses is the demonstration that nonlinearity and randomness have opposite influences on propagation. A discussion may be found in [77]. Randomness, as per the Anderson localization phenomenon, inhibits propagation by causing incoherent scattering in the medium. Nonlinearity, on the other hand, phase-locks and synchronizes the different frequencies and leads to a coherent response of the medium. Even in situations that are not exactly integrable, one should expect nonlinearity eventually to overcome the scattering effects of randomness.

- The picture of the solution is given in Figure 5.23.

A Mechanical Analogy for The Sine–Gordon Equation

Make the change of variables

$$z = \frac{T + Z}{2}, \quad \tau = \frac{T - Z}{2}$$

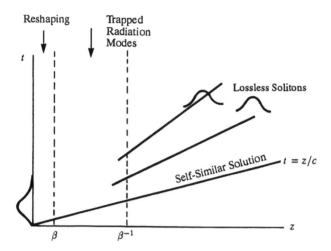

FIGURE 5.23   Picture of the solution in the (t, z) plane.

in (5.122) and obtain

$$\frac{\partial^2 u}{\partial T^2} - \frac{\partial^2 u}{\partial Z^2} + \sin u = 0. \tag{5.134}$$

As we discuss in Section 6c, Alwyn Scott proposed that we look at (5.134) as arising when we hang a sequence of pendula from a torsion wire so that the motion of any one pendulum will cause the torsion wire to twist and transfer some of its motion to a neighboring pendulum. We derive equation (5.134) from this picture in Section 6c. The soliton solution, which we called a 2π kink (see formula 5.125), corresponds to a 360° twist in the torsion wire so that even though the pendula ahead of and behind the twist may be at rest in the stable position, the state of the system as a whole is quite different from that state when there are no twists and *all* the pendula are in the stable position. Because of this, the 2π kinks are called topological solitons. The 0π soliton or breather (see formula (5.129)), on the other hand, does not involve a complete twist in the torsion wire but is merely a localized pulse in which several neighboring pendula swing synchronously with different amplitudes. The amplitude of the motion decreases exponentially (the envelope has the hyperbolic secant shape) away from the pendulum with maximum amplitude.

The attenuated medium problem we have been discussing corresponds to the case when, as $Z \to \pm\infty$, the pendula are all in the stable positions. The resulting motion of the chain of pendula can be described in terms of 2π kinks, breathers, and radiation-like modes.

On the other hand, what happens when, say at $T = 0$, the pendula are held in the unstable position $u = \pi$? It is clear that any perturbation (imagine that we give the first one in the chain a little kick) will cause the pendula to fall. If each pendulum were not connected to its neighbor, then a very small push would cause the pendulum to oscillate or overturn in a motion that in the phase plane of the pendulum (see Figure 6.14) follows very closely the separatrix joining the unstable position $u = -\pi$ with the unstable position $u = \pi$. However, because the pendula are coupled, some of the energy of the first pendulum is transferred to the second, causing it to fall, and the first pendulum has less than the energy it requires to reach $u = \pi$ again. It therefore oscillates about $u = 0$. At first the oscillations are large, but as more and more of the energy is transferred along the chain, the oscillations slowly become smaller and smaller in amplitude. The pulse that travels along the chain, causing the pendula to fall out of the unstable position $u = \pi$ and execute slowly decaying oscillations about the stable position $u = 0$, is well described by a special solution of (5.134) that we call a π pulse (not to be confused with a 0π pulse or breather) because it causes a given pendulum to make the transition from the unstable $u = \pi$ position to the stable $u = 0$ position. There is no damping in the system, but energy is lost from a given pendulum through radiation, as each pendulum transfers some of its motion to its neighbors. The π pulse solution is obtained by looking for a solution of (5.134) in self-similar form

$$u(T, Z) = f\left(\eta = \sqrt{T^2 - Z^2}\right), \tag{5.135}$$

and we shall meet these shapes in the next section when we discuss superfluorescence. Observe that the front (the transition from $u = \pi$ to $u = 0$), which occurs near 0, moves at the speed of light.

Superfluorescence, which we now discuss, is simply the coherent radiation emitted when the chain of pendula in the unstable state or the atoms in an amplifier (the excitable medium is prepared so that all the atoms are in the excited higher energy state) are perturbed by noise.

Superfluorescence

We address the problem of what happens where the medium is initially prepared in the excited state. It is clear that this state is unstable, and any perturbation will trigger a release of coherent electromagnetic radiation. This burst of radiation was first predicted by Dicke in 1954 and first observed in 1973 [78]. For simplicity we take the sharp line limit (in which case we can take λ real) and then the initial-boundary value problem for (5.133) is posed as follows:

$$e_z = \lambda$$

$$\lambda_\tau = en \tag{5.136}$$

$$n_\tau = -e\lambda$$

with

$$\lambda = \epsilon\rho(z), \quad n = (1 - \epsilon^2\rho^2(z))^2 \quad \text{on} \quad \tau = 0$$

and

$$e = 0, \quad \text{on} \quad z = 0.$$

The function $\rho(z)$ represents the initial noise in the system, and we include the small parameter ϵ in front of $\rho(z)$ to indicate that the noise is relatively small.

While inverse scattering theory again holds, the behavior of the system is best understood by appealing to a special solution to (5.134) found by making the transformation (5.121) so as to convert (5.134) to (5.122), namely

$$u_{z\tau} = -\sin u.$$

We look for solutions in terms of the variable $\eta = 2(z\tau)^{1/2}$ and obtain

$$u_{\eta\eta} + \frac{1}{\eta}u_\eta = -\sin u. \tag{5.137}$$

Equation (5.137) is a celebrated equation in mathematics because its solutions, called Painlevé transcendants of the third kind, have very special properties as functions of the complex variable η. We will not go into the details here except to say that one can solve the nonlinear ordinary differential equation (5.137) exactly and write its solution in terms of two free parameters (for example, the values of u and u_η at some nonsingular point $\eta = \eta_0$).

Now, near $\tau = 0$, we know $n = -\cos u$ is approximately unity (all the atoms are in the excited state) so that u is close to π. Let $u = \pi + \phi$ and linearize (5.122) to obtain

$$\phi_{\eta\eta} + \frac{1}{\eta}\phi_\eta - \phi = 0$$

so that near $\eta = \tau = 0$, ϕ is a linear combination of $I_0(\eta)$ and $K_0(\eta)$, the Bessel functions of the second kind of zero order. Since we want the fields $n, \lambda,$ and e to be bounded as $\eta \to 0$, we will choose only the one-parameter family of solutions of (5.137), which behave as $I_0(\eta)$ when η is small and positive. So, near $\eta = 0$,

$$u = \pi + A I_0(\eta)$$

where A is related to the mean squared average of the noise $\langle \rho^2(z) \rangle$, while near $\eta = +\infty$,

$$u = 0, \quad \frac{B}{\sqrt{\eta}} \cos(\eta + C)$$

with B and C given in terms of A. The graphs of u, the electric field $e = (2\pi/\eta)du/d\eta$ at a fixed value of τ, and $n = -\cos u$ are shown in Figure 5.24.

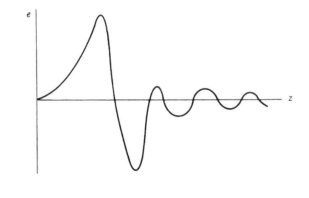

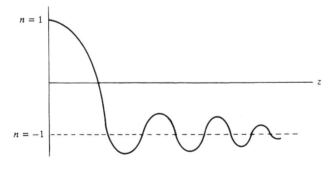

FIGURE 5.24

So what happens? A pulse that moves at the speed of light knocks the atoms from their excited states. Some of the photons are immediately reabsorbed, which means that locally the medium can return close to the excited state $n = +1$. But some of the photons are transferred forward and act to destabilize the atoms further down the medium. At a fixed position z, the medium is left in a ringing state that slowly (algebraically) decays as more and more of the photons travel to larger values of z.

The Mathematics: A Series of Exercises

These topics are treated further in references [76, 79, 80, 70, 71]. This section can be omitted on a first reading.

EXERCISE 5.10

Show that equations (5.133) are the solvability condition of

$$V_\tau = \begin{pmatrix} -i\zeta & \frac{1}{2}e \\ -\frac{1}{2}e^* & i\zeta \end{pmatrix} V, \quad V_z = \frac{i}{4} \begin{pmatrix} \langle \frac{n}{\zeta-\alpha} \rangle & -\langle \frac{\lambda}{\zeta-\alpha} \rangle \\ -\langle \frac{\lambda^*}{\zeta-\alpha} \rangle & -\langle \frac{n}{\zeta-\alpha} \rangle \end{pmatrix} V + cV \qquad (5.138)$$

where $\langle n/\zeta - \alpha \rangle = P \int (n(\alpha, z, \tau) g(\alpha) d\alpha / \zeta - \alpha)$ and P is the Cauchy principal value, and c is an arbitrary scalar constant. ☐

EXERCISE 5.11

The reader should consult Section 6k and recall that τ is x and z is t. Show that the Jost function

$$V = \begin{pmatrix} \phi_1(\tau, z, \zeta) \\ \phi_2(\tau, z, \zeta) \end{pmatrix},$$

defined as solutions to (5.138) by their asymptotic properties

$$\begin{pmatrix} \phi_1 \\ \phi_2 \end{pmatrix} \to \begin{pmatrix} 1 \\ 0 \end{pmatrix} e^{-i\zeta z} \quad \text{as} \quad \tau \to -\infty, \qquad (5.139)$$

satisfy (5.138) with the choice $c = (i/4)\langle 1/\zeta - \alpha \rangle$. (This choice is necessary so that (5.139) satisfy (5.138).) ☐

EXERCISE 5.12

From Section 6k, we see that as $\tau \to +\infty$,

$$\begin{pmatrix} \phi_1 \\ \phi_2 \end{pmatrix} \to \begin{pmatrix} a(\zeta, z)e^{-i\zeta\tau} \\ b(\zeta, z)e^{i\zeta\tau} \end{pmatrix} \qquad (5.140)$$

☐

EXERCISE 5.13

Show that we can identify

$$n(\alpha, \tau, z) = -\phi_1(\alpha, \tau, z)\phi_1^*(\alpha, \tau, z) + \phi_2(\alpha, \tau, z)\phi_2^*(\alpha, \tau, z)$$

$$\lambda(\alpha, \tau, z) = 2\phi_1(\alpha, \tau, z)\phi_2^*(\alpha, \tau, z) \tag{5.141}$$

by showing that the expressions on the right-hand side of (5.141) satisfy (5.133). The consequence of this exercise and the previous one is that as $\tau \to +\infty$,

$$n(\alpha, \tau, z) \to -1 + 2bb^*(\alpha, z)$$
$$\lambda(\alpha, \tau, z) \to 2ab^* e^{-2i\alpha\tau} \tag{5.142}$$

where we have used the fact that $aa^* + bb^* = 1$. □

EXERCISE 5.14

Show that from (5.133),

$$\frac{\partial}{\partial \tau}(n^2 + \lambda\lambda^*) = 0.$$

Evaluate n and λ as given by (5.141) using (5.139) and (5.140) and show that

$$(n^2 + \lambda\lambda^*)_{\tau=-\infty} = 1 \quad \text{and} \quad (n^2 + \lambda\lambda^*)_{\tau=+\infty} = aa^* + bb^*.$$

Therefore, $aa^* + bb^* = 1$. □

EXERCISE 5.15

Using (5.138) with $c = (i/4)\langle 1/\zeta - \alpha \rangle$, show that by taking the limit $\tau \to +\infty$, and using (5.142),

$$a_z = \frac{i}{4}\left\langle \frac{2bb^*}{\zeta - \alpha} \right\rangle a - \frac{i}{4}\lim_{\tau\to\infty}\left\langle \frac{2ab^* e^{2i(\zeta-\alpha)\tau}}{\zeta - \alpha} \right\rangle b$$

$$b_z = -\frac{i}{4}\lim_{\tau\to\infty}\left\langle \frac{2a^* b e^{-2i(\zeta-\alpha)\tau}}{\zeta - \alpha} \right\rangle a - \frac{i}{4}\left\langle \frac{-2 + 2bb^*}{\zeta - \alpha} \right\rangle b. \tag{5.143}$$

□

Note that the point $\alpha = \zeta$ is no longer singular when one combines the two integrals on the right-hand sides of the equations for a_z and b_z because the denominator also vanishes at $\alpha = \zeta$. Therefore we can change the Principal Value integral into either of the integrals C_u or C_A shown below in Figure 5.25, because

$$P \int \frac{f(\zeta) - f(\alpha)}{\zeta - \alpha} d\alpha = \int_{-\infty}^{\infty} \frac{f(\zeta) - f(\alpha)}{\zeta - \alpha} d\alpha$$

$$= \int_{C_u} \frac{f(\zeta) - f(\alpha)}{\zeta - \alpha} d\alpha = \int_{C_A} \frac{f(\zeta) - f(\alpha)}{\zeta - \alpha} d\alpha.$$

FIGURE 5.25

It is convenient to define $\langle \; \rangle$ on the right in the equation for a_z as $\int_{C_u}$ and $\langle \; \rangle$ on the left in the equation for b_z as $\int_{C_A}$, because for $\tau > 0$,

$$\int_{C_u} \frac{2ab^* e^{2i(\zeta - \alpha)\tau}}{\zeta - \alpha} d\alpha = 0$$

$$\int_{C_A} \frac{2a^*b \; e^{-2i(\zeta - \alpha)\tau}}{\zeta - \alpha} d\alpha = 0,$$

(5.144)

a result that is easy to verify by moving the integrals to semicircles at infinity in the lower and upper half-planes respectively ($e^{-2i\alpha\tau} \rightarrow 0$ for Im $\alpha < 0$ and $\tau > 0$, $e^{2i\alpha\tau} \rightarrow 0$ for Im $\alpha > 0$, $\tau > 0$). Hence

$$a_z = \frac{ia}{2} \int_{C_u} g(\alpha) \frac{bb^*(\alpha, z)}{\zeta - \alpha} d\alpha,$$

$$b_z = \frac{ib}{2} \int_{C_A} g(\alpha) \frac{bb^*(\alpha, z)}{\zeta - \alpha} d\alpha.$$

(5.145)

EXERCISE 5.16

Show that

$$b_z^* = -\frac{ib^*}{2} \int_{C_u} \frac{aa^*(\alpha, z)}{\zeta - \alpha} g(\alpha) \, d\alpha. \tag{5.146}$$

☐

EXERCISE 5.17

For ξ real, show that

$$\left(\frac{b^*}{a}(\xi, z)\right)_z = \frac{b^*}{a}(\xi, z) \left(-\frac{i}{2} \int_{C_u} \frac{g(\alpha)}{\zeta - \alpha} d\alpha\right). \tag{5.147}$$

☐

EXERCISE 5.18

Take $\zeta = \zeta_k$, Im $\zeta_k > 0$, and show that (5.144) becomes

$$\beta_{kz} = \beta_k \left(-\frac{i}{2} \int \frac{g(\alpha)}{\zeta - \alpha} d\alpha\right). \tag{5.148}$$

Since Im $\zeta_k > 0$, we can replace C_u by an integral along the real α axis.
☐

EXERCISE 5.19

Show that (5.144) integrates to

$$\frac{b^*}{a}(\xi, z) = \frac{b^*}{a}(\xi, 0) \exp\left(-\frac{\pi}{2} g(\xi) z - \frac{i}{2} z P \int_{-\infty}^{\infty} \frac{g(\alpha)}{\xi - \alpha} d\alpha\right). \tag{5.149}$$

Therefore, if $g(\alpha) = \beta/\pi(\alpha^2 + \beta^2)$, the energy $|b^*/a|^2$ in the radiation component of the initial pulse decays as

$$\left|\frac{b^*}{a}\right|^2 = \left|\frac{b^*}{a}\right|^2_{z=0} \exp(-\pi g(\xi) z) = \left|\frac{b^*}{a}\right|^2 \exp\left(-\frac{\beta z}{\xi^2 + \beta^2}\right). \tag{5.150}$$

For $\xi \gg \beta$, this energy decays in a distance β^{-1}. On the other hand,

$$\frac{b^*}{a}(0, z) = \frac{b^*}{a}(0, 0) \exp\left(-\frac{\pi}{2\beta} z\right) \tag{5.151}$$

since $P \int g(\alpha)/\alpha \, d\alpha = 0$ because $g(\alpha)$ is even, and so the energy in $\xi = 0$ decays in a distance $z = \beta$.   □

EXERCISE 5.20

The McCall–Hahn area theorem.
Suppose $e(0, \tau)$ is real. Write $e(0, \tau) = u_\tau$. Then from

$$\phi_{1\tau} + i\zeta\phi_1 = \tfrac{1}{2}u_\tau\phi_2$$

$$\phi_{2\tau} - i\zeta\phi_2 = -\tfrac{1}{2}u_\tau\phi_1$$

(5.152)

it is easy to see that at $\zeta = 0$,

$$\phi_1(0, \tau, z) = \cos \frac{u(\tau, z) - u(-\infty, z)}{2}$$

$$\phi_2(0, \tau, z) = -\sin \frac{u(\tau, z) - u(-\infty, z)}{2}.$$

Therefore

$$\frac{b^*}{a}(0, z) = \lim_{\tau \to \infty} \frac{\phi_2^*(0, \tau, z)}{\phi_1(0, \tau, z)} = -\tan \frac{u(\infty, z) - u(-\infty, z)}{2}.$$

(5.153)

However the area of the pulse is

$$\text{Area} = u(\infty, z) - u(-\infty, z) = \int_{-\infty}^{\infty} e(z, \tau) \, d\tau$$

and from (5.151) and (5.153),

$$\tan\left(\frac{\text{Area}}{2}\right) \to 0$$

for $z \gg \beta$. Thus for $z \gg \beta$

$$\text{Area} \to 2n\pi,$$

(5.154)

which is the McCall–Hahn area theorem.   □

The formulae for the solitons are as follows: (1)

$$\xi = \xi + i\eta : e(\tau, z) = 4\eta \operatorname{sech}\theta \, e^{-i\phi}$$

(5.155)

where

$$\theta = -2\eta\tau + \omega_2 z + \theta_0$$

$$\phi = -2\xi\tau + \omega_1 z + \phi_0 \tag{5.156}$$

$$\omega_1 + i\omega_2 = -\frac{1}{2}\int_{-\infty}^{\infty}\frac{g(\alpha)d\alpha}{\xi + i\eta - \alpha}. \tag{5.157}$$

(2) If $e(\tau, 0)$ is real, $\xi = \phi = 0$ and

$$e(\tau, z) = \frac{\partial u}{\partial \tau}(\tau, z),$$

then

$$u(\tau, z) = 4\tan^{-1}\exp\theta. \tag{5.158}$$

(3) If $e(\tau, 0)$ is real, solitons can occur in bound pairs called breathers with eigenvalues $-\xi + i\eta$ and $\xi + i\eta$,

$$u(\tau, z) = 4\tan^{-1}\frac{\eta}{\xi}\cos\phi\,\mathrm{sech}\theta \tag{5.159}$$

with ϕ and θ given by (5.156).
When

$$g(\alpha) = \frac{\beta}{\pi(\alpha^2 + \rho^2)},$$

$$\omega_1 = -\frac{1}{2}\frac{\xi}{\xi^2 + (\eta + \beta)^2} \tag{5.160}$$

$$\omega_2 = \frac{1}{2}\frac{\eta + \beta}{\xi^2 + (n + \beta)^2}.$$

5e. Stimulated Raman Scattering

An intense laser beam propagating through matter (solid, liquid, or gas) can scatter off of background fluctuations (noise) representing the quantized oscillations of the medium and generate an intense scattered optical wave frequency downshifted from the incident wave frequency by the material oscillation frequency. This downshifted Stokes wave may appear as a consequence of scattering from quantized electronic, vibrational, rotational, or optical lattice phonon modes of the material. These stimulated Raman scattering processes play an important role when intense laser pulses propagate in bulk materials over distances of the order of several centimeters or when much weaker optical pulses propagate in optical fibers over kilometer distances. Stimulated Raman scattering was encountered in the very early self-focusing experiments marking the advent of the field of nonlinear optics in the early 1960s and led to considerable confusion in the interpretation of early experiments. The scattered Stokes wave appears initially as an exponentially amplified wave when the incident laser intensity exceeds a certain well-defined threshold. We shall see below that there is a very simple explanation for this exponential gain. The sharp intensity threshold exists because, as discussed earlier for lasers, the gain must exceed all intrinsic losses arising due to a finite conductivity of the medium. The gain coefficient, to be derived explicitly below, depends on material parameters and on laser wavelength and typically is quoted in units of centimeters per megawatt to emphasize that peak intensities of many MW/cm^2 are needed to observe stimulated Raman scattering in bulk materials. For example, one of the strongest Raman emissions is observed in liquid carbon disulphide (CS_2), which has a gain coefficient of $g = 0.024$ cm/MW. The basic mathematical apparatus for the study of stimulated Raman emission was already derived from first principles in Section 4g via the effective two-level atom model. Before elaborating further on the theory of stimulated Raman scattering, let us first summarize what we can infer from our existing knowledge.

The directional characteristics of the Stokes wave are dictated by the material dispersion relation appearing in the resonance triad relation discussed in Section 6g and, importantly, by geometric constraints. As we shall see in the discussion of Chapter 6 where material dispersion relations for optical and acoustic lattice phonons are derived, there is a priori no preferred scattering direction for the former, which are the appropriate modes for Raman scattering. However, practically speaking, the collimated nature of laser beams causes Raman scattering to occur predominantly in the forward and/or backward directions relative to the incident laser pulse. Forward scattering is preferred for incident pulses of duration less than or comparable to the characteristic material oscillation damping time (γ_{12}^{-1} in equation (4.83) whereas backward scattering is preferred for long optical pulses. This can be understood intuitively on the following physical grounds. The Stoke scattered wavepacket, in order to grow, must maintain optimum overlap with the incident pump pulse from which it continually extracts its energy the three-wave interaction. For short pulses, it can do so by traveling forward with the incident pulse thereby maximizing the nonlinear interaction by maintaining overlap for distances of the order of the "walk-off" distance discussed in Section 4g. For long pulses, the Stokes wave by moving the backward direction can continuously feed off of the incident pulse with the result that its peak emission intensity can substantially exceed that of the incident wavepacket.

We will now analyze Raman scattering in some detail. If we wish understand the buildup of the Stokes emission from background noise from a long incident pulse, we can assume steady-state conditions and that the incident pump pulse remains undepleted ($A_1(z) = $ constant in equation (4.81) Setting time derivatives to zero in equations (4.81–4.83) and treating A_1 as a parameter in the equation for A_2, we obtain the following formula for the buildup of the Stokes intensity

$$I_s(z) = I_s(0)e^{gz}$$

where $I_s = (n\epsilon_0 c/2)|A_2|^2$ has been used. The gain coefficient g is given b

$$g = \frac{2\hbar\omega_2 n_a q_{12}^2 T_2}{n(\omega_2)\epsilon_0^2 c^2} \cdot I_p = g_0 I_p$$

where $I_p = (n\epsilon_0 c/2)|A_1|^2$. The intensity growth rate for the Stokes wave which is initially exponential, depends on the material parameters and the incident pulse intensity. Including background losses, this expression becomes

$$I_s(z) = I_s(0)e^{(g-\alpha_s)z} \tag{5.161}$$

where α_s is the linear absorption coefficient reflecting a finite conductivity of the medium.

While we could carry the analysis further based on equations (4.81–4.83), we choose instead to introduce the reader to a theoretical approach, specific to molecular Raman scattering, that is commonly encountered in the research literature. It is a particularly convenient framework within which to introduce the new ideas of anti-Stokes and higher-order Stokes/anti-Stokes scattering. The model, originally due to G. Plazcek, exploits the effective two-level atom quantum approach of Section 4g and introduces a phenomenological interaction term in the Hamiltonian based on the field-induced polarizability introduced in Section 4h in the discussion of molecular reorientation. This discussion should reinforce the ideas introduced throughout Chapter 4, in particular, the discrimination between various dynamical degrees of freedom of matter as discussed for the Born–Oppenheimer approximation in Section 4j. Formally, there is no difference between the phenomenological model that we now introduce and the effective two-level atom approach of Section 4g.

The induced polarization of a molecular system of a diatomic lattice has separate contributions due to high-velocity electrons and the sluggish motion of the relatively massive nuclei forming the molecule. An atomic system differs in that only the electronic component is present. The dipole moment induced in a molecule by an applied electric field depends on both the electronic and nuclear coordinates. Using the expression $\vec{P} = \epsilon_0 N \alpha \vec{E}$, introduced in Section 4h in the discussion of the molecular reorientation mechanism, it is reasonable to assume that the molecular polarizability α, induced by an applied optical field, is a function of the nuclear (normal) coordinate of vibration q.* Assuming small displacements of the vibrating nuclei from equilibrium, we expand $\alpha(q)$ in a Taylor series about $q = 0$,

$$\vec{P} = \epsilon_0 N \left(\alpha_0 + \left(\frac{\partial \alpha}{\partial q} \right)_0 \right) \langle q \rangle \vec{E}, \tag{5.162}$$

where $(\partial \alpha / \partial q)_0$ is called the differential polarizability and the definition of $\langle q \rangle$ will be given below. Consider the effective two-level molecular model sketched in Figure 5.26. Figure 5.26(a) depicts an incident pump wave (ω_p) exciting a vibrational oscillation at ω_v and generating a Stokes wave at ω_s so as to satisfy the conservation expression $\hbar\omega_p = \hbar\omega_s + \hbar\omega_v$. Figure 5.26(b) differs in that the material wave must be already excited, so that the pump wave can combine with it to produce an anti-Stokes frequency-upshifted wave satisfying $\hbar\omega_{as} = \hbar\omega_p + \hbar\omega_v$. In molecules it is often possible that the excited level $(v = 1)$ may be occupied at room temperature due to thermal fluctuations. The much smaller energy separation between vibrational (or rotational) levels compared with electronic levels often means that $\hbar\omega_v \leq k_B T$, and so the level occupancy is given by the Boltzmann distribution factor introduced in Section 4h (with $\delta V = \hbar\omega_v$).

The Hamiltonian appropriate to describe such interactions is

$$H = H_v - \frac{1}{2} \left(\frac{\partial \alpha}{\partial q} \right)_0 q E^2, \tag{5.163}$$

where the Hamiltonian H_v describes the vibrational spectrum in the absence the applied field (e.g., a harmonic oscillator). The reader can easily derive the two-level density matrix equations for this Hamiltonian by replacing the $\vec{p} \cdot$ interaction term in (4.5) by $\frac{1}{2}(\partial \alpha / \partial q)_0 q E^2$. Adding the phenomenological damping times T_1, T_2 and using the definitions $n = \rho_{22}$, $\langle q \rangle = q_{12}(\rho_{21} + \rho_{12})$

---

*$q = R - R_0$ where R_0 represents the equilibrium separation of atoms in a molecule (or nearest neighbor atoms in a lattice).

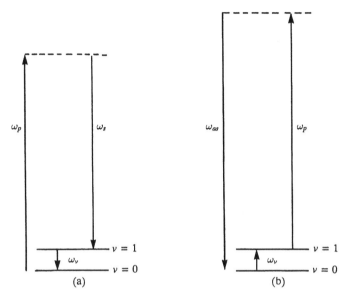

FIGURE 5.26 (a) A strong pump wave excites a downshifted Stokes optical wave an material vibration. (b) The strong pump wave combines with the already excited material vibration to produce a frequency-upshifted anti-Stokes wave.

and $\rho_{11} + \rho_{22} \simeq 1$ we obtain the material equations

$$\frac{\partial^2 \langle q \rangle}{\partial t^2} + \frac{2}{T_2} \frac{\partial \langle q \rangle}{\partial t} + \omega_0^2 \langle q \rangle = \frac{\omega_0 q_{12}^2}{\hbar} \left(\frac{\partial \alpha}{\partial q} \right)_0 E^2 (1 - 2n) \qquad (5.164)$$

$$\frac{\partial n}{\partial t} + \frac{1}{T_1}(n - \bar{n}) = \frac{1}{2\hbar\omega_0} \left(\frac{\partial \alpha}{\partial q} \right)_0 E^2 \frac{\partial \langle q \rangle}{\partial t}. \qquad (5.165)$$

In deriving these equations we have dropped small terms $\langle q \rangle / T_2^2$ in (5.164) by assuming small damping $1/T_2 \ll \omega_0$, and $\langle q \rangle / T_2$ in (5.165) relative to $\partial/\partial t \langle q \rangle$ for the same reason. The quantity n is the occupation number of the excited vibrational level and $\bar{n}$, its equilibrium value, is usually very small and given by a Boltzmann distribution. The quantity $\langle q \rangle$ is the quantum expectation (mean) value of the vibrational coordinate. The derivation of (5.164–5.165) is left as an exercise to the reader.

Written in this manner, the material equations illustrate clearly the physical origin of the stimulated Raman process depicted schematically in Figure 5.26(a). In many practical situations the excited level occupation number changes very little. If we drop it from (5.164–5.165) these become decoupled and we end up with the classical driven damped oscillator equation where the forcing term proportional to E^2 will be resonant if it contains a term oscillating at the vibrational frequency ω_0; this is simply the interference term $A_p A_s^*$ due to the presence of the pump and Stokes waves encountered in Section 4g. The quantum nature of the Raman interaction is contained in the coupling to the excited level occupation number n as before for which there is no classical analog. The coupled wave equations for stimulated Stokes emission in the forward z-direction, assuming infinite plane waves, are derived by substituting the following envelope expansions for the vibrational amplitude $\langle q \rangle = \frac{1}{2}\{Q e^{i(kz-\omega_0 t)} + (*)\}$, the total field $E = \frac{1}{2}\{A_L e^{i(k_L z - \omega_L t)} + A_s e^{i(k_s z - \omega_s t)} + (*)\}$ and the nonlinear polarization $P^{NL} = \frac{1}{2}\{P^{NL} e^{i(k_L z - \omega_L t)} + P_s^{NL} e^{i(k_s z - \omega_s t)} + (*)\}$ into the wave equation (2.9)

$$\left(\frac{\partial}{\partial z} + \frac{n(\omega_s)}{c}\frac{\partial}{\partial t} + \alpha_s\right) A_s = \frac{i\pi\omega_s}{cn(\omega_s)} N \left(\frac{\partial\alpha}{\partial q}\right)_0 A_L Q^* \qquad (5.166)$$

$$\left(\frac{\partial}{\partial z} + \frac{n(\omega_L)}{c}\frac{\partial}{\partial t} + \alpha_L\right) A_L = \frac{i\pi\omega_L}{cn(\omega_L)} N \left(\frac{\partial\alpha}{\partial q}\right)_0 A_s Q \qquad (5.167)$$

$$\frac{\partial}{\partial t} Q + \frac{i}{2\omega_0}\left(\omega^2 - \omega_0^2 - i\frac{2\omega_0}{T_2}\right) Q = \frac{i}{4m\omega_0} N \left(\frac{\partial\alpha}{\partial q}\right)_0 A_L A_s^*. \qquad (5.168)$$

It is easy to see that these equations are formally identical to the set (4.81–4.83) derived in Section 4g from a quantum description (i.e. $\omega_0^2 - \omega_v^2 \simeq 2\omega_v(\omega_0 - \omega_v)$, $\omega_0 \equiv \omega_1 - \omega_2$, $\omega_v = \omega_{21}$) allowing one to identify the correct expression for the differential polarizability $(\partial\alpha/\partial q)$ which is introduced here in an ad hoc fashion. The corresponding equations for backward scattering are simply obtained by replacing k_s by $-k_s$.

Let us now derive the steady-state Raman nonlinear susceptibility $\chi^{(3)}(\omega_s)$ at the Stokes frequency by assuming weak excitation, so that we can drop the n equation and set all time derivatives to zero. Furthermore, we assume for simplicity that the pump wave is undepleted ($|A_L|^2 = $ constant), so that we are interested in the amplification of a weak Stokes seed. The growth Stokes wave in z is given by

$$\frac{d}{dz} A_s = \frac{\pi\omega_s N^2 \left(\frac{\partial\alpha}{\partial q}\right)_0^2 \left(\frac{1}{T_2} + i(\omega_0 - \omega_v)\right) |A_L|^2}{4m\omega_0 nc} \frac{}{((\omega_0 - \omega_v)^2 + (1/T_2)^2)} A_s. \qquad (5.169)$$

This equation shows that the Stokes wave initially grows exponential from noise at a rate directly proportional to the pump intensity and more that this growth is maximal at the Stokes frequency ($\omega_0 \equiv \omega_1 - \omega_2$; $\omega_0 =$. The steady-state Raman susceptibility follows directly from the equation the induced polarization:

$$\vec{P} = \vec{P}_L + \vec{P}_{NL}$$

$$= \epsilon_0 N \alpha_0 \vec{E} + \epsilon_0 N \left(\frac{\partial \alpha}{\partial q}\right)_0 \langle q \rangle \vec{E}. \tag{5.170}$$

The nonlinear polarization $P_{NL}(\omega_s)$ induced at ω_s then defines $\chi^{(3)}(\omega_s)$ follows:

$$\vec{P}_{NL}(\omega_s) = \epsilon_0 \chi^{(3)}(\omega_s) \vec{E}(\omega_s)$$

$$= \epsilon_0 N \left(\frac{\partial \alpha}{\partial q}\right)_0 Q^* A_L e^{i[(k_L - k_v)z - (\omega_L - \omega_v)t]} \tag{5.171}$$

$$\chi^{(3)}(\omega_s) = \frac{N^2 \left(\frac{\partial \alpha}{\partial q}\right)_0^2}{4m\omega_0} \frac{\left((\omega_0 - \omega_v) - \frac{i}{T_2}\right)}{((\omega_0 - \omega_v)^2 + (1/T_2)^2)} \tag{5.172}$$

where we have substituted the expression for the vibrational amplitude Q. This is essentially the Raman response function graphed in Figure 4.5(b).

A proper treatment that accounts for both Stokes and anti-Stokes scattering (depicted pictorially in Figure 5.26) must include all three optical waves (the pump, Stokes, and anti-Stokes) in the expression for the total optical field. The reader is strongly encouraged to carry this analysis through by setting $E = A_L e^{i(k_L z - \omega_L t)} + A_s e^{i(k_s z - \omega_s t)} + A_{as} e^{i(k_{as} z - \omega_{as} t)} + (*)$ and repeating the derivation outlined above for the Stokes wave generation. It should be immediately evident the molecular vibration will be driven by a term $A_L^* A_{as}$ that is the anti-Stokes analog of the Stokes wave interaction. Carrying this analysis for the term through to the determination of the anti-Stokes Raman susceptibility $\chi^{(3)}(\omega_{as})$, one finds that the anti-Stokes wave is attenuated rather than amplified. Here $P^{NL}(\omega_{as}) \propto |A_L|^2 A_s$. There exists, however, another source of nonlinear polarization at ω_{as}. This term arises from the combination $P^{NL}(\omega_s) \propto A_L A_L A_s^* e^{i(2\omega_2 - \omega_1)t}$, which does not directly involve the anti-Stokes wave at all! Moreover, the wavevector conservation expression $\vec{k}_{as} = 2\vec{k}_L - \vec{k}_s$ does not involve the material dispersion at all, but instead, in an isotropic medium, the magnitudes of $\vec{k}_i$ are determined by their respective frequencies $|\vec{k}_i| = n_i(\omega_i)\omega_i/c, i = L, s$. For this reason anti-Stokes radiation is emitted in the form of a conical shell about the laser direction in experiments when one takes account of the finite divergence of the laser beam (recall the shape of the TEM$_{00}$ solution).

Finally, the real-life situation is often much more complicated! For example, if a sufficiently intense Stokes wave is generated, it can then act as a secondary pump wave and generate a higher-order Stokes-shifted wave at frequency $\omega_s^{(2)} = \omega_L - 2\omega_v$, and so on. A whole hierarchy of higher-order Stokes/anti-Stokes sidebands can be generated in this fashion. The reader who wishes to pursue this subject further will find a comprehensive review of the subject in reference [72]. More recent developments, particularly in optical fibers and counter-propagating beams, are appearing frequently in the nonlinear optics research literature.

Stimulated Raman Scattering with Small Damping

We now want to show that the problem of the excitation of a Stokes wave by a pump wave scattering off a natural vibration in the medium is exactly the same as the problem of superfluorescence discussed in Section 5d.

Consider (5.166–5.168) in the on-resonance limit ($\omega = \omega_0$), and make the change of variables

$$A_L = \sqrt{\omega_L} A_1, \quad A_s = \sqrt{\omega_s} A_2, \quad Q = i\sqrt{\frac{cn}{4\pi \omega_0 \pi}} X$$

$$\tau = N\left(\frac{\partial \alpha}{\partial q}\right)_0 \sqrt{\frac{\pi \omega_L \omega_s}{4m\omega_0 cn}} \left(t - \frac{n}{c}\right), \quad x = N\left(\frac{\partial \alpha}{\partial q}\right)_0 \sqrt{\frac{\pi \omega_L \omega_s}{4\pi \omega_0 cn}} z,$$

$$(5.173)$$

where we assume $n(\omega_L) = n(\omega_s) = n$ so that the group velocities of pump and Stokes waves are the same. We also neglect wave damping is, we set $\alpha_2 = \alpha_3 = 0$. We obtain

$$\frac{\partial A_1}{\partial z} = -A_2 X$$

$$\frac{\partial A_2}{\partial z} = A_1 X^* \qquad (5.174)$$

$$\frac{\partial X}{\partial \tau} + \gamma X = A_1 A_2^*$$

where

$$\gamma = \frac{1}{T_2}\sqrt{\frac{4m\omega_0 cn}{\pi \omega_L \omega_s}} \left(N\left(\frac{\partial \alpha}{\partial q}\right)_0\right)^{-1}.$$

The initial boundary value problem for (5.174) is defined by giving A_1, A_2 and functions of τ at $z = 0$ and X as a function of z at some time t (equals zero say) when the front of the pump wave first makes contact with the medium (see Figure 5.27).

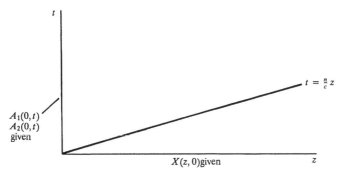

FIGURE 5.27

The Large Damping Limit

Observe that if $A_1(0, t)$ and $A_2(0, t)$ vary slowly with t so that $A_{j\tau} \ll \gamma A_j$, $j = 1, 2$ then we can adiabatically eliminate X from (5.174) by ignoring $\partial X / \partial \tau$ to obtain

$$X = \frac{1}{\gamma} A_1 A_2^*$$

$$\frac{\partial A_1}{\partial z} = -\frac{1}{\gamma} A_2 A_2^* A_1, \quad \frac{\partial A_2}{\partial z} = \frac{1}{\gamma} A_1 A_1^* A_2, \tag{5.175}$$

from which we find that

$$A_1 A_1^* + A_2 A_2^* = E(\tau)$$

and

$$\frac{\partial A_2 A_2^*}{\partial z} = \frac{1}{\gamma} A_2 A_2^* (E - A_2 A_2^*).$$

Any energy initially in the pump wave (imagine that all the energy was initially there so that $E = A_1 A_1^* (z = 0)$) goes into the Stokes wave in a distance of order $\gamma^{-1} c/n$.

The Small Damping Limit

We now look at the case for which we cannot adiabatically eliminate the material vibration. It proves to be convenient to set

$$n = A_1 A_1^* - A_2 A_2^*$$

$$\lambda = 2A_1 A_2^*,$$

whence (5.174) becomes

$$\frac{\partial e}{\partial \tau} + \gamma e = \lambda$$

$$\frac{\partial \lambda}{\partial z} = en \qquad\qquad (5.176)$$

$$\frac{\partial n}{\partial z} = -\frac{1}{2}(e\lambda^* + e^*\lambda).$$

These are exactly the equations for propagation of an electromagnetic wave envelope e in an excitable medium with polarization λ and population inversion n. The roles of the two independent variables τ (retarded time) and z (distance from the beginning of the medium) are switched. We list the correspondence below.

Symbol	Excitable Medium	Stimulated Raman Scattering
A_1	Amplitude of upper level state	Amplitude of pump wave
A_2	Amplitude of lower level state	Amplitude of Stokes wave
n	Population inversion	Difference in energies between the pump and Stokes wave
λ	Polarization	Quadratic product $2A_1 A_2^*$
e	Electric field envelope	Amplitude of optical mode of oscillation of the material
γ	Losses in the electric field due to finite conductivity	Decay rate of optical mode vibration in the material

Remark. In the Raman context, we choose not to normalize $A_1 A_1^* - A_2 A_2^*$ by dividing out the sum of the energies $E(\tau) = A_1 A_1^* + A_2 A_2^*$ contained in the pump and Stokes fields.

EXERCISE 5.21

Show that for stimulated Raman scattering,

$$\frac{\partial}{\partial z} E = 0$$

and that

$$n^2 + \lambda\lambda^* = E^2(\tau).$$

□

Remark. In deriving (5.174) from (5.166–5.168) we ignored damping. If we include these terms, we can see that they correspond to homogeneous broadening effects in the excitable medium.

EXERCISE 5.22

"Inhomogeneously broadened" stimulated Raman scattering. Imagine that there is a continuum of material states with which the pump and Stokes wave can almost resonate. In this case, we can treat $\omega - \omega_0$ as a nonzero random variable, and the equations for e, n, and λ would resemble closely the equations (5.133) for an inhomogeneously broadened excitable medium. The reader is invited to explore the consequences of this observation. □

We now address the question: what happens when a pump wave impinges on a medium in which stimulated Raman scattering can occur? How is the pump wave energy converted into that of the Stokes wave in the case where the damping γ of the material mode can be ignored? For simplicity, we will assume that all fields A_1, A_2, and X are real so that (5.176) can be written

$$e_\tau = \lambda, \quad \lambda_z = en, \quad n_z = -e\lambda. \tag{5.177}$$

By analogy with our previous work on the excitable medium in the sharp line limit, we set

$$e = u_z, \quad \lambda = -\sin u, \quad n = -\cos u \tag{5.178}$$

and obtain

$$u_{z\tau} = -\sin u, \tag{5.179}$$

the sine-Gordon equation. The relevant initial-boundary value problem can be posed as shown in Figure 5.28. At the beginning of the medium, there is only the pump wave, and we will take $E(\tau)$ to be unity for $\tau > 0$ (and zero for $\tau < 0$). The initial state of the material vibration is given as noise; that is, ϵ is small and $\rho(z)$ is a stationary random function of z of zero mean. Again, the relevant solution of (5.179) to which the actual solution tends is the self-similar solution $u = f(\eta = 2(z\tau)^{1/2})$ satisfying

$$f_{\eta\eta} + \frac{1}{\eta} f_\eta = -\sin f. \tag{5.180}$$

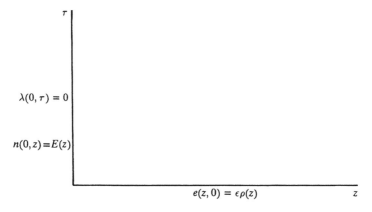

FIGURE 5.28

As in the superfluorescence case, we seek solutions $f(\eta)$ that are bounded near $\eta = 0$ and close to π, that is, near $\eta = 0$, $f = \pi + AI_0(\eta)$. Again A is related to the average strength of the noise $\langle \rho^2(z) \rangle$. For fixed z, the material vibration $e(z, \tau)$ and the energy difference $n(z, \tau)$ take the form shown in Figure 5.24.

Summary

When the material damping is large, the pump wave energy is monotonically transferred from the pump wave to the Stokes wave (i.e., $A_2 A_2^*$ increases and $A_1 A_1^*$ decreases monotonically until all the energy is in the Stokes wave). The material vibration plays the role of catalyst. When the material damping is small, there is a large transfer of energy into both the material vibrations and the Stokes wave. The medium continues the ring for a long time, and at a fixed location, the material vibration is seen to oscillate. As time goes on, the oscillations decrease slowly in amplitude (recall $f(\eta)$ behaves like $\eta^{-1/2} \cos \eta$ for large η), and eventually all the energy resides in the Stokes wave.

5f. Stimulated Brillouin Scattering

An intense propagating optical wave can couple, through the physical mechanism of electrostriction discussed in Section 4h, to background acoustic fluctuations and thereby generate a scattered optical wave whose frequency is downshifted by the acoustic frequency from the incident wave. This stimulated scattering phenomenon is called stimulated Brillouin scattering, and we see in Chapter 6, from the dispersion relation derived for acoustic modes in a simple diatomic lattice, that the scattering tends to occur predominantly in the backward direction. We shall now derive the equations describing this three-wave interaction process involving two electromagnetic waves and a single density (acoustic) wave. The acoustic wave equation (4.107) already shows how the electrostrictive force can resonantly excite acoustic waves when the total optical field contains two waves whose frequency difference $(\omega_1 - \omega_2)$ lies near the acoustic wave frequency $\omega_a = k_a v$. The acoustic wave frequency ω_a is many orders of magnitude less than the optical wave frequencies ω_1 and ω_2, as can be seen from the following estimate. Consider an incident pump wave

$$\vec{E}_1 = \hat{e} A_1 e^{i(k_1 z - \omega_1 t)} + (*)$$

propagating in the positive z direction and a backscattered optical wave

$$\vec{E}_2 = \hat{e} A_2 e^{-i(k_2 z + \omega_2 t)} + (*)$$

propagating in the negative z-direction. The resonance triad relation (6.152) shows that $k_a \simeq 2k_1$.

For a neodymium laser of wavelength $\lambda = 1.06~\mu$m incident on a medium of refractive index $n = 1.5$, we obtain $k_1 = 0.89 \times 10^7$ m^{-1} and at a typical sound velocity of approximately 10^3 m/s, we obtain $\omega_a = 1.8 \times 10^{10}$ rad/s. This acoustic wave frequency ($f_a = \omega_a/2\pi = 2.9 \times 10^9$ Hz), although many orders of magnitude less than the optical wave frequency $f_1 = 2.8 \times 10^{14}$ Hz, is referred to as hypersound and is characterized by an extremely short acoustic phonon lifetime. The phonon damping constant $\Gamma_B \simeq 10^9$ s^{-1} in many materials, so that the hypersound intensity damping is very high: $\alpha_s = (2\Gamma/v) \simeq 10^6$ to 10^5 m^{-1}. In other words, the hypersound excited by the incident optical wave E_1 has time to cover only an extremely short distance, $\sim 10^{-5}$ m from the excitation location, before being damped out. Substituting

$$\Delta \rho = \tilde{\rho} e^{i(k_a z - \omega_a t)} + (*)$$

and the above expressions for $\vec{E}_1$ and $\vec{E}_2$ into the acoustic wave equation and using $k_a = k_1 \pm k_2$, $\omega_a = \omega_1 - \omega_2$, we obtain an equation for the envelope of the acoustic wave $\tilde{\rho}$:

$$\frac{\partial \tilde{\rho}}{\partial t} + \Gamma_B \tilde{\rho} = i \frac{k_a}{v} \epsilon_0 n_0 \frac{dn}{d\rho} A_1 A_2^*. \tag{5.181}$$

In obtaining this equation we have invoked the slowly varying envelope approximation and used the above fact that the acoustic wave is damped out over very short distances. The induced nonlinear polarization for each optical wave $\vec{P}^{NL}(\omega_1) = 2\epsilon_0 n_0 \, (\partial n / \partial \rho) \, \Delta \rho \, \vec{E}(\omega_i)$ follows directly from the discussion in Section 4i. The coupled field–matter equations for stimulated Brillouin scattering follow by substituting the above expressions for $\vec{E}(\omega_i)$ and $\vec{P}^{NL}(\omega_i)$ in the Maxwell equations:

$$\left(\frac{\partial}{\partial z} + \frac{n_0(\omega_1)}{c} \frac{\partial}{\partial t} \right) A_1 = i \frac{\omega_1}{c} \left(\frac{dn}{d\rho} \right) \tilde{\rho} A_2 \tag{5.182}$$

$$\left(-\frac{\partial}{\partial z} + \frac{n(\omega_2)}{c} \frac{\partial}{\partial t} \right) A_2 = i \frac{\omega_2}{c} \left(\frac{dn}{d\rho} \right) \tilde{\rho}^* A_1 \tag{5.183}$$

$$\frac{\partial \tilde{\rho}}{\partial t} + \Gamma_B \tilde{\rho} = i \frac{k_a}{v} \epsilon_0 n_0 \left(\frac{dn}{d\rho} \right) A_1 A_2^*. \tag{5.184}$$

These equations describe transient stimulated Brillouin backward scattering and have the same general mathematical form as those for stimulated Raman scattering in the backward direction. The main physical difference between the two scattering events lies in the different magnitudes of the coupling parameters. In particular, the phonon damping time

$$\Gamma_B^{-1} \simeq 10^{-9} \text{ s}$$

for acoustic phonons contrasts with a value of

$$\Gamma^{-1} = T_2 \simeq 10^{-12} \text{ s}$$

for optical phonons. Moreover, the gain constant, which can be evaluated from the steady-state equation, as in the Raman case, by assuming a non-depl pump wave intensity ($|A|^2 =$ constant) and a very small Stokes wave

($A = 0$), turns out to be approximately three orders of magnitude larger than the stimulated Raman scattering. What this means is that incident optical pulse whose duration exceeds the phonon damping time (i.e., nanoseconds) under stimulated Brillouin scattering at significantly lower peak intensity thresholds. Of course, if the pump pulse duration is much less, say in the picosecond regime, there is no buildup of a Brillouin Stokes wave, and only stimulated Raman scattering is possible for sufficiently high intensities. We note that the general backward scattering problem, whether stimulated Raman or Brillouin is an initial boundary value problem with $A_1(0)$ and $A_2(L)$ specified at both ends of the scattering medium of length L. In fact, this problem is nontrivial as the Stokes wave A_2 grows from background noise with the large gain and resonance interaction selecting the frequency $\omega_2 = \omega_1 - \omega_a$ from a broad spectrum of noisy fluctuations. In solving this problem, one thinks of the backward wave $A_2(L)$ as a weak seed with a very small but well-defined amplitude and frequency ω_2 lying at the gain maximum. We can solve the boundary value problem exactly in steady state with $A_1(0)$ and $A_2(L)$ specified:

$$
\frac{d}{dz} A_1 = -g_B A_2 A_2^* A_1
$$

$$
\frac{d}{dz} A_2 = -g_B A_1 A_1^* A_2,
$$

(5.185)

where the linear gain coefficient

$$
g_B = \frac{\omega}{c} \left(\frac{dn}{d\rho} \right)^2 \frac{k_a \epsilon_0}{\Gamma_B v}
$$

and we have assumed $\omega = \omega_1 = \omega_2$ in g_B. These coupled o.d.e.'s can be solved exactly to yield (let $I_i = (n \epsilon_0 c / 2) |A_i|^2$)

$$
\frac{I_2(z)}{I_2(0)} = \frac{1 - I_2(0)/I_1(0)}{\exp[(I_1(0) - I_2(0))g_B z] - I_2(0)/I(0)}
$$

(5.186)

and

$$
I_2(0) = I_1(0) - I_1(z) + I_2(z).
$$

The solutions show that the difference of intensity of the pump and Stokes wave remains constant throughout the medium, at least in the lossless case considered here.

Chapter 6

MATHEMATICAL AND COMPUTATIONAL METHODS

Overview

THE READER IS ADVISED TO cover this material concurrently with that of the other chapters. Our goal here is to introduce the mathematical and computational methods that are important in the analysis of optics. Since we emphasized in Chapter 1 that a large part of present-day nonlinear optic can be considered as a weak coupling theory between the oscillators of light and matter, we first develop the ideas of perturbation theory and asymptotion expansions, then go on to the long-time behavior of solutions to linear wave equations. After that we introduce the method of multiple scales, the execution of which allows us to write down equations describing the long time and distance evolution of the complex amplitudes (measuring both energy and phase) of weakly coupled oscillators. We then go on to do the same thing for wavetrains and wavepackets and meet the nonlinear Schrödinger (NLS), coupled NLS, and three-wave (TWI) and four-wave (FWI) interaction equations. We cover some of the methods and ideas associated with bifurcation theory, solitons, and the inverse scattering transform, and we briefly discuss the notions of chaos and turbulence and the unpredictability associated with deterministic systems. Finally, we discuss computational algorithms.

6a. We introduce several motivating examples illustrating regular and singular perturbations.

6b. We meet several formal definitions of the big and little order relations, asymptotic sequences and series. Of particular importance to us is the notion of a *uniform asymptotic expansion.*

6c. We study the evolution of several types of solutions to linear wave equations, and we see how the notions of group velocity and dispersion emerge from an asymptotic (long time) analysis of these solutions using the method of steepest descent.

6d. We derive Snell's (Descartes') laws for the reflection and refraction of plane waves (wavetrains) at the interface between two dielectrics. This material is an important precursor to Chapter 3.

6e. The shapes of guided fields. We introduce the boundary value problem associated with electromagnetic fields that are confined by regions of elevated refractive index.

6f. We look at the spectrum of linear cavity modes important in lasers.

6g. Here we use the method of multiple time scales to find how the amplitudes of coupled oscillators evolve due to weak self and mutual interaction and to three- and four-wave resonant mixing processes. We also introduce the WKBJ method and show how it can be understood in terms of the multiple time scale method.

6h. The long time-distance evolution of wavepackets. The recurring theme in deriving these equations is as follows. We seek solutions to the electric field or other fields of interest as an asymptotic expansion

$$\vec{E}(x, y, z, t) = \epsilon \vec{E}_0(x, y, z, t) + \epsilon^2 \vec{E}_1(x, y, z, t) + \cdots, \quad 0 < \epsilon \ll 1,$$

Where $\vec{E}_0$ is a wavepacket (a wavetrain with slowly varying amplitude) or linear combinations of wavepackets. We choose the time or distance dependence of the slowly varying amplitudes in order to keep the asymptotic expansion above *uniformly* valid over space (x, y, z) and time t; that is, the successive ratios E_{n+1}/E_n remain bounded. The resulting amplitude equations are called asymptotic solvability conditions, and many of the interesting equations in both this book in particular and in mathematical physics in general are derived in exactly this way.

6i. Bifurcation theory and amplitude equations. Here we discuss systems like lasers that change from one state (nonlasing state) to another (the lasing state) as some external parameter is varied. We introduce the notion of amplitude or order parameter and show how one can understand the nature of the transition from the equations that the order parameter satisfies. We meet in this section both the Landau and complex Ginzburg-Landau equations. Some of the latter are new results.

6j. We derive the NLS equation for wavepackets in a planar waveguide. This calculation is a useful precursor to the derivation of the equation describing the propagation of solitons in optical fibers given in Chapter 3 because it has all the necessary ideas, but the computation is simpler as the electric field may be taken to be a scalar quantity.

6k. Solitons and the inverse scattering transform. We give a fairly comprehensive survey of the method used to investigate solutions of the NLS equation and show how to determine the effect of various perturbations that are important in the optical context.

6l. Chaos and turbulence. We give a brief overview of the origins of complicated and unpredictable behavior that solutions of certain deterministic systems (like the Maxwell–Bloch equations for a laser) can display.

6m. We discuss the numerical and graphical algorithms most commonly encountered in dealing with problems in nonlinear optics. We include specifically the coding of the NLS equation and strongly encourage the reader to do some computations. We also include the names of graphical package available on standard systems.

6a. Perturbation Theory

The presence of a small parameter in an equation often can be exploited to gain simple approximations to and information about solutions. But we must be careful, because it is not always the case that small events and disturbance have small effects and consequences. The goal of perturbation theory is to study the effects of small disturbances [13, 14, 82, 83]. If the effects are small, the perturbation is said to be *regular*; if not, it is said to be *singular*.

For example, let us compute the roots of the polynomial $x^2 - 2x + \epsilon$ by solving

$$x^2 - 2x + \epsilon = 0, \quad 0 < \epsilon \ll 1. \tag{6.1}$$

The exact solutions $x_{1,2}(\epsilon)$ can be written

$$x_{1,2}(\epsilon) = 1 \pm \sqrt{1 - \epsilon}, \tag{6.2}$$

which when expanded in a formal series is

$$x_{1,2}(\epsilon) = 1 \pm \left(1 - \frac{1}{2}\epsilon - \frac{1}{8}\epsilon^2 \cdots \right). \tag{6.3}$$

We call the infinite series *formal* because we have not yet analyzed for which values of ϵ, if any, it converges. (It converges for $-1 < \epsilon < 1$!) We can recover this formal series by looking for approximate solutions to (6.1) in the form (we will write $\simeq$ instead of $=$ to remind us that we have to say in which sense $x(\epsilon)$ is well represented by the infinite series)

$$x \simeq x_0 + \epsilon x_1 + \epsilon^2 x_2 + \cdots, \tag{6.4}$$

substituting (6.4) into (6.1), and identifying the coefficients of powers of ϵ. Do it. It works. So far, so good. Let us try another. Substitute (6.4) into

$$\epsilon x^3 - x + 1 = 0, \quad 0 < \epsilon \ll 1 \tag{6.5}$$

and find

$$x \simeq 1 + \epsilon + 3\epsilon^2 + \cdots$$

which does appear to be a close approximation to the root near one. But (6.5) is cubic, and a graph of $f(x, \epsilon) = \epsilon x^3 - x + 1$ will very quickly reveal that it has three real roots, not one. Where are the others? The first thing we must realize is that adding ϵx^3 to $-x + 1$ is only a small perturbation when we think of x as order one. But when x itself becomes large, ϵx^3 can be every bit as big as x itself and a lot bigger than 1. How can we get $\epsilon x^3 - x + 1$ to be zero? The first way we have already found. Treat x as order one, and then a balance is achieved between the second and third terms. How else might a balance be achieved? Let $x = y\epsilon^{-p}$. Then $f(x, \epsilon) = \epsilon^{1-3p} y^3 - \epsilon^{-p} y + 1 = 0$. We can balance the first and second terms by the choice $p = 1/2$, whence (6.5) becomes

$$\epsilon^{\frac{1}{2}} f(y\epsilon^{\frac{1}{2}}, \epsilon) = y^3 - y + \epsilon^{\frac{1}{2}} = 0. \tag{6.6}$$

Now let us look for a formal series

$$y \simeq y_0 + \epsilon^{\frac{1}{2}} y_1 + \cdots \tag{6.7}$$

and, bingo, we recover approximations for the three roots. Do this and use a Newton's method to check your approximate solutions for $\epsilon = 0.01, \epsilon = 0.04$.

EXERCISE 6.1

Find approximations for the solutions of $\epsilon x^3 - (x - 1)(x - 2) = 0$ and $\epsilon x^3 - (x - 1)^2 = 0$. The second one might be a bit challenging. As a hint, draw the graphs of ϵx^3 and $(x - 1)^2$.   □

Similar behavior occurs if we consider the ordinary differential equation (o.d.e.)

$$\epsilon y'' - y = 0, \quad y(0) = A, \quad y(1) = 0. \tag{6.8}$$

The trouble about approximating the solution $y(x, \epsilon)$ by $y = 0$ (set $y = y_0 + \epsilon y_1 + \cdots$ into (6.8) and find $y_0 = y_1 = \cdots = 0$) is that the approximation fails to take into account that in order for the exact solution

$$y(x, \epsilon) = \frac{A}{1 - e^{-2/\sqrt{\epsilon}}} \left(e^{-x/\sqrt{\epsilon}} - e^{-(2-x/\sqrt{\epsilon})} \right) \tag{6.9}$$

to reach the left boundary value $y = A$, it has to make a sharp transition in a layer of width $\sqrt{\epsilon}$ (called a boundary layer) very close to the left boundary. Indeed for small ϵ, we can neglect both the exponentially small terms $e^{-(2/\sqrt{\epsilon})}$ and $e^{-(2-x)/(\sqrt{\epsilon})}$ in the whole interval $0 \le x \le 1$ because they are smaller than any power of ϵ.

EXERCISE 6.2

Prove $\lim_{\mu \to 0} \mu^{-N} e^{-(1/\mu)} = 0$ for all $N > 0$. $\square$

Then $y(x, \epsilon)$ can be written as

$$y(x, \epsilon) = Ae^{-x/\sqrt{\epsilon}} + \text{exponentially small terms}. \tag{6.10}$$

For $x \gg \sqrt{\epsilon}$, say if $\epsilon = 10^{-4}$ and $x = 10^{-1}$, then $y(x, 0)$ is practically zero but y changes very rapidly in the layer $0 \le x \le \sqrt{\epsilon}$. The graph of $y(x, \epsilon)$ is shown in Figure 6.1.

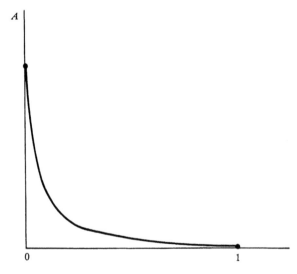

FIGURE 6.1

The singular behavior, namely the fact that $y(x, \epsilon)$ is not represented uniformly throughout the whole interval $0 \leq x \leq 1$ by $y = 0$, is brought about by the fact that near $x = 0$, the derivative y' of y is of order $1/\sqrt{\epsilon}$ and the second derivative is of order $1/\epsilon$ so that $\epsilon y''$ is also of order one and just as important as the term $-y$ in (6.8).

6b. Asymptotic Sequences, Expansions

We begin this section with the notion of ordering (see references [84–86]). There is no loss of simplicity in working with functions $\phi(z)$ of complex variables, so we will do that. In most applications of the ideas we will meet, z will be called ϵ or ϵ^{-1} and z will be zero or infinity. We say

$$\phi(z) = O(\psi(z)) \quad \text{as} \quad z \to z_0 \tag{6.11}$$

if $|\phi(z)| \leq \kappa |\phi(z)|$ for all z in $|z - z_0| < \delta$, where κ is a constant. The symbol simply means that near $z = z_0$, $\phi(z)$ is the same size as or smaller than $\psi(z)$. If $\psi(z)$ does not vanish in $0 < |z - z_0| < \delta|z - z_0| < \delta$, the ratio of $|\phi|$ to $|\psi|$ is bounded. For examples

$$\sin z = O(z) \quad \text{as} \quad z \to 0$$

$$\cos z = O(1) \quad \text{as} \quad z \to 0.$$

We say

$$\phi(z) = o(\psi(z)) \quad \text{as} \quad z \to z_0$$

if $|\phi(z)| < \mu |\psi(z)|$ for arbitrarily small, positive μ for any z in $0 < |z - z_0| < \delta$. We mean by this that near $z = z_0$, ϕ is much smaller than ψ. For examples,

$$e^{-1/\epsilon} = o(\epsilon^N) \qquad \text{as} \quad \epsilon \to 0$$

$$\epsilon = o(\log |\epsilon|) \qquad \text{as} \quad \epsilon \to 0$$

$$1 = o(\log |\epsilon|) \qquad \text{as} \quad \epsilon \to 0$$

$$\log t = o(t) \qquad \text{as} \quad t \to \infty.$$

From these definitions, it is clear that $\phi(z) = o(\psi(z))$ as $z \to z_0$ implies that $\phi(z) = O(\psi(z))$, $z \to z_0$. The converse need not be true, however. The big O relationship can often be proved by showing that the limit of the ratio $\phi(z)/\psi(z)$ has a finite value κ_0 as $z \to z_0$, although all one needs to show is that the ratio is bounded. For example, $\sin(1/x) = O(1)$ as $x \to 0$ through real values, although there is no limit. The little o relationship is proved by showing that in the limit $z \to z_0$, $(|\phi(z)|)/(|\psi(z)|) \to 0$.

The functions $\phi(z, x)$, $\psi(z, x)$ may depend on another variable x (in our applications x can be space or time) belonging to some domain D, and it is often important that the order relations (big O and little o) are uniform with respect to x in D. For example, for $0 < t < \infty$, it is not true that $\epsilon t = O(1)$ as $\epsilon \to 0$ because as soon as t is bigger than ϵ^{-1}, ϵt is no longer small compared to unity. We will meet this notion of uniformity again and again.

We say the sequence $\{\phi_n(z)\}_{n-1}^{\infty}$ is an *asymptotic sequence* if

$$\phi_{n+1}(z) = o(\phi_n(z)) \quad \text{for all } n, \quad \text{as} \quad z \to z_0. \tag{6.12}$$

We say the sequence $\{\phi_n(z, x)\}_{n-1}^{\infty}$ is a *uniform asymptotic sequence* with respect to x in D if

$$\phi_{n+1}(z, x) = o(\phi_n(z, x)) \quad \text{as } z \to z_0 \quad \text{for all } n \text{ and all } x \in D. \tag{6.13}$$

For examples, the sequence

$$1, \epsilon, \epsilon^2 \ln|\epsilon|, \epsilon^2, \epsilon^3, \ldots$$

is an asymptotic sequence as $\epsilon \to 0$. The sequence

$$1, \epsilon, \epsilon^2 t, \epsilon^3, \ldots$$

is not a uniform asymptotic sequence for all t. The sequence

$$1, \epsilon, \epsilon^2 \sin t, \epsilon^3, \ldots$$

is a uniform asymptotic sequence for all t. The fact that $\epsilon^2 \sin t$ is smaller than ϵ^3, ϵ^4 at a countable number of intervals in t which tend to zero as $\epsilon \to 0$ doesn't matter.

We say that the formal series $\sum_{n=1}^{\infty} a_n \phi_n(z)$ is an *asymptotic expansion* for $f(z)$ as $z \to z_0$, if, as $z \to z_0$

$$f(z) = \sum_{1}^{N} a_n \phi_n(z) + o(\phi_n(z)) \quad \text{for all } N. \tag{6.14}$$

We write $f(z) \simeq \sum_{1}^{\infty} a_n \phi_n(z)$ because at this point we know nothing about the convergence of the right-hand side. Once the sequence $\{\phi_n(z)\}$ is chosen, the coefficients $\{a_n\}$ are uniquely determined:

$$a_n = \lim_{z \to z_0} \frac{f(z) - \sum_{1}^{n-1} a_k \phi_k}{\phi_n}. \tag{6.15}$$

However, two different functions may have the same asymptotic expansion as $\epsilon \to 0$. For example, if $f(\epsilon) \simeq \sum a_n \phi_n(\epsilon)$ as $\epsilon \to 0$ and the $\phi_n(\epsilon)$ are powers of ϵ, then $g(\epsilon) = f(\epsilon) + e^{-(1/\epsilon)} \simeq \sum a_n \phi_n(\epsilon)$ because $e^{-(1/\epsilon)} \simeq 0$ as $\epsilon \to 0$.

The important thing to understand is that asymptotic expansions provide useful approximations, even though the expansion if carried out all the way to infinity is divergent. The reason is that we are really writing

$$f(z) = \sum_1^N a_n \phi_n(z) + R_N(z), \tag{6.16}$$

and as $z \to z_0$ we are approximating $f(z)$ by $\sum_1^N a_n \phi_n(z)$ by bounding $R_N(z)$ by some power of $|z - z_0|$. If the formal series $\sum a_n \phi_n$ is an asymptotic series for $f(z)$ as $z \to z_0$, then $R_N(z)$ is much smaller than the last term $\phi_N(z)$. In other words, we can approximate $f(z)$ to within a finite distance of its actual value. Convergence, on the other hand, guarantees us that for $|z - z_0| < \delta$, we can take N large enough that we can get arbitrarily close to the value of $f(z)$ at z. The difference is well illustrated by the following example. Consider, for $x > 0$,

$$F(x) = \sqrt{\pi}\, Erfcx = 2 \int_x^\infty e^{-t^2}\, dt = \int_{x^2}^\infty \frac{e^{-u}}{u^{\frac{1}{2}}}\, du. \tag{6.17}$$

Write $\int_{x^2}^\infty e^{-u} u^{-(1/2)}\, du$ as $-\int_{x^2}^\infty de^{-u} u^{-(1/2)} = (e^{-x^2})/x - 1/2 \int_{x^2}^\infty e^{-u} u^{-(3/2)}\, du$ and continue, so that formally,

$$F(x) \simeq \frac{e^{-x^2}}{x} \sum_0^\infty (-1)^n \frac{(2n-1)!!}{(2x^2)n} \tag{6.18}$$

which converges only at $x = \infty$. On the other hand, if we define $S_N = \sum_0^N (-1)^n ((2n-1)!!/(2x^2)^n)$, then

$$x e^{x^2} \sqrt{\pi}\, Erfcx - S_N = x e^{x^2} (-1)^{N+1} \frac{1 \cdot 3 \cdot 5 \cdots (2N+1)}{2^{N+1}}$$

$$\times \int_{x^2}^\infty e^{-u} u^{-N-(3/2)}\, du \tag{6.19}$$

$$= O\left(\frac{1}{2x^2}\right)^{N+1}$$

so that if $\phi_n(x) = (2x^2)^{-n}$, then $|x\, e^{x^2} F(x) - S_N| = O(\phi_{N+1}(x))$ (and therefore $o(\phi_N(x))$), and (6.18) is an asymptotic series. The values for $x = 2$ are $S_0 = 1$, $S_1 = 0.875$, $S_2 = 0.922$, $S_3 = 0.893$, $S_4 = 0.919$, $S_5 = 0.891, \ldots$ and thereafter the oscillations cause S_N to oscillate in an increasingly wild fashion. Nevertheless, the answer is 0.904, which is well approximated by S_4. Moreover, there are rules for telling us at which term in the asymptotic expansion to stop in order to achieve the best approximation.

So, asymptotic expansions are useful approximations even though the expansion when carried out to infinity may be divergent. In mathematical physics we are often interested in obtaining leading-order approximations, namely the first few terms in the expansion. However, in almost all cases, the asymptotic expansion in an asymptotic sequence $\phi_n(\epsilon)$ has coefficients that depend on space x and time t variables, and these variables can range over large sets of values. Our goal is to make sure that the asymptotic expansion for the solution $f(x, t, \epsilon)$ is uniformly valid for all x and t in their respective domains. We therefore introduce the notion of a uniform asymptotic expansion. We say $\sum a_n(x)\phi_n(x, \epsilon)$ is a uniform asymptotic expansion for $f(x, \epsilon)$ as $\epsilon \to 0$ for all $x \in D$ provided

$$\left| f(x, \epsilon) - \sum_{1}^{N} a_n(x)\phi_n(x, \epsilon) \right| = o(\phi_N(x, \epsilon)) \tag{6.20}$$

for all $x \in D$ as $\epsilon \to 0$.

Let us illustrate the problem with an example. We wish to calculate small amplitude solutions for the sample pendulum (g gravity, l pendulum length)

$$\frac{d^2\theta}{dt^2} + \omega^2 \sin\theta = 0, \quad \omega^2 = \frac{g}{l}, \quad \theta(0) = \epsilon, \quad \frac{d\theta}{dt}(0) = 0. \tag{6.21}$$

Let $\theta(t, \epsilon) = \epsilon\, x(t, \epsilon)$ and expand $\sin\theta$ in a Taylor series:

$$\frac{d^2x}{dt^2} + \omega^2 x = \epsilon^2 \frac{\omega^2}{6} x^3 + \text{higher order terms.} \tag{6.22}$$

Let

$$x(t, \epsilon) = x_0(t, \epsilon) + \epsilon x_1(t, \epsilon) + \epsilon^2 x_2(t, \epsilon) + \cdots \tag{6.23}$$

Substitute (6.23) into (6.22), identify powers of ϵ, solve and find ((*) is complex conjugate)

$$x_0(t) = ae^{i\omega t} + (*)$$

$$x_1(t) = 0 \tag{6.24}$$

$$x_2(t) = \frac{-a^3}{48}e^{3i\omega t} - \frac{i}{4}\omega t a^2 a^* e^{i\omega t} + (*),$$

where the complex coefficient a is determined by the initial conditions (it can and does depend on ϵ). The reason that x has a term proportional to t is that the equation

$$\frac{d^2 x_2}{dt^2} + \omega^2 x_2 = \frac{\omega^2}{6}a^3 e^{3i\omega t} + \frac{\omega^2}{2}a^2 a^* e^{i\omega t} + (*),$$

satisfied by $x_2(t)$, has forcing terms that are periodic with the same period as the solutions of the homogeneous equation. Such terms, $(\omega^2/2)a^2 a^* e^{i\omega t} + (\omega^2/2)a\, a^{*2} e^{-i\omega t}$, are called *secular*. Now look at (6.23). Because $x_2(t)$ is proportional to t, the asymptotic expansion (6.23) is not uniform for times $\epsilon^2 t$ of order one because then the third term is as big as the first. Moroever, if we were to continue the computation we would find x_4 would contain a term proportioned to $t^2 e^{i\omega t}$, x_6 would contain $t^3 e^{i\omega t}$, and so on.

What is "wrong" is that we assumed that for long times the weakly nonlinear pendulum would continue to have the basic frequency ω and period $2\pi/\omega$. As we will see, it does not. How do we correct this mistake and arrange things so that the asymptotic expansion (6.23) remains uniformly valid as $\epsilon \to 0$ for all time? We will learn about the method (the multiple scale method) to do this in Section 6g, and we will find there that in order to render (6.23) uniformly valid we must allow the complex amplitude coefficient $a(\epsilon)$, up to now a constant, to become a slowly varying function of time. We will choose its dependence on time

$$\frac{da}{dt} = \epsilon^2 f_1(a, a^*) + \epsilon^4 f_2(a, a^*) + \cdots, \tag{6.25}$$

namely, choose the $f_j(a, a^*)$, $j = 1, 2, \ldots$ in such a manner as to eliminate all secular terms such as $(\epsilon^2 t)^n$, $n = 1, 2, \ldots$, $(\epsilon^4 t)^n$, $n = 1, 2, \ldots$ from the formal expansion (6.23).

As we emphasize in Chapter 2, many of the canonical equations of mathematical physics are derived in precisely this way. They are often called *asymptotic solvability conditions*.

6c. The Propagation of Linear Dispersive Waves

In this section, we are interested in exploring the long-time behavior of solutions of dispersive wave equations (see also references [84, 85]). We will use two models:

(i) $$u_{tt} - c^2 u_{xx} + \omega_0^2 u = 0, \ c^2, \ \omega_0^2 \text{constants} \tag{6.26}$$

(ii) $$v_t = i v_{xx}. \tag{6.27}$$

One can think of the first equation as the manifestation of a mechanical model (first proposed by Alwyn Scott) to simulate the sine–ordon equation

$$u_{tt} - c^2 u_{xx} + \omega_0^2 \sin u = 0. \tag{6.28}$$

Suppose a series of pendula of length l are strung together on a torsion wire whose torque is proportional to the local twist. Then the equation (rate of change of angular momentum equals torque) for the angle u_n between the nth pendulum and the vertical is $ml^2(d^2 u_n/dt^2)$ or

$$ml^2 \ddot{u}_n = k(u_{n+1} - u_n) - k(u_n - u_{n-1}) - mgl \sin u_n.$$

The continuum limit (introduce a field $u(x, t)$ such that $u(x = nh, t) = u_n(t)$, in which we approximate $u_{n+1} - 2u_n + u_{n-1}$ as $h^2 u_{xx}$) of this equation is (6.28), and for $u(x, t)$ small, one can replace $\sin u$ by u to obtain (6.26). In Section 5d, we made further use of this analogy.

We will solve each model as an initial-value problem on the whole line $-\infty < x < \infty$ and then the second one as an initial-boundary value problem on the half-line $x > 0$ where the boundary $x = 0$ is forced in a sinusoidal manner. The purpose of these exercises is to show how wave propagation is affected by dispersion, the property that waves with different frequencies travel at different speeds, and how the notions of group velocity as measured by $\omega'(k)$ and dispersion as measured by $\omega''(k)$ are important. We also introduce the method of steepest descents, important in determining the long-time behavior of solutions represented as integrals, which often arise in the form of a Fourier transform. In each case $\omega(k)$ is the dispersion relation; it relates the frequency of propagation ω to the wavenumber k and vice versa. In the context of optics, as we have mentioned, it is more convenient to view k as a function of ω, and we follow that convention in most of this book. In this chapter, however, we adopt the more traditional view and think of ω as a function of k.

The dispersion relations for (i) and (ii) are found by looking for wavetrain solutions proportional to $\exp(ikx - i\omega t)$:

(i)
$$\omega^2 = \omega_0^2 + c^2 k^2$$

(ii)
$$\omega = k^2.$$

First let us solve (i) and (ii) on $-\infty < x < \infty$ with initial conditions in which the field is a weakly modulated plane wave, a wavepacket centered about a dominant wavenumber k_0. Let

(i)
$$u(x, 0) = Rl e^{-\alpha^2 x^2 + i k_0 x}, \quad u_t(x, 0) = 0, \quad 0 < \alpha \ll 1$$

(ii)
$$v(x, 0) = e^{-\alpha^2 x^2 + i k_0 x}, \quad 0 < \alpha \ll 1 \qquad .$$

Let

$$\begin{Bmatrix} u(x, t) \\ v(x, t) \end{Bmatrix} = \int_{-\infty}^{\infty} A(k, t) e^{ikx} dk, \quad A(k, t) = \frac{1}{2\pi} \int_{-\infty}^{\infty} \begin{Bmatrix} u(x, t) \\ v(x, t) \end{Bmatrix} e^{-ikx} dx,$$

and find for case (i) that

$$A_{tt} + \omega^2 A = 0, \ \omega_0^2(k) = \omega_0^2 + c^2 k^2, \ A(k, t) = (1/2) A(k, 0)(e^{-i\omega t} + e^{i\omega t})$$

and for case (ii) that

$$A_t + i\omega A = 0, \quad \omega(k) = k^2, \quad A(k, t) = A(k, 0) e^{-i\omega t}.$$

In both cases $A(k, 0) = 1/(2\alpha \sqrt{\pi}) \exp(-(k - k_0)^2 / 4\alpha^2)$, which means that, in wavenumber space, the power is located in a band of width α about k_0. Then

$$u(x, t) = \frac{Rl}{4\alpha \sqrt{\pi}} \int_{-\infty}^{\infty} \exp\left(-\frac{(k - k_0)^2}{4\alpha^2} + ikx - i\omega(k)t\right) dk + (\omega \to -\omega)$$

$$(6.29a)$$

and

$$v(x, t) = \frac{1}{2\alpha \sqrt{\pi}} \int_{-\infty}^{\infty} \exp\left(-\frac{(k - k_0)^2}{4\alpha^2} + ikx - i\omega(k)t\right) dk \qquad (6.29b)$$

We will now analyze what happens for x and t of the orders of no larger than α^{-1} and α^{-2} respectively. It is clear that for $0 < \alpha \ll 1$, the integrands are dominated by the behavior near $k = k_0$. Set $k = k_0 + \alpha \kappa$ and obtain

$$v(x, t) = \frac{1}{4\sqrt{\pi}} e^{i(k_0 x - i\omega(k_0)t)}$$

$$\times \int_{-\infty}^{\infty} \exp -\frac{\kappa^2}{4}(1 + 2i\omega''(k_0)\alpha^2 t) + i\kappa\alpha(x - \omega'(k_0)t)d\kappa \quad (6.30)$$

where we have expanded $\omega(k)$ up to order α^2 in a Taylor series about $k = k_0$. For (6.29b), (6.30) is exact. For (6.29a), and in general, there are additional terms such as $\omega'''(k_0)\alpha^3 t$ and $\omega^{iv}(k_0)\alpha^4 t_0$, in the exponent of the integrand. Since we are limiting ourselves to times $\alpha^3 t \ll 1$, we can safely ignore them. Equation (6.30) can be evaluated exactly using

$$\int_{-\infty}^{\infty} e^{-\beta x^2} dx = \sqrt{\frac{\pi}{\beta}} \quad (6.31)$$

which is valid for complex β as long as $-(\pi/2) \le \text{Arg}\beta \le \pi/2$. We obtain

$$v(x, t) = e^{i(k_0 x - \omega(k_0)t)} \sqrt{\frac{1}{1 + 2i\alpha^2\omega''(k_0)t}} \exp \frac{-\alpha^2(x - \omega'(k_0)t)^2}{1 + 2i\alpha^2 t\omega''(k_0)}, \quad (6.32)$$

which shows that the envelope of the wavepacket moves with speed $\omega'(k_0)$, called the group or packet velocity, for times $\alpha^2 t$ less than one. When $\alpha^2 t$ becomes of order unity and larger, the solution loses the form of a carrier wave $e^{ik_0 x - i\omega(k_0)t}$ multiplied by the smooth envelope shape $e^{-\alpha^2(x-\omega'(k_0)t)^2}$. Rather the latter begins to behave as $\exp(i(x - \omega'(k_0)t)^2)/(\alpha^2 t\omega''(k_0))$, and the amplitude decays as $(\alpha^2\omega''t)^{1/2}$. In short, after a time of the order of the inverse of the square of the width of the original wavepacket, different waves have spread out so far that the wavepacket no longer exists.

Next, we turn to the more general initial value problem

(i) $u(x, 0) = f(x), \quad u_t(x, 0) = 0$

(ii) $v(x, 0) = g(x)$,

and if $F(k)$ and $G(k)$ are the Fourier transforms of $f(x)$ and $g(x)$ respectively, the solutions $u(x, t)$, $v(x, t)$ are

$$u(x, t) = \tfrac{1}{2} \int_{-\infty}^{\infty} F(k)e^{ikx - i\omega(k)t} dk + \tfrac{1}{2} \int_{-\infty}^{\infty} F(k)e^{ikx + i\omega(k)t} dk \quad (6.33a)$$

and

$$v(x, t) = \int_{-\infty}^{\infty} G(k) e^{ikx - i\omega(k)t} \, dk \qquad (6.33b)$$

respectively. We analyze the long-time behavior of (6.33b) for $t \gg 1$ and for x/t finite. We write (6.33b) as

$$I = \int_C G(k) e^{th(k)} \, dk, \qquad (6.34)$$

where in this case C is the contour to $-\infty$ along the real k axis (see Figure 6.2) and $h(k) = ik(x/t) - i\omega(k)$. The method we are about to discuss is called the method of *steepest descents* for a reason you will soon appreciate. We assume that both $G(k)$ and $h(k)$ are analytic in some neighborhood D of C.

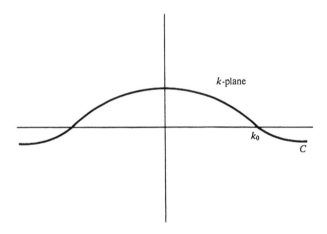

k-plane

k_0

C

FIGURE 6.2

Let $k = \xi + i\eta$, and then

$$h(k) = u(\xi, \eta) + iv(\xi, \eta) = 2\eta \left(\xi - \frac{x}{2t} \right) - i \left(\left(\xi - \frac{x}{2t} \right)^2 - \eta^2 + \frac{x^2}{4t^2} \right) \qquad (6.35)$$

for the case $\omega(k) = k^2$.* Let us plot the equal value contours of the function $u(\xi, \eta)$ in the complex k plane in Figure 6.3. Observe that u is negative for $\xi < x/2t, \eta > 0$ and $\xi > x/2t, \eta < 0$ and positive in the other quadrants. It

---

*We have run out of notation; our apologies. Please do not confuse the solutions $u(x, t)$, $v(x, t)$ of (6.29) with the real $u(\xi, \eta)$ and imaginary $v(\xi, \eta)$ parts of $h(k)$.

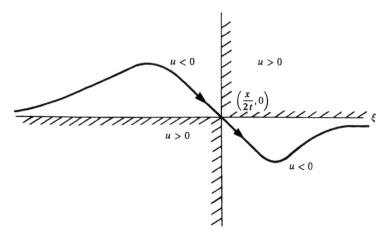

FIGURE 6.3   Path of steepest descent in the (ξ, η) plane.

has a critical point $u_\xi = u_\eta = 0$ at $\xi = x/2t, \eta = 0$, which is a saddle point. Also $u(\xi, \eta)$ has the steepest gradients in the directions $\theta = \pi/4, -3\pi/4$ (steepest ascent) and $\theta = -\pi/4, 3\pi/4$ (steepest descent). If we can arrange that C goes through $k_0 = x/2t$, the saddle point of $u(\xi, \eta)$ along C, the path of steepest descent, then the integral I will be dominated by its contribution from the neighborhood of $k = k_0$ because the modulus $e^{tu(\xi,\eta)}$ of $e^{th(k)}$ will be much bigger (exponentially bigger) there than at any other point on C. In fact we have, setting $k = k_0 + re^{i\phi}$ with ϕ chosen to be the path of steepest descent,

$$2\phi + \text{Arg}h''(k_0) = \pm\pi. \tag{6.36}$$

The choice of sign corresponds with the direction of the contour through k_0.

$$I \simeq \int_{-\infty}^{\infty} G(k_0) \exp\left(th(k_0) - \frac{t|h''(k_0)|}{2}r^2\right) dr$$

$$\times \exp i\left(-\frac{\pi}{2} - \tfrac{1}{2}\text{Arg}h''(k_0)\right) \tag{6.37}$$

$$= G(k_0) \exp\left(th(k_0) + i\left(-\frac{\pi}{2} - \tfrac{1}{2}\text{Arg}h''(k_0)\right)\right)\sqrt{\frac{2\pi}{|h''(k_0)|t}}.$$

The expression (6.37) is the leading term in an asymptotic expansion for I in inverse powers of $t^{1/2}$ obtained as follows.

Step 1. Split C into two paths, C_- which goes from $-\infty$ to k_0 and C_+ which goes from k_0 to ∞.

Step 2. Let $h(k_0) - h(k) = \lambda$ for λ real and positive. The inverse of this relation has two roots, one on C_+ and the other on C_-. In our case, we obtain $k - x/2t = \pm\sqrt{\lambda}e^{-i(\pi/4)}$, the local paths of steepest descent away from $k_0 = x/2t$.

Step 3. Write

$$I = \int_0^\infty G\left(\frac{x}{2t} + \sqrt{\lambda}e^{-(i\pi/4)}\right) e^{th(k_0)-\lambda t}\frac{1}{2\sqrt{\lambda}}d\lambda e^{-i(\pi/4)}$$
$$ \; {}^{C_+^0}$$

$$+ \int_0^\infty G\left(\frac{x}{2t} - \sqrt{\lambda}e^{-i\pi/4}\right) e^{th(k_0)-\lambda t}\left(-\frac{1}{2\sqrt{\lambda}}\right) d\lambda e^{-i(\pi/4)}$$
$$ \; {}^{C_-}$$

$$= \frac{1}{2}\int_0^\infty \left(G\left(\frac{x}{2t} + \sqrt{\lambda}e^{-i(\pi/4)}\right)\right.$$

$$\left. + G\left(\frac{x}{2t} - \sqrt{\lambda}e^{-i(\pi/4)}\right)\right) e^{th(k_0)-\lambda t}\frac{d\lambda}{\sqrt{\lambda}}e^{-i(\pi/4)}$$

Let $\lambda = (u/t)$, expand G in a Taylor series, and find $(\Gamma(z) = \int_0^\infty e^{-u}u^{z-1}du)$

$$I = e^{th(k_0)-i(\pi/4)}\int_0^\infty du \sum_{n \text{ even}} \frac{1}{n!}G^{(n)}\left(\frac{x}{2t}\right)\frac{1}{t^{\frac{(n+1)}{2}}}e^{-u}u^{\frac{(n-1)}{2}}$$

$$= e^{th(k_0)-i(\pi/4)} \sum_{n \text{ even}} \frac{1}{n!}G^{(n)}\left(\frac{x}{2t}\right)\frac{1}{t^{\frac{(n+1)}{2}}}\int_0^\infty e^{-u}u^{\frac{(n-1)}{2}}du \qquad (6.38)$$

$$= e^{th(k_0)-i(\pi/4)} \sum_{n \text{ even}} \frac{1}{n!}G^{(n)}\left(\frac{x}{2t}\right)\cdot\frac{1}{t^{\frac{(n+1)}{2}}}\Gamma\left(\frac{n+1}{2}\right).$$

Observe that the first term $n = 0$ in (6.38) is (6.37) because $h''(k_0) = 2i$, $\Gamma(1/2) = \sqrt{\pi}$.

In applying the method of steepest descent, the important points to make are these.

1. Since $h(k)$ is an analytic function of k for $k \in D$, its real and imaginary parts, $h(k) = u(\xi, \eta)+iv(\xi, \eta)$, satisfy the Cauchy-Riemann conditions

$\partial u/\partial \xi = \partial v/\partial \eta$, $\partial u/\partial \eta = -(\partial v/\partial \xi)$ and therefore are also harmonic functions, namely $(\partial^2/\partial \xi^2 + \partial^2/\partial \eta^2)(u, v) = 0$. Therefore $u(\xi, \eta)$ cannot have a maximum or a minimum in D. If it does have a critical point, that is, a point $k_0 = \xi_0 + i\eta_0$ where

$$h'(k_0) = \partial u/\partial \xi(\xi_0, \eta_0) + i(\partial v/\partial \xi)(\xi_0, \eta_0)$$

$$= -i(\partial u/\partial \eta)(\xi_0, \eta_0) + \partial v/\partial \eta(\xi_0, \eta_0) = 0,$$

this point must be a saddle point.

2. We must make sure that we can deform the original contour C into one that connects up with the paths of steepest descent from k_0 and does not go through any points $k = \xi + i\eta$ where $u(\xi, \eta) > u(\xi_0, \eta_0)$. This crucial point means that in applying the method one must have an idea of the global topography of $u(\xi, \eta)$, even though the actual calculation of the asymptotic expansion for I is local. We will illustrate the importance of this point shortly.

The direction for the path of steepest descent was found locally by choosing the argument of $k - k_0$ according to (6.36). There is, however, a more global result. Write the directional derivative of $u(\xi, \eta)$ in the direction given by $\hat{s} = (\cos \theta, \sin \theta)$:

$$\nabla u \cdot \hat{s} = \frac{\partial u}{\partial \xi} \cos \theta + \frac{\partial u}{\partial \eta} \sin \theta$$

and choose so that this gradient is maximal or minimal. We find

$$\frac{d}{d\theta} \nabla u \cdot \hat{s} = 0 = -\frac{\partial u}{\partial \xi} \sin \theta + \frac{\partial u}{\partial \eta} \cos \theta$$

which, by the Cauchy–Riemann conditions, gives

$$0 = -\frac{\partial v}{\partial \eta} \sin \theta - \frac{\partial v}{\partial \xi} \cos \theta = -\nabla v \cdot \hat{s}.$$

Therefore the path of steepest descent is the contour along which the imaginary part of $h(k)$ is constant, namely $v(\xi, \eta) = v(\xi_0, \eta_0)$. In our case, this curve is the degenerate conic

$$\left(\xi - \frac{x}{2t}\right)^2 = \eta^2,$$

which splits into a pair of two straight lines, the path of steepest descent $\xi - \frac{x}{2t} = -\eta$ and the path of steepest ascent $\xi - \frac{x}{2t} = \eta$.

Remark. The alert reader will argue that any path along which $u(\xi, \eta) < u(\xi_0, \eta_0)$ will give an approximation to the integral. That is true, but the steepest descent path gives the best approximation for the fewest number of terms.

We find then that for $t \gg 1$,

$$v(x,t) \simeq \sqrt{\frac{\pi}{t}} G\left(\frac{x}{2t}\right) \exp\left(i\frac{x^2}{4t} - i\frac{\pi}{4}\right), \tag{6.39}$$

which bears little resemblance to the initial profile $g(x)$ whose Fourier transform is $G(k)$. What it means and what is remarkable is that, at a fixed location x in the far field (remember x/t was finite), we see at that point x the power $G(x/2t)$ associated with the wavenumber $k = x/2t$; namely we see that wavenumber whose group velocity is such that the waves with this wavenumber would be carried to the position x in time t. Observe that the local wavenumber k, defined as the gradient $\partial\theta/\partial x$ of the phase $\theta = x^2/4t - \pi/4$, is precisely $x/2t$. Therefore an observer who sits at a point x in the far field will see pass by the different wave groups with amplitudes $G(x/2t)$, and as time progresses the observer finds the waves becoming longer and longer since the shorter waves have the larger group velocity. Because the packets have spread out, the local amplitude is reduced by $1/\sqrt{t}$ or $1/\sqrt{x}\sqrt{t/x}$. The time for dispersion to occur is clearly of the order of $L^2|\omega''(k_0)|^{-1}$, where L is a typical pulse width (proportional to α^{-1} in the earlier discussion). As we pointed out in Chapter 2, for light beams whose waist size is of the order of 1 mm, the dispersion time corresponds to a propagation distances of about a meter. In light fibers, which we discussed in Chapter 3, the light is confined in the transverse direction and only longitudinal dispersion is important. In that context, the dispersion is extremely small and, for pulses of picosecond widths, corresponds to a propagation distance of tens of kilometers.

EXERCISE 6.3

The following integral rises when we are considering the propagation of linear waves in inhomogeneously broadened media (Section 5d).

$$F(z,t) = \frac{1}{2\pi} \int_{-\infty}^{\infty} \hat{F}(\Omega) e^{z(ik(\Omega) - i\Omega t/z)} d\Omega$$

where $k(\Omega) = \Omega/c - \gamma\Omega/(4\beta^2 + \Omega^2) + i\gamma(2\beta/(4\beta^2 + \Omega^2))$. Find the first term in the asymptotic expansion for $F(z, t)$ as $z \to \infty$ and t/z is held finite. Write $h = i(\Omega/c + (i\gamma/2\beta - i\Omega)) - i\Omega t/z$. The critical points $h'(\Omega_c) = 0$ are given by $\Omega_c = -2i\beta \pm (\gamma z/(t - z/c))^{1/2}$ for $ct > z$ (inside the light cone). The asymptotic state is

$$F(z, t) \simeq \sqrt{\frac{\pi}{\gamma z}} \left(\frac{\gamma z}{t - z/c}\right)^{3/2} e^{-2\beta|t - z/c|}$$

$$(F(\Omega_c^+)e^{-2i(\gamma z(t - z/c))^{1/2}} + F(\Omega_c^-)e^{2i(\gamma z(t - z/c))^{1/2}}).$$

Thus along the ray $t = (1/v)z, v < c$, the amplitude decays as $e^{-2\beta z}|1/v - 1/c|$. $\qquad\square$

We finally turn to the initial-boundary value problem,

$$v_t = iv_{xx}, \quad 0 < x < \infty$$

$$v(0, t) = e^{ivt}, \quad v > 0, \quad v(x, t) \to 0 \quad \text{as} \quad x \to \infty, \tag{6.40}$$

$$v(x, 0) = 0.$$

Using the Fourier integral sine transform,

$$v(x, t) = \int_0^\infty B(k, t) \sin kx \, dk,$$

$$\tag{6.41}$$

$$B(k, t) = \frac{2}{\pi} \int_0^\infty v(x, t) \sin kx \, dx,$$

we find, multiplying $v_t = iv_{xx}$ by $\sin kx$ and integrating over $(0, \infty)$,

$$B_t + i\omega B = \frac{2i}{\pi}ke^{ivt}, \quad \omega(k) = k^2. \tag{6.42}$$

(One cannot substitute the expansion for $v(x, t) = \int_0^\infty B(k, t) \sin kx \, dk$ into $v_t = iv_{xx}$ and differentiate because the convergence is not uniform at $x = 0$.)
One obtains

$$B(k, t) = \frac{2}{\pi}k\frac{e^{-ivt} - e^{-i\omega t}}{\omega - v}$$

and

$$v(x, t) = \int_0^\infty \frac{2}{\pi} k \frac{e^{-ivt} - e^{-i\omega t}}{\omega - v} \sin kx \, dk \tag{6.43}$$

It is convenient to write $\sin kx = (e^{ikx} - e^{-ikx}/2i)$, write v as the sum of two integrals and change $k \to -k$ in the second integral ($\omega(k)$ does not change sign) and find

$$v(x, t) = \frac{1}{i\pi} \int_{-\infty}^\infty k \frac{e^{ikx - ivt} - e^{ikx - i\omega(k)t}}{\omega - v} dk \tag{6.44}$$

The integrand is an entire function (analytic for all finite complex k) because the apparent pole at $k = \pm\sqrt{v}$ is canceled by a zero residue as the numerator is also zero there. We can therefore deform the contour from the real k axis, and since the second integral will be evaluated using the method of steepest descent, we know we should choose a contour that cuts the real k axis at $k_0 = x/2t$ at an angle of $-(\pi/4)$. We therefore choose a contour that passes over the apparent pole at $k = -\sqrt{v}$ and under $k = +\sqrt{v}$ (Figure 6.4). With this choice, the highest point of the real part of $ikx - i\omega(k)t$ occurs at $k_0 = x/2t$. The rest of C lies in regions where $Re(ik_0 x - i\omega(k_0)t) < 0$. Now if $x/2t < \sqrt{v}$, then (6.44) is evaluated by using (6.39) for the second integral and by closing the first integral along the semicircle contour C_∞, $k = Re^{i\theta}$, $R \to \infty$, $0 < \theta < \pi$, at $k \to \infty$ in the upper half-plane. Since the integral on C is zero, $|e^{ikx}| = e^{-R \sin \theta}$ on $k = Re^{i\theta}$, the only contribution comes from the pole at $k = \sqrt{v}$:

$$v(x, t) \simeq e^{i\sqrt{v}x - ivt} - \sqrt{\frac{1}{\pi t} \frac{2xt}{x^2 - 4vt^2}} \exp i \left(\frac{x^2}{4t} - \frac{3\pi}{4} \right) + \cdots . \tag{6.45}$$

For $x > 2\sqrt{vt}$, on the other hand, we must take the contour on the second integral across its pole at $k = +\sqrt{v}$ in order to have it go over the saddle at

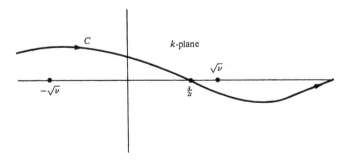

FIGURE 6.4

$x/2t$, which is now to the right of $\sqrt{v}$. This gives us an extra contribution that exactly cancels $e^{i\sqrt{v}x-ivt}$. Therefore we have

$$v(x,t) \simeq e^{i\sqrt{v}x-ivt} - \sqrt{\frac{1}{\pi t}\frac{2xt}{x^2 - 4vt^2}}\, \exp i\left(\frac{x^2}{4t} - \frac{3\pi}{4}\right),$$

$$0 < x < 2\sqrt{vt} \tag{6.46a}$$

and

$$v(x,t) \simeq \frac{1}{\sqrt{\pi t}}\frac{2xt}{x^2 - 4vt^2}\, \exp i\left(\frac{x^2}{4t} - \frac{3\pi}{4}\right),$$

$$x > 2\sqrt{vt}. \tag{6.46b}$$

This means that the different wavenumbers contained in the wavepacket that is excited at $x = 0$ begin to spread out and interfere at distances larger than that to which the group velocity would carry the information. The cancellation effects of dispersion are again evident.

Observe that at $x = 2\sqrt{vt}$, the function $G(k) = (-1/i\pi)(k/k^2 - v)$ has a pole at the saddle point $k = x/2t$ of the phase. In this case, a slightly more sophisticated analysis [85] is required from which one finds that, as $x/2t$ increases from values less than to values greater than k, a smooth transition between (6.46a) and (6.46b) is achieved.

EXERCISE 6.4

Carry out this example when $v < 0$. Also do the initial-boundary value problem for example (i) with $u(x, 0) = u_t(x, 0) = 0$ and $u(0, t) = \sin vt$.
□

6d. Snell's Laws

We now discuss the laws of reflection and refraction at an interface between two different dielectrics for two types of electromagnetic waves, called the TE and TM modes respectively. Consider Figure 6.5. The fields consist of three plane waves, an incoming wave $\exp i(\vec{k}_1 \cdot \vec{x} - \omega t)$, a reflected wave $\exp i(\vec{k}_2 \cdot \vec{x} - \omega t)$, and a transmitted wave $\exp i(\vec{k}_3 \cdot \vec{x} - \omega t)$, with wavevectors $\vec{k}_j(l_j, m_j, k_j)$, $j = 1, 2, 3$, and wavenumbers $|\vec{k}_1| = |\vec{k}_2| = n_0(\omega/c)$, $|\vec{k}_3| = n_1(\omega/c)$. On the plane $x = 0$, the continuity of the fields demands that the arguments of the exponentials are equal, which means that

$$m_1 = m_2 = m_3 = m, \quad k_1 = k_2 = k_3 = k. \tag{6.47}$$

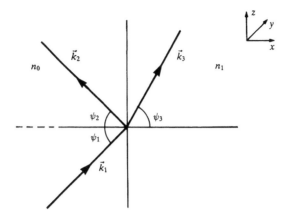

FIGURE 6.5   Interface between two dielectrics.

From these relations one finds

$$l_1 = \sqrt{\frac{n_0^2\omega^2}{c^2} - m^2 - k^2},$$

$$l_2 = -\sqrt{\frac{n_0^2\omega^2}{c^2} - m^2 - k^2} = -l_1 \qquad (6.48)$$

$$l_3 = \sqrt{\frac{n_1^2\omega^2}{c^2} - m^2 - k^2}$$

EXERCISE 6.5

Prove that for some α, β, γ not all zero, $\alpha\vec{k}_1 + \beta\vec{k}_2 + \gamma\vec{k}_3 = 0$. Therefore, since we can express one of the wavevectors as a linear combination of the other two, it must lie in the plane defined by them, and the three vectors are coplanar. Since all the three vectors lie in a plane, without loss of generality we choose coordinates so that $m = 0$. $\square$

We look at two electromagnetic wave types, the TE mode and the TM mode.

TE mode:

$$E_x = E_z = 0$$

$$E_y = \sum_{j=1}^{3} E_j \exp i(\vec{k}_j \cdot \vec{x} - \omega t) + (*), \qquad \begin{matrix} E_1, E_2 = 0 & x > 0 \\ E_3 = 0, & x < 0 \end{matrix} \qquad (6.49)$$

$$\vec{H} = \sum_{j=1}^{3} \frac{\vec{k}_j \times \vec{E}_j}{\mu_0 \omega} \exp i(\vec{k}_j \cdot \vec{x} - \omega t) + (*), \; \vec{E}_j = (0, E_j, 0) \tag{6.50}$$

where (6.50) comes from (2.1), $\partial \vec{H}/\partial t = -(1/\mu_0)\nabla \times \vec{E}$.

TM mode:

$$H_x = H_z = 0$$

$$H_y = \sum_{j=1}^{3} H_j \exp i(\vec{k}_j \cdot \vec{x} - \omega t) + (*), \quad \begin{matrix} H_1, H_2 = 0, & x > 0 \\ H_3 = 0, & x < 0 \end{matrix} \tag{6.51}$$

$$\vec{E} = \sum_{j=1}^{3} -\frac{\vec{k}_j \times \vec{H}_j}{n_0^2 \omega \epsilon_0} \exp i(\vec{k}_j \cdot \vec{x} - \omega t) + (*) \tag{6.52}$$

$$\vec{H}_j = (0, H_j, 0), n_j = n_j^2 = 1, 2; n_j^2 = n_1^2, j = 3,$$

where (6.52) comes from $(\partial \vec{E}/\partial t) = (1/\epsilon_0 n_0^2)\nabla \times \vec{H}$.

For both waves we demand the continuity of the tangential and normal components of the $\vec{E}$ and $\vec{H}$ fields at the interface $x = 0$. There are no surface charges or currents. For the TE mode, this gives (recall $l_2 = -l_1$) goes after gives

$$E_1 + E_2 = E_3$$
$$E_1 l_1 - E_2 l_1 = E_3 l_3 \tag{6.53}$$

and for the TM mode,

$$H_1 + H_2 = H_3$$
$$\frac{H_1 l_1}{n_0^2} - \frac{H_2 l_1}{n_0^2} = \frac{H_3 l_3}{n_1^2} \tag{6.54}$$

It is convenient to introduce the angles $\psi_1, \psi_2,$ and ψ_3 as shown in Figure 6.5, from which $\vec{k}_1 = (n_0 \omega/c)(\cos \psi_1, 0, \sin \psi_1)$, $\vec{k}_2 = (n_0 \omega/c)(-\cos \psi_2, \cos \psi_2, 0, \sin \psi_2)$, $\vec{k}_3 = (n_1 \omega/c)(\cos \psi_3, 0, \sin \psi_3)$. Equation (6.47) tells us that

$$\psi_1 = \psi_2, \quad n_0 \sin \psi_1 = n_1 \sin \psi_3, \tag{6.55}$$

and equations (6.53), (6.54) tell us that for the TE mode

$$\frac{E_2}{E_1} = \frac{\cos \psi_1 - (n_1/n_0) \cos \psi_3}{\cos \psi_1 + (n_1/n_0) \cos \psi_3}, \quad \frac{E_3}{E_1} = \frac{2 \cos \psi_1}{\cos \psi_1 + (n_1/n_0) \cos \psi_3} \quad (6.56)$$

and for the TM mode

$$\frac{H_2}{H_1} = \frac{\cos \psi_1 - (n_0/n_1) \cos \psi_3}{\cos \psi_1 + (n_0/n_1) \cos \psi_3}, \quad \frac{H_3}{H_1} = \frac{2 \cos \psi_1}{\cos \psi_1 + (n_0/n_1) \cos \psi_3} \quad (6.57)$$

We observe several features:

1. The angle of reflection equals the angle of incidence.

2. The angle of refraction ψ_3 is greater than or less than the angle of incidence ψ_1 depending on whether n_0 is greater than or less than n_1. Again we see that light bends towards regions of greater refractive index. In particular, if $n_0 > n_1$, $\sin \psi_3 = (n_0/n_1) \sin \psi_1$ and thus $\psi_3 > \psi_1$. Observe that if the angle of incidence ψ_1 is such that $\sin \psi_1 = n_1/n_0$, then $\psi_3 = \pi/2$ and $E_2/E_1 = 1$. (See Figure 6.6.)

3. For $n_0 < n_1$, the reflected field can be up to $180°$ out of phase with the incident field. In particular, for $\psi_1 = \psi_3 = 0$, $E_2/E_1 = -(n_1 - n_0)/(n_1 + n_0)$.

4. For the TM mode, the amplitude of the reflected mode can be zero. Suppose $n_0 > n_1$; then $\sin \psi_3 > \sin \psi_1$ and $\cos \psi_3 < \cos \psi_1$. Thus there is a value of the incident angle such that $H_2/H_1 = 0$. This is called the Brewster angle.

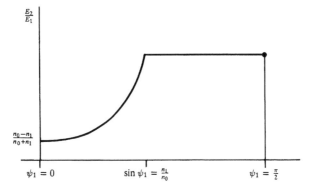

FIGURE 6.6   Graph of reflection coefficient as function of incident angle when $n_0 > n_1$.

6e.   Waveguides

The main idea of a waveguide is to guide a beam of light by employing a variation in refractive index in the transverse direction so as to cause the light to travel along a well-defined channel [3]. The dependence of refractive index on the transverse direction, the direction perpendicular to that in which the wave propagates, can be continuous or discontinuous. The essential element, however, is that the refractive index is maximal in the channel along which one wishes the light to be guided. In order to present the idea in concrete form, we introduce a simple model for which all details can be explicitly worked out with a minimum of effort. This model treats a TE mode $\vec{E} = (0, E(x, z, t), 0)$ which depends only on one of the two transverse directions. While this example is sufficient to guide most of our thinking, we will also include the second transverse dimension, which will be relevant when we discuss the propagation of light along light fibers with approximately circular cross-section.

Consider Figure 6.7, which depicts a center region called the core (or guiding layer) and an outside region called the cladding. For the planar waveguide depicted in the figure, we allow for the fact that the cladding may consist of two distinct regions called the cladding and substrate. We assume that both regions are identical in the following analysis ($n_1 = n_2$). The refractive indices of these two regions are n_0 and n_1 respectively with $n_0 > n_1$, and we take the thickness of the cladding to be so large compared to that of the core that it may be considered infinite. We wish to find the TE field

$$\vec{E}(x, z, t) = (0, E(x), 0) \exp i\omega(\beta(z/c) - t)$$

$$\vec{H}(x, z, t) = \left(-\frac{\beta}{\mu_0 c} E, 0, -\frac{i}{\omega \mu_0} \frac{dE}{dx}\right) \exp i\omega(\beta(z/c) - t)$$

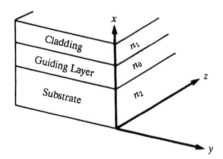

FIGURE 6.7   Model waveguide.

where $E(x)$ satisfies

$$\frac{d^2 E}{dx^2} - (\beta^2 - n^2)\frac{\omega^2}{c^2}E = 0 \tag{6.58}$$

for $|x| > L$ and $|x| < L$. The boundary conditions (continuity of transverse and normal electric and magnetic fields) are that both E and dE/dx are continuous at $x = \pm L$ and that E tends to zero as x tends to $\pm \infty$. The constant β is to be determined. It is the *effective refractive index* (sometimes $\beta(\omega/c)$ is called the mode propagation constant) of the waveguide, which consists of a combination of the core and cladding. We can see immediately that in order that the electric field decays as $x \to \pm \infty$, the effective refractive index β must be greater than n_1, the refractive index in the cladding, as otherwise the solutions of (6.58) in the cladding would be periodic sines and cosines. What is not quite so obvious here is that it also must be less than n_0. This we shall now assume and verify a posteriori. The fields in the three regions $x < -L, -L < x < L$, and $x > L$ are, respectively,

$$x < L: \quad E(x) = A \exp\sqrt{\beta^2 - n_1^2}\,\frac{\omega}{c}(x + L) \tag{6.59}$$

$$-L < x < L: \quad E(x) = B \cos\sqrt{n_0^2 - \beta^2}\,\frac{\omega}{c}x + C \sin\sqrt{n_0^2 - \beta^2}\,\frac{\omega}{c}x \tag{6.60}$$

$$x > L: \quad E(x) = D \exp\sqrt{\beta^2 - n_1^2}\,\frac{\omega}{c}(L - x). \tag{6.61}$$

Applying continuity of E and dE/dx at $x = -L$ and $x = +L$ gives four linear homogeneous algebraic equations for A, B, C, D. A nontrivial solution demands that the determinant of the matrix of these equations is zero, and this condition gives us a transcendental equation for the effective refractive index β. We are going to leave the details as an exercise for the reader. We also leave it as an exercise for the reader to show that if $\beta^2 > n_0^2$, the only solution that satisfies equations and boundary conditions is the trivial one $E = 0$. Defining the angle ϕ by

$$\beta^2 - n_1^2 = (n_0^2 - n_1^2)\cos^2\phi, \tag{6.62}$$

then one finds the equation that determines β:

$$\sin 2\left(\frac{\omega}{c}L\sqrt{n_0^2 - n_1^2}\sin\phi + \phi\right) = 0. \tag{6.63}$$

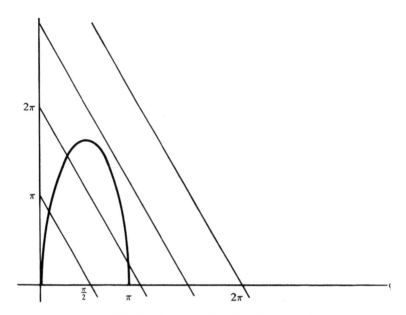

FIGURE 6.8   Graphs of $y_1(\phi)$, $y_2(m, \phi)$, $m = 1, 2, \ldots$.

The solutions $\phi(\beta)$ are given by the intersections of the graphs (see Figure 6.8)

$$y_1 = 2\frac{\omega}{c}L\sqrt{n_0^2 - n_1^2}\sin\phi, \quad y_2 = m\pi - 2\phi, \quad m \text{ an integer.}$$

The allowed solutions lie in the range $0 < \phi < \pi/2$. Observe following:

1. The number of solutions depends on the magnitude of the *waveguide parameter*

$$V = 2\pi\frac{(2L)}{\lambda}\sqrt{n_0^2 - n_1^2}, \tag{6.64}$$

   which is the ratio of core thickness multiplied by the square root of the electric susceptibility mismatch $n_0^2 - n_1^2$ divided by the wavelength of the light.

2. There is always one solution no matter how small or large V is.

3. If V is chosen sufficiently small, then there is only one solution. In this case we say that the waveguide is a single-mode waveguide. In a cylindrical geometry, an exactly analogous situation obtains, and it is possible to design optical fibers so that they support only one transverse mode.

4. As V becomes very large, the solutions get very close together near $\phi = 0$. For V very large and for $\phi \simeq 0$, $y_1 = V\phi_m = m\pi - 2\phi_m$ whence $\phi_m = (m/(V+2))\pi$.

5. Note that β in general is a function of ω even though n_1 and n_2 need not be.

EXERCISE 6.6

How should we choose V so that the second line $y_2 = 2\pi - 2\phi$ does not intersect $V \sin \phi$ before $\phi = \pi/2$? (Answer: choose $V < \pi$.) □

To each solution of (6.63) for ϕ and therefore β, there is a solution $E(x)$ of the boundary value problem. These solutions are called cavity modes. We can write them (up to an overall multiplicative constant):

$$m \text{ odd: } \quad x < -L, \quad E(x) = (-1)^{(m+1)/2} \sin \phi \exp \left(\frac{V}{2} \left(\frac{x}{L} + 1 \right) \cos \phi \right),$$

$$-L < x < L, \quad E(x) = \cos \left(\frac{Vx}{2L} \sin \phi \right),$$

$$x > L, \quad E(x) = (-1)^{(m+1)/2} \sin \phi \exp \left(\frac{V}{2} \left(1 - \frac{x}{L} \right) \cos \phi \right).$$

$$m \text{ even}: x < -L, \quad E(x) = (-1)^{m/2} \sin \phi \exp \left(\frac{V}{2} \left(\frac{x}{L} + 1 \right) \cos \phi \right),$$

$$-L < x < L, \quad E(x) = \sin \left(\frac{Vx}{2L} \sin \phi \right),$$

$$x > L, \quad E(x) = \sin \phi \exp \left(\frac{V}{2} \left(1 - \frac{x}{L} \right) \cos \phi \right). \tag{6.65}$$

They are even and odd functions of x respectively, as shown in Figure 6.9.

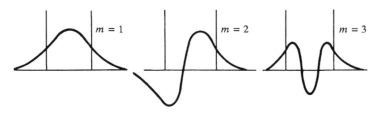

FIGURE 6.9

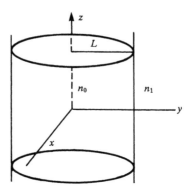

FIGURE 6.10

We next turn to the cylindrical geometry shown in Figure 6.10. In this geometry, it is normally best to use cylindrical coordinates for both the dependent $\vec{E}(E_r, E_\theta, E_z)$, $\vec{H}(H_r, H_\theta, H_z)$ and independent (r, θ, z) variables. The reason for this is that the continuity conditions at the boundary $r =$ are that the tangential components E_ϕ, E_z, H_ϕ, and H_z are continuous. Als at $r = L$, ϵE_r and H_r are continuous. Assuming all the fields have the tim dependence $e^{-i\omega t}$, $\vec{E}$ and $\vec{H}$ satisfy

$$\vec{H} = -\frac{i}{\omega \mu_0} \nabla \times \vec{E},$$

$$\vec{E} = \frac{i}{\omega \epsilon} \nabla \times \vec{H}$$

(6.66)

when $\epsilon = \epsilon_0 n^2$, n is the refractive index, and $\mu_0 \epsilon_0 = c^{-2}$. The equation for $\vec{E}$ is found by taking the curl of $\vec{H}$ whence

$$-\nabla \times \nabla \times \vec{E} + \frac{n^2 \omega^2}{c^2} \vec{E} = 0.$$

(6.67)

The operators curl, divergence, gradient in cylindrical coordinates are ($\hat{r}, \hat{\theta}$, and $\hat{z}$ are unit vectors in the directions r, θ, z)

$$\nabla \times \vec{E} = \frac{1}{r} \begin{pmatrix} \hat{r} & r\hat{\theta} & \hat{z} \\ \frac{\partial}{\partial r} & \frac{\partial}{\partial \theta} & \frac{\partial}{\partial z} \\ E_r & rE_\theta & E_z \end{pmatrix}$$

$$\nabla \cdot \vec{E} = \frac{1}{r}\frac{\partial}{\partial r} r E_r + \frac{1}{r}\frac{\partial}{\partial \theta} E_\theta + \frac{\partial E_z}{\partial z},$$

$$\nabla \phi = \frac{\partial \phi}{\partial r}, \frac{1}{r}\frac{\partial \phi}{\partial \theta}, \frac{\partial \phi}{\partial z}.$$

A little calculation shows that both E_z and H_z satisfy the wave equation

$$\left(\nabla^2 + \frac{n^2\omega^2}{c^2}\right)\left(\begin{matrix}E_z\\H_z\end{matrix}\right) = 0,$$

and if each of the fields has a z-dependence given by $e^{i\beta(\omega/c)z}$, then from (6.66), we can write all components $E_r, E_\theta, H_r, H_\theta$ in terms of E_z and H_z. In this way, we can find the general solutions of (6.66), and it is also clear from the form of (6.66) that $\nabla \cdot \epsilon \vec{E} = \nabla \cdot \vec{H} = 0$. For details of these calculations, we suggest Yariv's book [3]. The normal modes of propagation can be divided into two classes, HE_{lm} and EH_{lm} modes.

The analysis can be greatly simplified, however, if the difference in refractive indices $n_0 - n_1$ is small because in that case E_r is also "almost" continuous. Then since $E_r = E_x \cos\theta + E_y \sin\theta$, $E_\theta = -E_x \sin\theta + E_y \cos\theta$, E_x and E_y are both continuous at $r = L$. Hence we can use Cartesian coordinates $\vec{E}(E_x, E_y, E_z)$ and $\vec{H}(H_x, H_y, H_z)$ for both the electric and magnetic fields, the dependent variables, and each scalar component $\psi(E_x, E_y, E_z, H_x, H_y, H_z)$ satisfies the Helmholtz equation

$$\left(\nabla_1^2 - \frac{\omega^2}{c^2}(\beta^2 - n^2)\right)\psi = 0, \qquad (6.68)$$

where $\nabla_1^2 = (\partial^2/\partial r^2) + 1/r(\partial/\partial r) + (1/r^2)(\partial^2/\partial\theta^2)$, and we have removed the z dependence $\exp i\beta(\omega/z)$. The components must still relate through (6.66) so that both fields $\epsilon\vec{E}$ and $\vec{H}$ are divergence-free. Inside the cylinder $0 \le r < L$, the solution of (6.68), which is bounded, is

$$\psi(r, \theta) = J_l((\omega/c)\sqrt{n_0^2 - \beta^2}r)e^{il\theta}$$

while outside, $r > L$, it is

$$\psi(r, \theta) = K_l((\omega/c)\sqrt{\beta^2 - n_1^2}r)e^{il\theta}.$$

As in the plane case, $n_0 > \beta > n_1$ and because $n_0 - n_1$ is small, so are both $\sqrt{n_0^2 - \beta^2}$ and $\sqrt{\beta^2 - n_1^2}$. Therefore derivatives of E_x with respect to z are much greater than derivatives with respect to x and y because $\beta \gg \sqrt{n_0^2 - \beta^2}, \sqrt{\beta^2 - n_1^2}$. This is called the weak guiding assumption.

We can write all the fields E_z, H_x, H_y, and H_z in terms of two independent fields E_z and E_y. The y-polarized modes are given by

$$E_x = 0$$

$$E_y = A_l J_l \left(\frac{\omega}{c} \sqrt{n_0^2 - \beta^2} r \right) e^{il\theta + i\beta(\omega/c)z - i\omega t}, \quad r < L$$

$$= B_l K_l \left(\frac{\omega}{c} \sqrt{\beta^2 - n_1^2} r \right) e^{il\theta + i\beta(\omega/c)z - i\omega t}, \quad r > L$$

$$H_x = -\frac{\beta}{\mu_0 c} E_y - \frac{i}{\omega \mu_0} \frac{\partial E_z}{\partial y} \simeq -\frac{\beta}{\mu_0 c} E_y$$

$$H_y = \frac{i}{\omega \mu_0} \frac{\partial E_x}{\partial x} \simeq 0$$

$$H_z = -\frac{i}{\omega \mu_0} \frac{\partial E_y}{\partial x}$$

$$E_z = \frac{ic}{\omega \beta} \frac{\partial E_y}{\partial y}, \tag{6.69}$$

where we have neglected terms proportional to $\sqrt{n_0^2 - n_1^2}$ compared to those proportional to β. The dominant components E_y and H_x are transverse. The x-polarized modes are given by

$$E_x = C_l J_l \left(\frac{\omega}{c} \sqrt{n_0^2 - \beta^2} r \right) e^{il\theta + i\beta(\omega/c)z - i\omega t}, \quad r < L$$

$$= D_l K_l \left(\frac{\omega}{c} \sqrt{\beta^2 - n_1^2} r \right) e^{il\theta + i\beta(\omega/c)z - i\omega t}, \quad r > L$$

$$E_y = 0$$

$$H_x = \frac{-i}{\omega \mu_0} \frac{\partial E_z}{\partial y} \simeq 0$$

$$H_y = \frac{\beta}{\mu_0 c} E_x + \frac{i}{\omega \mu_0} \frac{\partial E_z}{\partial x} \simeq \frac{\beta}{\mu_0 c} E_x$$

$$H_z = \frac{i}{\omega \mu_0} \frac{\partial E_x}{\partial y}$$

$$E_z = \frac{ic}{\omega \beta} \frac{\partial E_x}{\partial x}. \tag{6.70}$$

Note that in both cases, the divergences of $\vec{E}$ and $\vec{H}$ ($\nabla \cdot \vec{E}$ and $\nabla \cdot \vec{H}$) are zero. The dominant components are E_x and H_y.

The continuity of $E_x(E_y)$ at $r = L$ gives

$$\begin{Bmatrix} B_l \\ D_l \end{Bmatrix} = \frac{J_l(\frac{\omega}{c}\sqrt{n_0^2 - \beta^2}L)}{K_l(\frac{\omega}{c}\sqrt{\beta^2 - n_1^2}L)} \begin{Bmatrix} A_l \\ C_l \end{Bmatrix}.$$

The continuity of E_z is calculated as follows. For the y-polarized mode,

$$E_z = -(ic/\omega\beta)(\partial E_y/\partial y)$$

and

$$(\partial E_y/\partial y) = (\partial E_y/\partial r)(\partial r/\partial y) + (\partial E_y/\partial \theta)(\partial \theta/\partial y)$$

$$= (\partial E_y/\partial r)\sin\theta + il E_y \cos\theta,$$

so that the continuity of E_z is equivalent to the continuity of $(\partial E_y/\partial r)$, which gives

$$\sqrt{n_0^2 - \beta^2} J_l' \left(\frac{\omega}{c}\sqrt{n_0^2 - \beta^2}\right) K_l \left(\frac{\omega}{c}\sqrt{\beta^2 - n_1^2}L\right)$$

$$= \sqrt{\beta^2 - n_1^2} J_l \left(\frac{\omega}{c}\sqrt{n_0^2 - \beta^2}L\right) K_l' \left(\frac{\omega}{c}\sqrt{\beta^2 - n_1^2}L\right). \qquad (6.71)$$

The same expression holds for the continuity of E_z for the x-polarized mode. Write $\beta^2 - n_1^2 = \alpha_l^2(n_0^2 - n_1^2)$ and obtain

$$\sqrt{1 - \alpha_l^2} J_l' \left(\frac{V}{2}\sqrt{1 - \alpha_l^2}\right) K_l \left(\frac{V}{2}\alpha_l\right) = \alpha_l J_l \left(\frac{V}{2}\sqrt{1 - \alpha_l^2}\right) K_l' \left(\frac{V}{2}\alpha_l\right)$$

$$\qquad (6.72)$$

where

$$V = \frac{2\pi}{\lambda}2L\sqrt{n_0^2 - n_1^2}, \quad \lambda = \frac{2\pi c}{\omega},$$

is the waveguide parameter.

For each l, there may be a discrete set of solutions $\alpha_{lm}, m = 1, 2, \dots$ to (6.72). The propagation constants β_{lm} for the $e^{il\theta}$ are thus determined from (6.72). In particular, we are most interested in situations where $V < 2.405$, for which there is only one propagation mode $l = 0$ and one solution to (6.72) with $l = 0$. The corresponding eigenmode then is the transverse structure of the fields in a single-mode optical fiber.

Note, however, that the solution is degenerate. The general solution is

$$\vec{E} = (A\vec{E}_1 + B\vec{E}_2)e^{i(\beta\omega/c)z - i\omega t}, \tag{6.73}$$

a linear combination of the x and y polarizations.

6f. TEM$_{rs}$ Cavity Modes

The open cavity resonator shown in Figure 6.11 was the brainchild of Schawlow and Townes and marked the critical breakthrough in the design of working lasers [3]. The reason is that its geometry can be used to select the fundamental TEM$_{00}$ mode because the higher-order transverse modes can be made to suffer significantly greater losses.

FIGURE 6.11   Fabry–Perot resonator with spherical mirrors.

The structure of these modes can be found directly from the paraxial equation (2.67)

$$\frac{\partial A}{\partial z} - \frac{i}{2k}\nabla_\perp^2 A = 0. \tag{6.74}$$

In Section 2g, you were asked to show that a Gaussian beam with shape

$$A(0, x, y) = \exp - \left(\frac{x^2 + y^2}{\omega_0^2}\right) \tag{6.75}$$

at its narrowest point $z = 0$ takes on the shape

$$A(z, x, y) = \frac{1}{\left(1 + \frac{2iz}{k\omega_0^2}\right)} \exp - \left(\frac{x^2 + y^2}{\omega_0^2 + \frac{2iz}{k}}\right) \tag{6.76}$$

for arbitrary z. The quantity ω_0 is called the beam waist.

EXERCISE 6.7

Take the Fourier transform

$$A(z, x, y) = \int \hat{A}(z, l, m) l^{ilx+imy} dl dm \qquad (6.77)$$

of (6.74) and find $\hat{A}(z, l, m) = \hat{A}(0, l, m) \exp(-i(l^2 + m^2)z)/2k$. Using the facts that

$$\int_{-\infty}^{\infty} \exp\left(-\frac{x^2}{\omega_0^2} - ilx\right) dx = \sqrt{\pi}\omega_0 \exp\left(\frac{-\omega_0^2 l^2}{4}\right),$$

and

$$\int_{-\infty}^{\infty} \exp\left(-\left(\frac{\omega_0^2}{4} + \frac{iz}{2k}\right) l^2 + ilx\right) = \frac{\sqrt{\pi}}{\sqrt{\frac{\omega_0^2}{4} + \frac{iz}{2k}}} \exp\frac{-x^2}{\omega_0^2 + \frac{2iz}{2k}}$$

show that (6.76) is correct. □

Now, we rewrite (6.76) by noting that

$$\frac{1}{\omega_0^2 + \frac{2iz}{k}} = \frac{1}{\omega_0^4 + \frac{4z^2}{k^2}}\left(\omega_0^2 - \frac{2iz}{k}\right)$$

and defining

$$w^2(z) = w_0^2\left(1 + \left(\frac{z}{z_0}\right)^2\right), \quad R(z) = z + \frac{z_0^2}{z}, \quad z_0 = \frac{1}{2}w_0^2 k \qquad (6.78)$$

then finding

$$A(z, x, y) = \frac{w_0}{w(z)} \exp\left(-\frac{x^2 + y^2}{w^2(z)} + \frac{ik(x^2 + y^2)}{2R(z)} - i\tan^{-1}\frac{z}{z_0}\right). \qquad (6.79)$$

The parameter z_0 is the distance over which the width of the pulse increases by a factor of $\sqrt{2}$. For a beam waist w_0 of 1 mm, a light beam with vacuum wavelength $\lambda = (2\pi c/\omega) = (2\pi n/k)$ equal to 1 μm in a medium with refraction index one, z_0 is about 5 m.

The fundamental TEM mode is

$$\vec{E}_{00} = \hat{e}\left(\frac{w_0}{w}\exp\left(-\frac{x^2+y^2}{w^2(z)}+ikz\right.\right.$$

$$\left.\left.+\frac{ik(x^2+y^2)}{2R(z)}-i\tan^{-1}\frac{z}{z_0}-i\omega t\right)+(*)\right). \qquad (6.80)$$

The higher-order modes are given by similar expressions:

$$\vec{E}_{rs} = \hat{e}\frac{w_0}{w}H_r\left(\sqrt{2}\frac{x}{w(z)}\right)H_s\left(\sqrt{2}\frac{y}{w(z)}\right)\exp-\left(\frac{x^2+y^2}{w^2(z)}\right)$$

$$\left(\exp i\left(kz+\frac{k(x^2+y^2)}{2R(z)}-(r+s+1)\tan^{-1}\frac{z}{z_0}-\omega t\right)+(*)\right) \qquad (6.81)$$

where H_r is the rth Hermite polynomial. Pictures of the intensities of the low-order modes seen in cross sections are given in Figure 6.12.

Observe that if w_0 is sufficiently large so that $z_0 = \frac{1}{2}w_0^2 k$ is much larger than the distance between the mirrors, we can ignore the beam spreading (i.e., $w(z) \simeq w_0$) and write

$$\vec{E}_{rs} = \hat{e}A(x,y)e^{ikz-i\omega t}+(*) \qquad (6.82)$$

with $k = n\omega/c$, namely as a slowly varying wavetrain in which we leave the shape $A(x,y)$ unspecified and where k is quantified as $q\pi/d$, q an integer and d is the intermirror distance. It turns out to be useful to express the electric field in a laser in this way in circumstances where the gain band of the laser covers many of the frequencies associated with the higher transverse modes. In that case we allow the shape $A(x,y,t)$ to evolve in time. We discuss the laser in this case in Section 6i.

Now what have these modes got to do with lasers? Imagine that there is an excitable medium placed somewhere in the cavity space between two spherical mirrors placed at $z = z_2 > 0$ and $z = z_1 < 0$, as shown in Figure 6.11. Choose their radii of curvature to be $R_2 = R(z_2)$ and $R_1 = -R(z_1)$ respectively. Observe that a point (x,y,z) located on the mirror at R_2 satisfies the equation

$$x^2 + y^2 + (z - z_2 + R_2)^2 = R_2^2,$$

which can be rewritten as

$$\frac{x^2+y^2}{2R_2}+z = z_2 + \frac{(z-z_2)^2}{2R_2}.$$

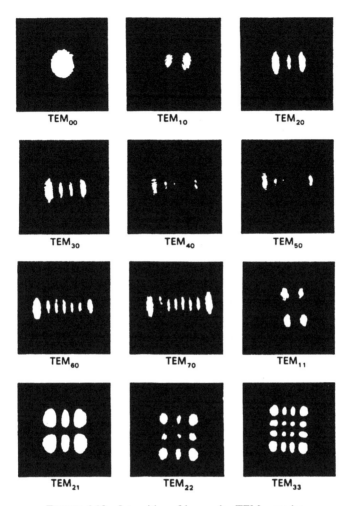

FIGURE 6.12   Intensities of low-order TEM$_{rs}$ modes.

Therefore the surface of constant phase θ of the wavefronts,

$$\theta = kz + \frac{k(x^2 + y^2)}{2R(z)} + (r + s + 1)\tan^{-1}\frac{z}{z_0} = \omega t,$$

when evaluated on the mirror surface, is equal to

$$kz_2 - (r + s + 1)\tan^{-1}\frac{z}{z_0} - \omega t$$

$$+ \frac{k(x^2 + y^2)}{2}\left(\frac{1}{R(z)} - \frac{1}{R(z_2)}\right) + \frac{k}{2R_2}(z - z_2)^2$$

and therefore coincides almost exactly with the mirror at z_2. The rays strike the mirrors perpendicularly and propagate back and forth across the cavity. In order for lasing action to occur, the amount of energy received by the wave in one round trip must exceed the losses due to finite conductivity of the medium and imperfect reflection of the mirrors. The losses due to the phase mismatch can be removed by tuning the wavenumber $k = n\omega/c$ so that the total phase change of the ray traveling between the two mirrors is an integer q times π, namely

$$\frac{n\omega}{c}d - (r+s+1)\left(\tan^{-1}\frac{z_2}{z_0} - \tan^{-1}\frac{z_1}{z_0}\right) = q\pi \qquad (6.83)$$

where $d = z_2 - z_1$. The frequency ω satisfying (6.83) for given values of q, r, and s is called ω_{qrs}. The corresponding electric field,

$$\vec{E}_{rs} = \hat{e}\frac{w_0}{w(z)}H_r\left(\sqrt{2}\frac{x}{w(z)}\right)H_y\left(\sqrt{2}\frac{y}{w(z)}\right)\exp\left(-\frac{x^2+y^2}{w^2(z)}\right)$$

$$\left(\exp i\left\{\frac{n\omega_{qrs}}{c}z + \frac{x^2+y^2}{2R(z)} - \omega_{qrs}t - (r+s+1)\tan^{-1}\frac{z}{z_0}\right\} + (*)\right),$$

$$(6.84)$$

is called the TEM$_{rs}$ cavity mode. In particular we note that the larger r and s are, the less the wavefronts and the mirror coincide, and so the arrangement selectively favors the TEM$_{00}$ mode. Observe that the change in frequency between two longitudinal modes

$$\Delta_{qrs} = \omega_{q+1rs} - \omega_{qrs} = \frac{c\pi}{nd}. \qquad (6.85)$$

The frequency difference between two transverse modes is

$$\Delta\omega_{qrs} = \frac{c}{nd}\Delta(r+s)\left(\tan^{-1}\frac{z_2}{z_0} - \tan^{-1}\frac{z_1}{z_0}\right), \qquad (6.86)$$

where $\Delta(r+s)$ is the change in $(r+s)$. If the mirrors are confocal, namely

$$R_2 = -R_1 = R, \qquad -z_1 = z_2 = \frac{d}{2},$$

then, since $\tan^{-1}x \simeq x$ for x small,

$$\Delta\omega_{qrs} = \frac{c}{nz_0}\Delta(r+s). \qquad (6.87)$$

Observe that when the diffraction distance z_0 is much larger than the distance d between the mirrors,

$$\frac{(\Delta\omega)\ \text{transverse}}{(\Delta\omega)\ \text{longitudinal}} = \frac{d}{\pi z_0} \tag{6.88}$$

is much less than one. The spectrum of cavity modes is shown in Figure 6.13.

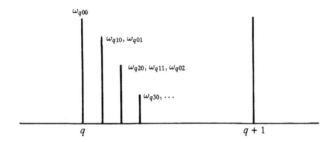

FIGURE 6.13

As discussed, we can design the optical resonator by using two spherical mirrors, one at $z_2 > 0$ and one at $z_1 < 0$, chosen so that their radii of curvature R_2 and $-R_1$ are the same as that of the beam front at the two locations. The values of z_2 and z_1 are chosen by solving

$$-R_2 = R(z_1) = z_1 + \frac{z_0^2}{z_1}$$

$$R_2 = R(z_2) = z_2 + \frac{z_0^2}{z_2}, \tag{6.89}$$

from which

$$z_2 = \frac{R_2}{2} \pm \frac{1}{2}\sqrt{R_2^2 - 4z_0^2}$$

$$z_1 = -\frac{R_1}{2} \pm \frac{1}{2}\sqrt{R_1^2 - 4z_0^2}. \tag{6.90}$$

Given $d = z_2 - z_1$, we can choose the minimum spot size w_0, its positions in relation to the mirrors by subleading the two equations in (6.90) and finding

$$z_0^2 = \frac{d(R_1 - d)(R_2 - d)(R_2 + R_1 - d)}{(R_2 + R_1 - 2d)^2},$$

from which we can find z_2, z_1, and $w_0 = (2z_0/k)^{1/2}$.

6g.   Nonlinear Oscillators, Wavetrains, and Three- and Four-Wave Mixing, the Method of Multiple Scales, and WKBJ Expansions

Let us begin (see also references [13, 14, 82]) by discussing the simple pendulum

$$\frac{d^2u}{dt^2} + \omega^2 \sin u = 0, \tag{6.91}$$

where $u(t)$ is the angle a pendulum of length l makes with the vertical, $\omega^2 = g/l$, and g is gravitational acceleration. Multiplying by du/dt and integrating gives

$$\frac{1}{2}\left(\frac{du}{dt}\right)^2 + V(u) = E, \tag{6.92}$$

where $K = 1/2(du/dt)^2$ may be considered to be the kinetic energy, E the constant total energy, and

$$V(u) = \omega^2(1 - \cos u)$$

is the potential energy, normalized to be zero at $u = 0$. We get a good idea about the nature of the solution by plotting the graph of du/dt versus u. The $((du/dt), u)$ plane is called the *phase* plane of this dynamical system. In a conservative Hamiltonian system of $2N$ degrees of freedom, the $2N$-dimensional space of N momentum $(p_1, \ldots p_n)$ and N position coordinates $(q_1, \ldots qN)$ are often referred to as the phase space of the system. Here $p = du/dt, q = u$.

Contours corresponding to different values of the total energy are given in Figure 6.14. For E small, both K and V are small and (6.92) is an ellipse,

$$\frac{1}{2}\left(\frac{du}{dt}\right)^2 + \frac{\omega^2}{2}u^2 = E, \tag{6.93}$$

the solution for $u(t)$ is sinusoidal,

$$u(t) = \sqrt{\frac{2E}{\omega}} \cos(\omega t - \phi),$$

and the period is $2\pi/\omega$ independent of the energy and amplitude of the motion. For larger $E = \omega^2(1 - \cos A)$, the general solution can be given in terms of

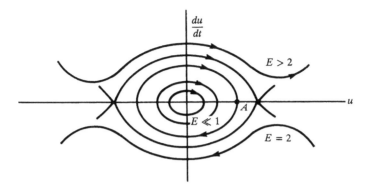

FIGURE 6.14   Phase plane for pendulum.

elliptic functions, and the period is

$$\frac{4}{\omega} \int_0^{\frac{\pi}{2}} \frac{d\phi}{\sqrt{1 - m^2 \sin^2 \phi}}, \quad m^2 = \sin^2 \frac{A}{2} \tag{6.94}$$

where A is the maximum amplitude of the motion. The period is now amplitude dependent. For $A = \pi$, $E = 2\omega^2$ the trajectory is the separatrix $u(t) = 2 \sin^{-1} \tanh \omega t$, which takes the orbit from the saddle point at $du/dt = 0$, $u = -\pi$ to the saddle at $du/dt = 0$, $u = \pi$, namely one complete revolution from the unstable configuration back to itself. The period of this orbit is ∞. Observe that it is also very unstable. A slight push causes the pendulum to continue rotating; a slight pull causes it to reverse its motion when it makes the top. More about this sensitivity in subsection 1. For $E > 2\omega^2$, the orbit is nonperiodic in u, and the pendulum continues to rotate with monotonically increasing or decreasing angle u. It is clear, therefore, that as the amplitude A (energy E) of the pendulum motion increases from 0 to π (0 to $2\omega^2$), the period also increases monotonically from $2\pi/\omega$ to ∞.

We now introduce a method that allows one to calculate the period of the pendulum as a function of amplitude when the latter is small. This method is called the method of multiple scales, and it can be used to determine not only how nonlinearity can change the fundamental frequency of a single oscillator, but also to describe the exchange of energy between different oscillators that are weakly coupled through nonlinear terms in the equation. All the cases we discuss involve oscillators which, to leading order, are uncoupled so that the field can be approximated to first order as a linear combination

$$\sum_{j=1} \left(a_j e^{i\omega j t} + (*) \right).$$

However, when we solve for corrections to the field as we did in Section 6b, when we examined the weakly nonlinear pendulum, we find that these corrections can exhibit an unbounded growth in time. In order to suppress this secular behavior we allow the oscillator amplitudes a_j be slowly varying functions of time. The method of multiple time scales shows us how to make these choices.

What we want to stress is that the presence of secular terms (terms that lead to unbounded growth in time) means that the effect of a small perturbation is no longer small if one waits long enough. Our goal is to attempt to understand the exchange of energy between the oscillators over long times by constructing nonlinear differential equations for the oscillator amplitude a_j. To understand what will happen for all time is often very difficult, so in most cases we settle for knowing what happens over long ($t = O(\epsilon^{-N})$, ϵ the weak coupling parameter, N an integer) but still finite times.

In particular, we examine the following situations, which serve as prototypes for all the interactions we meet in nonlinear optics.

1. The modulation of a single oscillator due to nonlinear forces. From this result we discover how to calculate the amplitude dependent frequency and/or wavenumber modulation (the nonlinear refractive index) of single wavetrains.

2. The modulation of the frequencies and/or wavenumbers of two weakly coupled nonlinear oscillators.

3. The sharing of energy between two resonant oscillators, a result important in understanding linear and nonlinear birefringence and in understanding switches.

4. The sharing of energy between oscillators due to indirect rather than direct resonant interactions. This section introduces three- and four-wave mixing processes important in Raman and Brillouin scattering and in phase conjugation.

5. The evolution of wavepackets rather than wavetrains and the derivation of equations of nonlinear Schrödinger (NLS) type which describe the propagation and interaction of pulses in fires, light beams in nonlinear dielectrics, and the coupling between copropagating and counter propagating beams. The wavepacket analysis is given in Section 6h.

The Self-Interaction of a Single Oscillator

Consider the equation for the pendulum with initial condition

$$u(0) = \epsilon A, \quad 0 < \epsilon \ll 1, \quad \frac{du}{dt}(0) = 0. \tag{6.95}$$

Write $u(t) = \epsilon y(t)$. Then

$$\frac{d^2 y}{dt^2} + \omega^2 y = \frac{\epsilon^2 \omega^2}{6} y^3 - \frac{\epsilon^4 \omega^2}{120} y^5 + \cdots \tag{6.96}$$

with initial conditions

$$y(0) = A, \quad \frac{dy}{dt}(0) = 0. \tag{6.97}$$

Let us look for solutions $y(t, \epsilon)$ in the form

$$y(t, \epsilon) = y_0(t) + \epsilon y_1(t) + \epsilon^2 y_2(t) + \cdots \tag{6.98}$$

with

$$y_0(t) = ae^{i\omega t} + a^* e^{-i\omega t}. \tag{6.99}$$

From our earlier calculation, we know that the second iterate y_2 contains terms proportional to $te^{\pm i\omega t}$, which causes the asymptotic expansion (6.98) to fail to be uniformly valid in time because after a time $\epsilon^2 t = O(1)$, the third term in (6.98) is as big as the first. Moreover, continuing the series we would find $t^n e^{\pm i\omega t}$ terms in y_{2n}, $n \geq 2$. The cause of the secular behavior is that we have fixed the frequency of the nonlinear oscillator incorrectly. It is close to but not equal to ω. To give ourselves flexibility, we allow the amplitude a, heretofore constant, to be slowly varying in time and given by the asymptotic series

$$\frac{da}{dt} = \epsilon f_1(a, a^*) + \epsilon^2 f_2(a, a^*) + \cdots . \tag{6.100}$$

We now find that extra terms involving $f_n(a, a^*)$, $n \geq 1$, occur in the equations for the determination of the iterates $y_1, y_2, \ldots$ and by appropriately choosing $f_n(a, a^*)$, we can eliminate the secular terms, namely those terms that lead to nonbounded behavior in the iterates.

In the literature, this method is often called the multiple time scale method because instead of using the ansatz (6.100) one can imagine a to be a function of the slow times

$$a = a(T_1 = \epsilon t, \quad T_2 = \epsilon^2 t, \ldots) \tag{6.101}$$

whence, using the chain rule,

$$\frac{da}{dt} = \epsilon \frac{\partial a}{\partial T_1} + \epsilon^2 \frac{\partial a}{\partial T_2} + \cdots \tag{6.102}$$

and we can identify $\partial a/\partial T_n$ with $f_n(a, a^*)$. However, if we formally use more than one time scale in $a(T_1, T_2, \ldots)$, we must remember that the T_n's are not independent variables, and the validity of familiar operations such as the commutation of derivatives may not hold. Further, the interpretation of (6.100) and (6.102) is as follows. One obtains a uniformly valid asymptotic expansion (6.98) for $y(t, \epsilon)$ for times up to $\epsilon t = O(1)$, $\epsilon^n t = O(1)$ by solving (6.100) to order ϵ (order ϵ^n). It is not necessarily true that if one finds the solution to

$$\frac{\partial a}{\partial T_1} = f_1(a, a^*)$$

to be $a(T_1) = a(0) \exp(ia(0)a^*(0)T_1)$, for example, that the behavior of $a(t)$ for times $\epsilon^2 t = O(1)$ will have the form

$$a(T_1, T_2) = b(T_2) \exp ib(T_2)b^*(T_2)T_1.$$

For times $\epsilon^2 t = O(1)$, one must solve

$$\frac{da}{dt} = \epsilon f_1 + \epsilon^2 f_2$$

and you will see in Exercise 6.9 below that the solution may not contain the time scale $T_1 = \epsilon t$ at all! So be careful, especially if you are using the method to find the behavior of the oscillator amplitudes beyond that time at which the first nontrivial change in the amplitude $a(t)$ occurs.

Substituting (6.54) and (6.56) into (6.52), we find at order ϵ^0:

$$\frac{d^2 y_0}{dt^2} + \omega^2 y_0 = 0 \tag{6.103}$$

whose solution is

$$y_0 = a(t)e^{i\omega t} + a^*(t)e^{-i\omega t}$$

with $a(t)$ slowly varying in time. The equation for y_1 is

$$\frac{d^2 y_1}{dt^2} + \omega^2 y_1 = -2i\omega f_1(a, a^*)e^{i\omega t} + (*). \tag{6.104}$$

To eliminate terms that resonate with the natural frequency, choose $f_1(a, a^*) = 0$ whence $y_1 = 0$. Since the general solution of the homogeneous equation

(6.104) has exactly the same form as that of y_0, it can be incorporated in the latter. The amplitude a can and usually will depend on ϵ. Next, at order ϵ^2,

$$\frac{d^2 y_2}{dt^2} + \omega^2 y_2 = -2i\omega f_2(a, a^*)e^{i\omega t} + a^3 \frac{\omega^2}{6}e^{3i\omega t} + \frac{a^2 a^*}{2}e^{i\omega t} + (*). \quad (6.105)$$

It is at this point that we make the first nontrivial choice

$$f_2(a, a^*) = \frac{\partial a}{\partial T_2} = -\frac{i\omega a^2 a^*}{4}, \quad (6.106)$$

and then

$$y_2(t) = \frac{-a^3}{48}e^{3i\omega t} + (*), \quad (6.107)$$

which is bounded. What is remarkable (and what we will not prove here) is that the choice (6.106) eliminates all terms proportional to $(\epsilon^2 t)^n$ that would otherwise have arisen in the asymptotic series (6.98). Continuing, we find that $f_3(a, a^*) = 0$ but that

$$f_4(a, a^*) = \frac{\partial a}{\partial T_4} = \frac{15i\omega}{192}a^3 a^{*2}. \quad (6.108)$$

This choice removes all terms proportional to $(\epsilon^4 t)^n e^{\pm i\omega t}$, $n \geq 1$, from (6.98). In determining (6.108), in addition to the $2i\omega f_4 e^{i\epsilon t}$ from $2i\omega (da/dt)e^{i\omega t}$ in $(d^2 y_0/$
$dt^2)$, one must also include the contribution from $(d^2 a/dt^2)e^{i\omega t}$, which to this order is $-(\omega^2/16)a^3 a^{*2}$.

The solution is then as follows. We solve (6.100),

$$\frac{da}{dt} = -\frac{i\omega}{4}\epsilon^2 a^2 a^* + \frac{15i\omega\epsilon^4}{192}a^3 a^{*2} + \cdots, \quad (6.109)$$

which choice guarantees that (6.98) is uniformly valid in time for times up to and including

$$\epsilon^4 t = O(1). \quad (6.110)$$

The solution of (6.109) is (aa^* is constant)

$$a = a_0 \exp\left(-\frac{i\omega}{4}\epsilon^2 |a_0|^2 + \frac{15}{192}\epsilon^4 |a_0|^4 + \cdots +\right)(*). \quad (6.111)$$

The nonlinear frequency is

$$\omega\left(1 - \frac{1}{4}\epsilon^2|a_0|^2 + \frac{15}{192}\epsilon^4|a_0|^4 + \cdots\right),\tag{6.112}$$

where a_0 is related to A. It is equal to $A/2$ to leading order, but in general is calculated as a power series. From the initial conditions we have

$$y(0) + \epsilon^2 y_2(0) + \epsilon^4 y_4(0) + \cdots = A$$

$$\frac{dy(0)}{dt} + \epsilon^2\frac{dy_2(0)}{dt} + \cdots = 0.\tag{6.113}$$

from which a_0 is found in terms of A.

Now that we have reached an understanding of what we are after, the reader can use the method of multiple scales directly by replacing the operator

$$\frac{d}{dt} \rightarrow \frac{\partial}{\partial t} + \epsilon\frac{\partial}{\partial T_1} + \epsilon^2\frac{\partial}{\partial T_2} + \cdots\tag{6.114}$$

in (6.96) directly and treating t, T_1, T_2, as formally independent. This allows us to keep track of the various powers of ϵ in a very convenient fashion. Remember, however, not to commute derivatives unless you have satisfied yourself that they do indeed commute!

In executing the method of multiple scales, we have effectively ignored all nonresonant higher harmonic terms like $\exp(\pm in\omega t), n > 1$. We could look at this in another way. Set

$$y = \sum_{n=-\infty}^{\infty} a_n(t)e^{in\omega t}, \quad a_{-n} = a_n^*,$$

in (6.96) and obtain

$$\sum_n e^{in\omega t}\left((1 - n^2)a_n + 2in\omega\frac{da_n}{dt} - \frac{d^2 a_n}{dt^2}\right)$$

$$-\frac{\epsilon^2}{6}\sum_{n_1,n_2,n_3} \omega^2 e^{i(n_1+n_2+n_3)\omega t}a_{n_1}a_{n_2}a_{n_3} + \cdots = 0.\tag{6.115}$$

We can obtain equations for the coefficients a_m by multiplying across by $e^{-im\omega t}$ and averaging over the period $2\pi/\omega$. Because $a_n(t)$ is slowly varying, all the terms that contain fast oscillations "average" out to zero because

$$\frac{\omega}{2\pi} \int_0^{(2\pi/\omega)} e^{ir\omega t} \, dt = 0$$

for all nonzero integers r.

For $n = 1$, we find

$$2i\omega\frac{da_1}{dt} + \frac{d^2a_1}{dt^2} = \omega^2 \frac{\epsilon^2}{2}(a_1^2 a_1^* + 2a_1 a_1^* a_3 + \cdots), \tag{6.116}$$

where we include all terms in the second sum in (6.115) for which $n_1 + n_2 + n_3 = 1$. For $n^2 \neq 1$, we would get an algebraic equation for a_n because $(1 - n^2)a_n$ is the biggest term in the bracket of the first sum in (6.115). For example,

$$-8a_3 = \frac{\epsilon^2}{6}a_1^3 + \cdots .$$

Since $a_3 = O(\epsilon^2)$, to leading order

$$2i\omega\frac{da_1}{dt} = \omega^2 \frac{\epsilon^2}{2}a_1^2 a_1^*,$$

which is precisely what we obtained by the method of multiple scales. Therefore, the method of multiple time scales is equivalent to an *averaging method* that

- ignores all higher oscillations

- solves the remaining equations iteratively, namely

$$2i\omega\frac{da_1}{dt} = \frac{\epsilon^2}{2}a_1^2 a_1^* + \text{terms of higher order in } \epsilon,$$

$$-8a_3 = \frac{\epsilon^2}{6}a_1^3 + \text{terms of higher order in } \epsilon.$$

The neglect of terms containing fast oscillations that disappear on averaging is based on the idea that in an equation such as

$$2i\omega\frac{da_1}{dt} = \frac{\epsilon^2}{2}a_1^2 a_1^* + \frac{\epsilon^2}{6}a_1^3 e^{2i\omega t} + \frac{\epsilon^2}{2}a_1 a_1^2 e^{-2i\omega t} + \cdots ,$$

the only cumulative influence on da_1/dt is the nonoscillating term $(\epsilon^2/2)a_1^2 a_1^*$. Since a_1 changes slowly, all the oscillating terms are alternatively pushing and pulling a_1 and more or less acting like random fluctuations. Can we always ignore the fluctuations? The answer is strictly no if we want to talk about what happens over all times $t \to \infty$, especially if the solutions to the averaged equations are separatrices or homoclinic orbits (see Section 61 on chaos and turbulence) that join the unstable and stable manifolds of saddle points (unstable equilibria), in which case the solutions are ultrasensitive to fluctuations. However, for finite but long times, $\epsilon^m t = O(1), m = 1, 2, \ldots,$ there is a precise sense in which the neglect of the fast oscillations is OK. In optics, this neglect is called the *rotating wave approximation* because the terms ignored here have phases that rotate very fast.

The following examples are suggested as exercises.

EXERCISE 6.8

Given

$$\frac{d^2 y}{dt^2} + y = \epsilon \frac{dy}{dt}(1 - y^2).$$

$$\left(\frac{\partial^2}{\partial t^2} + 2\epsilon \frac{\partial^2}{\partial t \partial T_1} + 1\right)(y_0 + \epsilon y_1) = \epsilon \frac{\partial y_0}{\partial t}(1 - y_0^2) + 0(\epsilon^2)$$

$$\frac{\partial^2 y_0}{\partial t^2} + y_0 = 0, \quad y_0 = a(T_1, \ldots)e^{it} + (*).$$

$$\frac{\partial^2 y_1}{\partial t^2} + y_1 = -2\frac{\partial^2}{\partial y \partial T_1}y_0 + \frac{\partial y_0}{\partial t}\left(1 - y_0^2\right)$$

$$= \left(-2i\frac{\partial a}{\partial T_1} + i(1 - aa^*)a\right)$$

$$\times e^{it} - ia^3 e^{3it} + (*).$$

Choose $\dfrac{\partial a}{\partial T_1} = \dfrac{a}{2}(1 - aa^*).$

                                                      □

EXERCISE 6.9

Given

$$\frac{d^2 y}{dt^2} + y = -\epsilon y^3 + \epsilon^2 \frac{dy}{dt}(1 - y^2).$$

Find

$$\frac{\partial a}{\partial T_1}, \frac{\partial a}{\partial T_2}$$

and solve

$$\frac{da}{dt} = \epsilon \frac{\partial a}{\partial T_1} + \epsilon^2 \frac{\partial a}{\partial T_2}.$$

Observe the time scale $T_1 = \epsilon t$ does not appear in the solution. Observe also that $(\partial/\partial T_1) \cdot (\partial a/\partial T_2) \neq (\partial/\partial T_2) \cdot (\partial a/\partial T_1)$.

Answer:

$$\frac{\partial a}{\partial T_1} = \frac{3i}{2}a^2 a^*, \quad \frac{\partial a}{\partial T_2} = \frac{-15i}{16}a^3 a^{*2} + \frac{a}{2}(1 - aa^*).$$

Solution:

$$a = re^{i\phi}, \quad r(t) = \sqrt{\frac{r_0^2 e^{\epsilon^2 t}}{1 - r_0^2 + r_0^2 e^{\epsilon^2 t}}}.$$

$$\phi(t) = \phi_0 + \left(\frac{3}{2\epsilon} - \frac{15}{16}\right) \ln(1 + r_0^2(e^{\epsilon^2 t} - 1))$$

$$+ \frac{15}{16} \frac{r_0^2(1 - r_0^2)(e^{\epsilon^2 t} - 1)}{1 + r_0^2(e^{\epsilon^2 t} - 1)}$$

Show that as $\epsilon^2 t \to 0$, ϵt finite, $r \to r_0$, $\phi \to \phi_0 + 3/2r_0^2\epsilon t$, and that as $\epsilon^2 t \to \infty$, $r \to 1$,

$$\phi \to \phi_0 + \frac{3}{2}\epsilon t - \frac{15}{16}\epsilon^2 t + \left(\frac{3}{2\epsilon} - \frac{15}{16}\right)\ln r_0^2 + \frac{15}{16}(1 - r_0^2).$$

Notice the transition in ϕ_t from $3/2r_0^2\epsilon$ to $\frac{3}{2}\epsilon - \frac{15}{16}\epsilon^2$, reflecting the change in r^2 from r_0^2 to 1. □

EXERCISE 6.10

Given

$$\frac{d^2 y}{dt^2} + (\omega^2 + \epsilon \cos t)y = -\mu \frac{dy}{dt} \cdot y^2, \quad 0 < \epsilon, \mu \ll 1.$$

This is a very useful example to study parametric resonance and understand phase-locking. To leading order $y_0 = ae^{i\omega t} + a^* e^{-i\omega t}$ so that the right-hand side of the equation for y_1 contains terms $-(\epsilon/2)(e^{it} + e^{-it})(ae^{i\omega t} + a^* e^{-i\omega t})$ as well as $-i\omega\mu a^2 a^* e^{i\omega t} + i\omega\mu aa^* e^{-i\omega t} - i\omega\mu a^3 e^{3i\omega t} + i\omega\mu a^{*3} e^{-3i\omega t}$. Resonance occurs if any of the forcing frequencies $\pm 1 \pm \omega$ is equal to the frequency of the homogeneous equation $y_{1tt} + \omega^2 y_1 = 0$ for y_1. Note that this happens when ω is equal to 1/2 because then $-\omega + 1 \simeq \omega$. Let $\omega = 1/2 + \epsilon\delta$ and $y_0 = a^{(1/2)t} + a^* e^{-(1/2)t}$ and find $(T = \epsilon t)$,

$$
y_{1tt} + \frac{1}{4}y_1 = \left(-i\frac{d}{dT}a - \delta a - \frac{1}{2}a^* + \frac{-i\mu}{2\epsilon}a^2 a^* \right) e^{it} + (*)
$$

$$
+ \text{ resonant terms.}
$$

$$
\left(\frac{d}{dT} - i\delta \right) a = \frac{i}{2}a^* - \frac{\mu}{2\epsilon}a^2 a^*.
$$

Observe that if we neglect the nonlinear term (i.e., μ/ϵ is small), then

$$
\left(\frac{d}{dT} + i\delta \right)\left(\frac{d}{dT} - i\delta \right) a = \frac{i}{2}\left(\frac{d}{dT} + i\delta \right) a^* = \frac{1}{4}a
$$

so that

$$
\left(\frac{d^2}{dT^2} + \delta^2 - \frac{1}{4} \right) a = 0.
$$

For $\delta^2 > 1/4$, the response is oscillatory. For $\delta^2 < 1/4$, on the other hand, the frequency of y_0 is locked to the exact subharmonic 1/2 of the frequency of the periodic coefficient $\cos t$ multiplying y in the original equation. The amplitude a grows until saturated by the nonlinear term. Therefore within the window $1/2 - \epsilon/2 < \omega < 1/2 + (1/2)\epsilon$, the frequency of y is locked to the first subharmonic 1/2 of driving frequency 1. It reaches a saturated state such that $(\mu/2\epsilon)r^2 = 1/2 \sin 2\theta$ and where θ is the root of $\cos 2\theta = -2\delta$, for which $\sin 2\theta > 0$. For resonance with higher-order subharmonics, $\omega = m/n, m < n$, one has to continue the expansion to higher order in ϵ. Therefore the windows of phase-locking become progressively narrower $O(\epsilon^2)$, ... and so on as n increases.   □

The Modal Interaction of Two Oscillators without Energy Exchange

This time we will look at the interaction of two right-traveling wavetrains

$$u_0(x, t) = \sum_{j=1}^{2} a_j(T_2 = \epsilon^2 t) \exp i(k_j z - \omega_j t) + (*),$$

$$\omega_j = \omega(k_j) = \sqrt{\omega_0^2 + c^2 k_j^2}$$

of the partial differential equation

$$\frac{\partial^2}{\partial t^2} - c^2 \frac{\partial^2 u}{\partial z^2} + \omega_0^2 u = \frac{\epsilon^2 \omega_0^2}{6} u^3.$$

Let

$$u(z, t) = u_0(z, t) + \epsilon^2 u_2(z, t) + \cdots \tag{6.117}$$

and obtain

$$\frac{\partial^2 u_2}{\partial t^2} - c^2 \frac{\partial^2 u_2}{\partial z^2} + \omega_0^2 u_2 = RHS,$$

when the *RHS* contains the secular terms

$$\frac{\omega_0^2}{6}(3a_1^2 a_1^* + 6a_2 a_2^* a_1)e^{i(k_1 z - \omega_1 t)} + \frac{\omega_0^2}{6}(3a_2^2 a_2^* + 6a_1 a_1^* a_2)e^{i(k_2 z - \omega_2 t)}$$

plus their complex conjugates and the terms

$$-2i\omega_1 \frac{\partial a_1}{\partial T_2} e^{i(k_1 z - \omega_1 t)} - 2i\omega_2 \frac{\partial a_2}{\partial T_2} e^{i(k_2 z - \omega_2 t)}$$

arising from the mixed partial derivative $2(\partial^2 u_0)/(\partial t \partial T_2)$, $T_2 = \epsilon^2 t$. (Here the derivatives $(\partial/\partial t) \cdot (\partial/\partial T_2)$ do commute because u_0 is a product of a function of t and a function of T_2.) Choose

$$\frac{da_1}{dt} = \epsilon^2 \frac{\partial a_1}{\partial T_2} = \frac{i\omega_0^2}{4\omega_1} \epsilon^2 (a_1 a_1^* + 2a_2 a_2^*) a_1$$

$$\frac{da_2}{dt} = \epsilon^2 \frac{\partial a_2}{\partial T_2} = \frac{i\omega_0^2}{4\omega_2} \epsilon^2 (2a_1 a_1^* + a_2 a_2^*) a_2 \tag{6.118}$$

and then $u_2(x, t)$ is bounded and the asymptotic expansion (6.117) uniformly valid for times $\epsilon^2 t \leq O(1)$. From (6.118), we see that the individual energies $\omega_1 a_1 a_1^* = E_1$ and $\omega_2 a_2 a_2^* = E_2$ are conserved and that

$$\frac{da_1}{dt} = \frac{i\omega_0^2}{2\omega_1}\epsilon^2 \left(\frac{E_1}{\omega_1} + \frac{2E_2}{\omega_2}\right) a_1$$

$$\frac{da_2}{dt} = \frac{i\omega_0^2}{2\omega_2}\epsilon^2 \left(\frac{2E_1}{\omega_1} + \frac{E_2}{\omega_2}\right) a_2,$$

so that the result of the nonlinear interaction is simply to modify the two frequencies

$$\omega_1 \rightarrow \omega_1 - \frac{\omega_0^2}{2\omega_2}\epsilon^2 \left(\frac{E_1}{\omega_2} + \frac{2E_2}{\omega_2}\right)$$

$$\omega_2 \rightarrow \omega_2 - \frac{\omega_0^2}{2\omega_2}\epsilon^2 \left(\frac{2E_1}{\omega_1} + \frac{E_2}{\omega_2}\right).$$

(6.119)

This interaction between the two modes is called a modal interaction. For wavetrains (this is not the case for wavepackets!), it merely adjusts the frequencies of each and no energy is exchanged.

Exchange of Energy between Resonant Oscillators Due to Weak Linear and Nonlinear Coupling

Consider the following simple mechanical model of two coupled pendula shown in Figure 6.15, which displays behavior similar to two coupled modes of polarization in linearly and nonlinearly birefringent media. The pendula are each of length l, and the restoring force of the spring that couples the two masses is taken to be $k'\Delta x$ where Δx is the amount of extension or compression. For small deflections of order ϵ, $0 < \epsilon \ll 1$ from the vertical axes, the equations, for the angular displacements $\phi_j = \epsilon x_j$, $j = 1, 2$ are

$$\frac{d^2 x_1}{dt^2} + \frac{g}{l} x_1 = \frac{g}{6l}\epsilon^2 x_1^3 + k'(x_2 - x_1)$$

$$\frac{d^2 x_2}{dt^2} + \frac{g}{l} x_2 = \frac{g}{6l}\epsilon^2 x_2^3 - k'(x_2 - x_1).$$

(6.120)

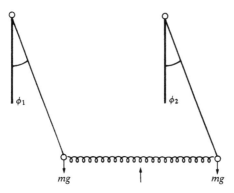

FIGURE 6.15 Coupled pendula.

We take k' to be of the same order as the nonlinear terms and set $k' = \epsilon^2 k$. Write

$$x_1 = A(T = \epsilon^2 t)e^{-i\omega t} + A^*(T)e^{i\omega t} + \epsilon^2 x_1^{(1)} + \cdots$$
$$x_2 = B(T = \epsilon^2 t)e^{-i\omega t} + B^*(T)e^{i\omega t} + \epsilon^2 x_2^{(1)} + \cdots \tag{6.121}$$

where $\omega^2 = (g/l) + \epsilon^2 k > 0$. To remove secular terms from $x_1^{(1)}$ and $x_2^{(1)}$, we must choose

$$\frac{dA}{dT} = i\Delta k B + i\beta A^2 A^*$$
$$\frac{dB}{dT} = i\Delta k A + i\beta B^2 B^* \tag{6.122}$$

where $\Delta k = (k/2\omega)$ and $\beta = (g/4\omega l)$. If the two pendulum lengths were slightly different so that $g/l_1 = \omega_1^2 = (\omega + \epsilon^2 \delta/2)^2$ and $g/l_2 = \omega_2^2(\omega - \epsilon^2 \delta/2)^2$, then phase mismatch factors $e^{+i\delta T}$ and $e^{-i\delta T}$ would multiply $i\Delta k B$ and $i\Delta k A$ respectively. In what follows we take $\delta = 0$. These equations may be solved exactly by introducing

$$E = AA^* + BB^*$$
$$N = AA^* - BB^*$$
$$P_= = AB^* + A^*B$$
$$P_- = i(AB^* - A^*B). \tag{6.123}$$

From (6.122), we find

$$\frac{dE}{dT} = 0$$

$$\frac{dN}{dT} = -2\Delta k P_-$$

$$\frac{dP_+}{dT} = \beta N P_- \tag{6.124}$$

$$\frac{dP_-}{dT} = 2\Delta k N - \beta N P_+.$$

Eliminate P_- and obtain

$$\frac{dP_+}{dT} = -\frac{\beta}{2\Delta k} N \frac{dN}{dT}$$

or

$$P_+ + \frac{\beta}{4\Delta k} N^2 = \frac{\beta}{4\Delta k} N_0^2 + P_{+0} \tag{6.125}$$

and then

$$\frac{d^2 N}{dT^2} = -\frac{\partial U}{\partial N} \tag{6.126}$$

where

$$U(N) = \left(2(\Delta k)^2 - \beta \Delta k P_{+0} - \frac{\beta^2}{4} N_0^2 \right) N^2 + \frac{\beta^2}{8} N^4. \tag{6.127}$$

Let us assume all the energy starts out in one mode, either A or B is initially zero. Then $P_{+0} = P_{-0} = (dN/dT)(0) = 0$ and $N_0 = N(0) = \pm E$ depending on whether the energy is initially in A or B respectively. If the square of the linear frequency mismatch $4(\Delta k)^2$ is less than the square of the nonlinear frequency corrections $(\beta^2/2)E^2$, then the potential $U(N)$ has one stable minimum at $N = 0$ and energy is shared equally between the two modes. On the other hand, if $4(\Delta k)^2 < (\beta^2/2)E^2$, then if all the energy is in A, say, and if

$$U(0) = \left(2(\Delta k)^2 - \frac{\beta^2}{8} E^2 \right) E^2 < 0,$$

N oscillators about the minimum

$$N = \sqrt{E^2 - (8(\Delta k)^2/\beta^2)}$$

and the energy remains largely in A. If $U(0) > 0$ (i.e., $(\beta^2/4)E^2 > 2(\Delta k)^2 > (\beta^2/8)E^2$), then the energy is shared. The reader is asked consider what happens when the frequency mismatch is reintroduced. Set $A \to Ae^{i\delta/2T}$, $B \to Be^{-i\delta/2T}$, whereupon one gets the extra terms $-i\delta/2A$ and $-i\delta/2B$ in (6.122). Repeat the above analysis and show that the exchange of energy is inhibited as δ becomes large.

Resonant Interactions between Three and Four Waves

The goal of this section is to introduce the reader to strong or resonant interactions that take place between three or four weakly coupled wavetrains

$$A_j e^{i\vec{k}\cdot\vec{x}-i\omega_j t} + (*) \tag{6.128}$$

with dispersion relation $\omega_j = \omega(\vec{k}_j) j = 1, 2, 3, 4$ whose frequencies and wavevectors satisfy the resonance conditions

$$\sum_{j=1}^{N} \vec{k}_j = 0, \quad N = 3, 4$$

$$\tag{6.129}$$

$$\sum_{j=1}^{N} \omega_j = 0.$$

If ω is a multivalued function of $\vec{k}_j$, then any of the roots can be used in satisfying (6.129). Such interactions give rise to Brillouin and Raman scattering. In order to become comfortable with the mathematical treatment, we start by considering a simple model of Rossby waves, which the long (> 1000 kms) waves as in the earth's atmosphere and oceans and in the atmospheres of other rotating planets. We then look at the vibrational modes in a diatomic crystal lattice. Finally, we look at the four-wave mixing associated with the NLS equation.

Rossby Waves

Consider,

$$\frac{\partial \psi}{\partial t}(\nabla^2 - \alpha^2)\psi + \beta \frac{\partial \psi}{\partial x} = J(\psi, \nabla^2 \psi), \tag{6.130}$$

an equation that expresses the conservation of the total vorticity of a vertical column of the atmosphere.

$$J(\psi, \nabla^2\psi) = \frac{\partial\psi}{\partial x}\frac{\partial}{\partial y}\nabla^2\psi - \frac{\partial\psi}{\partial y}\frac{\partial}{\partial x}\nabla^2\psi.$$

The linear wavetrain solutions

$$\psi(x, y, t) = Ae^{i\vec{k}\cdot\vec{x}-i\omega t} + (*), \quad \vec{k} = (k_x, k_y), \tag{6.131}$$

are also exact solutions of the nonlinear equation, because for ψ given by (6.131), $J(\psi, \nabla^2\psi) = 0$. They have the dispersion relation

$$\omega(\vec{k}) = -\frac{\beta k_x}{\alpha^2 + k^2}, \quad k^2 = k_x^2 + k_y^2. \tag{6.132}$$

The phase velocity ω/k_x in the x (or east–west direction) is in the westward or negative x direction. On the other hand, the group velocity

$$\nabla_{\vec{k}}\omega = \beta\frac{k_x^2 - k_y^2 - \alpha^2}{(\alpha^2 + k^2)^2}, \quad 2\beta\frac{k_x k_y}{\alpha^2 + k^2} \tag{6.133}$$

can be in either direction. Let us look for a solution that to first order is a linear superposition of low-amplitude wavetrains

$$\psi(x, y, t) = \epsilon\sum_{j=1}^{3}\left(A_j(T = \epsilon t)e^{i\theta_j} + (*)\right) + \epsilon^2\psi_1\cdots+\cdots \tag{6.134}$$

where

$$\theta_j = \vec{k}_j\cdot\vec{x} - \omega_j t, \quad \omega_j = \omega(k_j) \tag{6.135}$$

and $(\vec{k}_j, \omega_j)_{j=1}^{3}$ satisfy the resonance relations (6.129) with $N = 3$. The equation for ψ_1 is

$$\frac{\partial}{\partial t}(\nabla^2 - \alpha^2)\psi_1 + \beta\frac{\partial\psi_1}{\partial x} = J(\psi_0, \nabla^2\psi_0) - \frac{\partial}{\partial T}(\nabla^2 - \alpha^2)\psi_0 \tag{6.136}$$

where $\psi_0 = \sum_{j=1}^{3}(A_j\exp i\theta_j + (*))$. Because of the resonance conditions (6.129), the quadratic products of $e^{-i\theta_2}$ and $e^{-i\theta_3}$ in $J(\psi_0, \nabla^2\psi_0)$ gives rise to the term $e^{i\theta_1}$, which is a solution of the homogeneous equation for ψ_1. Consequently, these terms give rise to a resonant contribution in ψ_1 proportional to $te^{i\theta_1}$ that renders the asymptotic expansion (6.134) nonuniform. We avoid this difficulty and remove these terms by choice of (dA_1/dT); similar choices of (dA_2/dT) and (dA_3/dT) remove the resonant terms $e^{i\theta_2}$ and $e^{i\theta_3}$.

Another way to look at this is as follows. Suppose we consider the quadratic product of two different wavetrains $\exp(-i\vec{k}_2 \cdot \vec{x} + i\omega_2 t)$ and $\exp(-i\vec{k}_3 \cdot \vec{x} + i\omega_3 t)$ in $J(\psi_0, \nabla^2 \psi_0)$. We obtain

$$-(\vec{k}_2 \times \vec{k}_3)(k_2^2 - k_3^2)e^{-i(\vec{k}_2+\vec{k}_3)\cdot\vec{x}+i(\omega_2+\omega_3)t} \tag{6.137}$$

Now look for a solution $\psi_1 = C(t)\exp -i(\vec{k}_2 + \vec{k}_3) \cdot x$ and we find

$$(\alpha^2 + |\vec{k}_2 + \vec{k}_3|^2)\frac{dC}{dt} + i\beta(k_{2x} + k_{3x})C = (\vec{k}_2 \times \vec{k}_3)(k_{2x} - k_3^2)e^{i(\omega_2+\omega_3)t}$$

whose particular solution is

$$C(t) = \frac{-i(\vec{k}_2 \times \vec{k}_3)(k_2^2 - k_3^2)e^{i(\omega_2+\omega_3)t}}{(\alpha^2 + |\vec{k}_2 + \vec{k}_3|)(\omega_2 + \omega_3) + \beta(k_{2x} + k_{3x})}. \tag{6.138}$$

The denominator of (6.138) is zero when

$$\omega(\vec{k}_2) + \omega(\vec{k}_3) = \omega(\vec{k}_2 + \vec{k}_3), \tag{6.139}$$

which is precisely the resonance condition (6.129) when $\vec{k}_1$ is eliminated.

The removal of the zero dominator terms requires the three complex equations

$$\frac{dA_l}{dT} = K_l A_m^* A_n^*,$$
$$K_l = \frac{(\vec{k}_m \times \vec{k}_n)(k_m^2 - k_n^2)}{\alpha^2 + k_l^2} \tag{6.140}$$

when l, m, n is cycled over 1, 2, 3. Observe that there are several conserved quantities

$$\frac{d}{dT}((\alpha^2 + k_1^2)^j A_1 A_1^* + (\alpha^2 + k_2^2)^j A_2 A_2^* + (\alpha^2 + k_j^2)^j A_3 A_3^*)$$

$$= 0, j = 1, 2 \tag{6.141}$$

because $\vec{k}_1 \times \vec{k}_2 = k_{1x}k_{2y} - k_{1y}k_{2x} = \vec{k}_2 \times \vec{k}_3 = \vec{k}_3 \times \vec{k}_1$. Also,

$$\frac{\alpha^2 + k_1^2}{(\vec{k}_2 \times \vec{k}_3)(k_1^2 - k_3^2)}A_1 A_1^* - \frac{\alpha^2 + k_2^2}{(\vec{k}_3 \times \vec{k}_1)(k_3^2 - k_1^2)}A_2 A_2^* = \text{constant} \tag{6.142}$$

as well as the same relations found by cycling over 1, 2, and 3. (In physics, (6.142) are called the Manley–Rowe equations.) The net result is that over time scales $t = O(1/\epsilon)$, there is an order one exchange of energy between the three modes in the resonant triad.

Given a weakly nonlinear system with quadratic nonlinearity, the locus of resonant waves can be found by solving (6.129) or alternatively (6.139), which, if $\vec{k}_2$ is given, is a one-dimensional manifold (a curve) in $\vec{k}_3(l_3, m_3)$ space.

We also observe that the resonance condition can be interpreted as the conservation of (wave) momentum and (wave) energy in a scattering process in which an incoming wave with momentum $-\vec{k}_1$ and energy $-\omega_1$, say, scatters into two outgoing waves with momenta $\vec{k}_2$ and $\vec{k}_3$ and energies ω_2 and ω_3, as depicted in Figure 6.16.

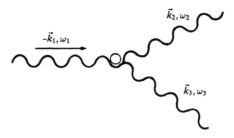

FIGURE 6.16

Vibrational Modes of a Diatomic Crystal Lattice: Brillouin and Raman Scattering

We now consider the interaction between light and matter in which indirect or triad resonances play an important role. In particular, we consider two electromagnetic waves, the *pump* wave

$$\vec{E}_p = \hat{e}(A_p e^{i\vec{k}_p \cdot \vec{x} - i\omega_p t} + (*)) \tag{6.143}$$

and the Stokes wave

$$\vec{E}_s = \hat{f}(A_s e^{i\vec{k}_s \vec{x} - i\omega_s t} + (*)) \tag{6.144}$$

and a vibrational mode

$$\vec{Q}_m = \hat{g}(Q e^{i\vec{K} \cdot \vec{x} - i\Omega t} + (*)) \tag{6.145}$$

of the medium. We assume that the light and matter fields are weakly and quadratically coupled. The precise nature of the coupling term is given in

Chapter 4. What we are interested in here are the possible resonances in which an electromagnetic pump wave $\vec{k}_p, \vec{\omega}_p$ can scatter into an electromagnetic Stokes wave $\vec{k}_s, \omega_s$ and material vibration $\vec{K}, \Omega$. The difference with the previous example is simply that the electromagnetic and material waves satisfy different equations and have different dispersion relations. The pump and Stokes wave $\vec{k}_p, \omega_p$ and $\vec{k}_s, \omega_s$ satisfy the dispersion relation

$$D(|\vec{k}|, \omega) = 0 \tag{6.146}$$

for electromagnetic waves; the material vibration satisfies a material dispersion relation

$$D_M(\vec{K}, \Omega) = 0. \tag{6.147}$$

The material dispersion relation often has several branches $\Omega = \Omega(\vec{K})$, corresponding to the various possibilities. In the illustrative example of a diatomic lattice, there are two branches $\Omega_{\text{optic}}(K = |\vec{K}|)$ and $\Omega_{\text{acoustic}}(K = |\vec{K}|)$. The three-wave resonant scattering of the pump wave off the former gives rise to *Roman scattering* in which the frequency Ω_{optic} of the scattered wave can be of the same order of magnitude as that of the electromagnetic waves. The scattering of the pump wave off the latter, the acoustic modes with much lower frequencies, is called *Brillouin scattering*.

If the scattered electromagnetic wave has a higher frequency than the incoming pump wave, it is called an anti-Stokes wave.

We now calculate the material dispersion relation for the case of the diatomic lattice and develop a geometrical picture of the resonant wavevectors. The model is simple and consists of two interpenetrating face-centered-cubic lattices, one of which is displaced along the body diagonal of the other by one-quarter of its length. We may picture the model (Figure 6.17) as a chain of atoms of interspersed masses M and m connected by springs with spring constant μ.

The transverse displacement of the nth mass is measured by y_n. The equations are

$$M\frac{d^2 y_{n+1}}{dt^2} = \mu(y_{2n+2} - 2y_{2n+1} + y_{2n})$$

$$m\frac{d^2 y_{2n}}{dt^2} = \mu(y_{2n+1} - 2y_n + y_{2n-1}), \tag{6.148}$$

M	m	M	m	M
$2n+1$	$2n$	$2n+2$	$2n+2$	$\cdots$

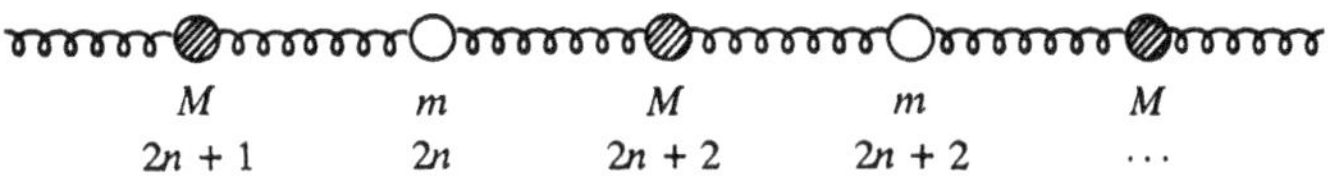

FIGURE 6.17 Model of resonant wavevectors.

which admit wavetrain solutions,

$$y_{2n+1} = \alpha e^{ik(2n+1)a - i\omega t} + (*)$$

$$y_{2n} = \beta e^{ik2na - i\omega t} + (*)$$

(6.149)

where

$$\begin{pmatrix} 2\mu - \omega^2 M & -2\mu \cos ka \\ -2\mu \cos \mu a & 2\mu - \omega^2 m \end{pmatrix} \begin{pmatrix} \alpha \\ \beta \end{pmatrix} = 0.$$

(6.150)

The dispersion relation is

$$\omega^2 = \mu \left(\frac{1}{M} + \frac{1}{m} \right) \pm \mu \left(\left(\frac{1}{M} + \frac{1}{m} \right)^2 - \frac{4 \sin 2ka}{Mm} \right)^{1/2}$$

(6.151)

and has two branches, the optical branch ω_+^2 and the acoustic branch ω_-^2. Observe that for $ka \ll 1$,

$$\omega_+^2 \to 2\mu \left(\frac{1}{M} + \frac{1}{m} \right), \quad \text{optical branch}$$

$$\omega_-^2 \to \frac{4a^2 \mu}{Mm} k^2, \quad \text{acoustical branch.}$$

The graphs of the two branches are shown in Figure 6.18. We now discuss the existence of triad resonances.

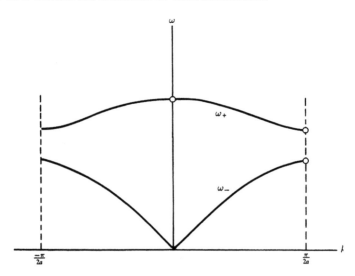

FIGURE 6.18   Optic (ω_+) and acoustic (ω_-) branches of dispersion relation.

Consider the acoustic branch (Brillouin scattering). Observe in Figure 6.19 that in ω, $\vec{k}$ space, the vectors $\vec{01}(\omega_p, \vec{k}_p)$, $\vec{02} = \vec{31}(\omega_s, \vec{k}_s)$, an $\vec{03}(\omega_a, \vec{k}_a)$ form a triangle and therefore satisfy the resonance relations

$$\vec{k}_a + \vec{k}_s = \vec{k}_p$$

$$\omega_a + \omega_s = \omega_p. \tag{6.152}$$

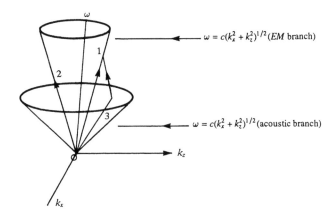

FIGURE 6.19

Notice that since $\omega_a \ll \omega_s$, ω_p, then $\omega_s \simeq \omega_p$. Also observe that $\vec{k}_s$ and $\vec{k}_p$ point in opposite directions so that $k_{sz} \simeq -k_{pz}$, which means that $k_{az} \simeq 2k_{pz}$. Therefore in Brillouin scattering we expect that the dominant scattered wave is backwards and propagates in the direction opposite to that of the incoming pump wave.

Consider the optical branch (Raman scattering), depicted in Figure 6.20. Since the optical mode has very little dispersion, we simply choose

$$\omega_0 + \omega_s = \omega_p$$

and then any $\vec{k}_s$ belonging to the intersection of the two surfaces. Raman scattering can occur equally in all directions. In both cases, the equations are

$$\frac{dA_p}{dt} = K_p A_s Q \tag{6.153a}$$

$$\frac{dA_s}{dt} = K_s A_p Q^* \tag{6.153b}$$

$$\frac{dQ}{dt} + \gamma Q = K_m A_p A_s^* \tag{6.153c}$$

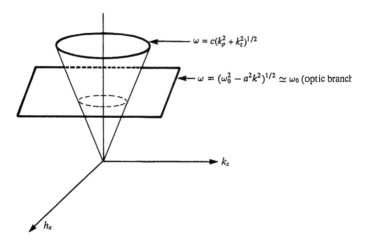

FIGURE 6.20

where K_p, K_s, and K_m are the coupling coefficients and γ is the damping rate of the material mode. Often it is the case that the time scale on which A_p and A_s vary is longer than γ^{-1} so that one can adiabatically eliminate Q from (6.153c) whence (6.153a) and (6.153b) become

$$\frac{dA_p}{dt} = \frac{K_p K_M}{\gamma} A_s A_s^* A_p$$

$$\frac{dA_s}{dt} = \frac{K_s K_M^*}{\gamma} A_p A_p^* A_s. \tag{6.154}$$

The rate at which the Stokes wave gains energy from the pump wave is given by

$$g = \text{Re}\frac{K_s K_M^*}{\gamma},$$

which we will find, when we calculate the coefficients, is positive.

Four-wave resonant processes

An excellent and appropriate model for four-wave resonant processes is the two- or three-dimensional NLS equation

$$\frac{\partial \psi}{\partial t} - i\nabla^2 \psi - i\alpha\psi^2\psi^* = 0 \tag{6.155}$$

with cubic nonlinearity. Take

$$\psi = \epsilon \psi_0 + \epsilon^3 \psi_2 + \cdots \tag{6.156}$$

where ψ_0 consists of four wavetrains $(\vec{k}_j, \omega_j)$,

$$\psi_0 = \sum_{j=0}^{3} A_j(T) e^{i(\vec{k}_j \cdot \vec{x} - \omega_j t)}, \quad T = \epsilon^2 t, \tag{6.157}$$

where

$$\vec{k}_0 + \vec{k}_1 = \vec{k}_2 + \vec{k}_3$$

$$\omega_0 + \omega_1 = \omega_2 + \omega_3 \tag{6.158}$$

and where $\omega(\vec{k})$ is given by the linear dispersion relation

$$\omega = k^2. \tag{6.159}$$

The equation for ψ_2 is

$$\frac{\partial \psi_2}{\partial t} - i \nabla^2 \psi_2$$

$$= i\alpha (A_0 e^{i(\vec{k}_0 \cdot \vec{x} - \omega_0 t)} + A_1 e^{i(\vec{k}_1 \cdot \vec{x} - \omega_1 t)} + A_2 e^{i(\vec{k}_2 \cdot \vec{x} - \omega_2 t)}$$

$$+ A_3 e^{i(\vec{k}_3 \cdot \vec{x} - \omega_3 t)})^2$$

$$\times (A_0^* e^{-i(\vec{k}_0 \cdot \vec{x} - \omega_0 t)} + A_1^* e^{-i(\vec{k}_1 \cdot \vec{x} - \omega_1 t)} + A_2^* e^{-i(\vec{k}_2 \cdot \vec{x} - \omega_2 t)}$$

$$+ A_3^* e^{-i(\vec{k}_3 \cdot \vec{x} - \omega_3 t)})$$

$$- \frac{dA_0}{dT} e^{i(\vec{k}_0 \cdot \vec{x} - \omega_0 t)} - \frac{dA_1}{dT} e^{i(\vec{k}_1 \cdot \vec{x} - \omega_1 t)} - \frac{dA_2}{dT} e^{i(\vec{k}_2 \cdot \vec{x} - \omega_2 t)}$$

$$- \frac{dA_3}{dT} e^{i(\vec{k}_3 \cdot \vec{x} - \omega_3 t)} \tag{6.160}$$

All terms proportional to $e^{+i(\vec{k}_j \cdot \vec{x} - \omega_j t)} j = 0, 1, 2, 3$ on the right-hand side of (6.160) give rise to a resonant response in ψ_2. These are removed by the appropriate choices of (dA_j/dT), $j = 0, 1, 2, 3$, which are

$$\frac{dA_0}{dT} = i\alpha A_0^2 A_0^* + 2i\alpha \sum_{j\neq 1}^{3} A_j A_j^* A_0 + i\alpha A_2 A_3 A_1^*$$

$$\frac{dA_1}{dT} = i\alpha A_1^2 A_1^* + 2i\alpha \sum_{j\neq 1} A_j A_j^* A_1 + i\alpha A_2 A_3 A_0^*$$

$$\frac{dA_2}{dT} = i\alpha A_2^2 A_2^* + 2i\alpha \sum_{j\neq 2} A_j A_j^* A_2 + i\alpha A_0 A_1 A_3^*$$

$$\frac{dA_3}{dT} = i\alpha A_3^2 A_3^* + 2i\alpha \sum_{j\neq 3} A_j A_j^* A_3 + i\alpha A_0 A_1 A_2^*. \tag{6.161}$$

Note that $\sum_0^3 A_j A_j^*$, $\sum_0^3 \vec{k}_j A_j A_j^*$ and $\sum_0^3 \omega_j A_j A_j^*$, corresponding to conservations of power, momentum and energy are all conserved. Also note that $A_0 A_0^* - A_1 A_1^*$, $A_0 A_0^* + A_2 A_2^*$, $A_0 A_0^* + A_3 A_3^*$, $A_1 A_1^* + A_2 A_2^*$, $A_1 A_1^* + A_3 A_3^*$ and $A_2 A_2^* - A_3 A_3^*$ are also conserved. This is the analog to (6.142) and tells us how energy is shared. If $A_0 A_0^*$ gains, $A_2 A_2^*$ loses and so on. As a special application of four-wave mixing, we now take a second look at the Benjamin–Feir instability.

EXERCISE 6.11

The focusing or Benjamin–Feir instability, revisited.
Consider the four waves $\vec{k}_0, \vec{k}_0, \vec{k}_0 + \epsilon\vec{K}, \vec{k}_0 - \epsilon\vec{K}$ satisfying

$$\vec{k}_0 + \vec{k}_0 = \vec{k}_0 + \epsilon\vec{K} + \vec{k}_0 - \epsilon\vec{K}$$

$$\omega(\vec{k}_0) + \omega(\vec{k}_0) = \omega(\vec{k}_0 + \epsilon\vec{K}) + \omega(\vec{k}_0 - \omega\vec{K}) - \epsilon^2 \omega_0'' K^2,$$

namely the resonance conditions (6.158) to order ϵ^2.

In this case, the $\vec{k}_0$ and $\vec{k}_1$ waves are exactly the same, so the sum in (6.161) omits $j = 1$ and the last term in the first equation is multiplie by $\exp(i\omega_0'' K^2 t)$ and the last term in the last two by $\exp(-i\omega_0'' K^2 t)$. Now look at the stability of the exact solution $A_0 = a_0 \exp i\alpha |a_0|^2 T$. Set $A_2, A_3 = (a_2, a_3) \exp i(\alpha |a_0|^2 - (1/2)\omega_0'' K^2)T$, linearize for small a_2, a_3, and obtain

$$\left(\frac{d}{dT} - i\frac{\omega_0''}{2}K^2 - i\alpha|a_0|^2\right)a_2 = i\alpha a_0^2 a_3^*$$

$$\left(\frac{d}{dT} - i\frac{\omega_0''}{2}K^2 - i\alpha|a_0|^2\right)a_3 = i\alpha a_0^2 a_2^*.$$

Set $a_2, a_3 \propto e^{\sigma T}$ and find

$$\sigma^2 = \alpha|a_0|\omega_0''K^2 - \frac{\omega_0''^2 K^4}{4},$$

which is exactly the result obtained in Exercise (2.13) with $\beta = \alpha$ and $\gamma = 1$ and $\omega_0'' = 2$. Thus the Benjamin–Feir or focusing instability results from an energy transfer to the sidebands $\vec{k}_0 \pm \epsilon\vec{K}$ through an "almost" resonant four-wave mixing process. $\qquad\square$

Geometric Optics and the WKBJ Approximation

The basic idea of geometric optics is that in a medium where the refractive index changes significantly over many wavelengths of light, $(1/n)\nabla n \ll 1/\lambda$, the wavetrains are almost planar almost everywhere in space. I do not think the reader finds this statement unreasonable. The goal of this section is to illustrate with simple examples exactly what we mean by "almost" and the methods one uses to handle such situations. Let us begin with a pendulum in which the length is being slowly shortened so that the rate of shortening is much slower than the local pendulum period $2\pi\sqrt{l(\epsilon t)/g}$. The equation of motion is

$$\frac{d^2x}{dt^2} + \omega^2(\epsilon t)x = 0 \tag{6.162}$$

where $0 < \epsilon \ll 1$. At first one might be attempted to such approximate solutions to (6.162) in the form

$$x \simeq ae^{i\omega(\epsilon t)t} + (*), \tag{6.163}$$

but a little calculation shows this must be wrong because the derivative of $x(t)$ contains terms proportional to $\omega'(\epsilon t)\epsilon t$, which clearly cannot be balanced in the equation for times ϵt of order one. The best thing therefore is to leave the phase undetermined for the moment and write it as

$$\theta = \frac{1}{\epsilon} h(T)$$

(6.164)

$$T = \epsilon t.$$

Then, according to the chain rule of multiple scales,

$$\frac{d}{dt} = h'(T) \frac{\partial}{\partial \theta} + \epsilon \frac{\partial}{\partial T}$$

(6.165)

where $h' = (dh/dT) = (d\theta/dt)$. Substitute (6.165) into (6.162) and let

$$x = x_0 + \epsilon x_1 + \cdots$$

(6.166)

to obtain

$$\left(h'^2 \frac{\partial^2}{\partial \theta^2} + \omega^2 + \epsilon \left(2h' \frac{\partial^2}{\partial T \partial \theta} + h'' \frac{\partial}{\partial \theta} \right) + \epsilon^2 \frac{\partial^2}{\partial T^2} \right) (x_0 + \epsilon x_1 + \cdots) = 0.$$

(6.167)

At the first balance, we obtain:

$$\epsilon^0 : \left(h'^2 \frac{\partial^2}{\partial \theta^2} + \omega^2 \right) x_0 = 0.$$

(6.168)

Choose $h'^2 = \omega^2$. This choice is necessary because otherwise $x_0 \simeq ae^{i(\omega/h')\theta}$ and each T derivative would bring down a factor θ. Thus $h(T) = \int^T \omega(T)dT$. The phase θ is the average of the slowly varying frequency ω and the instantaneous frequency is $d\theta/dt = dh/dT = \omega(T)$. Then

$$x_0 = a(T)e^{i\theta} + (*).$$

At order ϵ,

$$\omega^2 \left(\frac{\partial^2}{\partial \theta^2} + 1 \right) x_1 = - \left(2\omega \frac{\partial^2}{\partial T \partial \theta} + \frac{d\omega}{dT} \frac{\partial}{\partial \theta} \right) x_0$$

$$= - \left(2\omega \frac{da}{dT} + \frac{d\omega}{dT} a \right) i e^{i\theta} + \left(2\omega \frac{da^*}{dT} + \frac{d\omega}{dT} a^* \right) i e^{-i\theta}.$$

(6.169)

Removing secular terms means we choose

$$2\omega\frac{da}{dT} + \frac{d\omega}{dT}a = 0$$

or

$$\omega a a^* = \text{constant}. \tag{6.170}$$

The relation (6.170) is called *conservation of action*. To first approximation, then, the solution of (6.162) is

$$x \simeq \frac{r_0}{\sqrt{\omega(T)}} \exp i \int^t (\omega(\epsilon t)dt + i\theta_0) + (*). \tag{6.171}$$

The reader should recognize the similarity of results contained here with Section 2f. In that case, the effective wavenumber

$$l = \sqrt{n^2(x)\omega^2/c^2 - k^2}$$

can become zero at the caustic, near which point the WKBJ approximation breaks down.

6h. Wavepackets, the NLS Equation, and the Three (TWI) and Four (FWI) Wave Interaction Equations

In Section 6g, we discussed the notion of wavepackets in the optical context. We pointed out that it is often convenient to think of optical fields as functions of time t and transverse coordinates x and y evolving in z, the direction of propagation of the carrier wave. Here we adopt the more conventional approach of taking the time t to be the evolution variable. Our goal in this section is to consider the effects of spatial modulation in addition to and combined with the effects of the nonlinear interactions considered in the last section. We now want to deal with self, mutual, and resonant interactions of *wavepackets* rather than wavetrains. We will find that the wavepacket analogs of the self-modal, mutual modal and three- and four-wave resonant interactions of wavetrains to be described by the NLS, coupled NLS, and TWI and FWI equations respectively (see references [23, 29]).

We begin by analyzing small-amplitude wavepackets of the sine–Gordon equation

$$\frac{\partial^2 u}{\partial t^2} - c^2\frac{\partial^2 u}{\partial z^2} + \omega_0^2 \sin u = 0, \tag{6.172}$$

which, we saw in Chapter 5, is the governing equation for the electric field in an excitable medium of two-level atoms when the frequency ω of the external wave that excites the medium is the same as the frequency $\omega_{12} = (E_1 - E_2)/h$ of a photon emitted by an atom that makes a transition from the higher E_1 to lower energy state E_2. We want to look at the evolution of a group of wavetrains $\exp(ikz - i\omega(k)t)$ with wavenumbers k lying in a small band of order $\mu, 0 < \mu \ll 1$ about k_0. The amplitudes of each of these wavetrains is $\epsilon, 0 < \epsilon \ll 1$. Each wavetrain satisfies the linearized equation (6.172) at the leading order, but because of nonlinear coupling, the different wavetrains in the wavepacket may exchange energy over long times, and so we let each of the wavetrain amplitudes $A(\vec{k}, t)$ depend weakly on time. We will discuss what size $(\partial A/\partial t)$ must be shortly. The method used follows [23] closely.

Accordingly, we approximate the field $u(z, t)$ by

$$u(z, t) = \epsilon \sum_j \left(A_j(t) e^{i(k_0 + \mu K_j)z - i\omega(k_0 + \mu K_j)t} + (*) \right) + \epsilon^2 u_2 + \epsilon^3 u_3 + \cdots$$

$$(6.173)$$

which can be written as

$$u(z, t) = \epsilon \left(A(Z = \mu z, T = \mu t) e^{ik_0 z - i\omega(k_0)t} + (*) \right) + \epsilon^2 u_1 + \epsilon^3 u_2 + \cdots$$

$$(6.174)$$

with

$$\omega(k_0) = \sqrt{\omega_0^2 + c^2 k_0^2}. \qquad (6.175)$$

We write

$$\frac{\partial A}{\partial T} = \frac{\partial A}{\partial T_1} + \mu \frac{\partial A}{\partial T_2} + \cdots \qquad (6.176)$$

where $T_1 = \mu t, T_2 = \mu^2 t, \ldots$. As always, our goal is to choose $(\partial A/\partial T_1)$, $(\partial A/\partial T_2)$, and so on, so that (6.174) remains a uniform asymptotic expansion in time.

How do we choose the relation between μ and ϵ? Let us first suppose that the wavepacket is very narrow so that $\mu = \epsilon^2$. Then to order ϵ^3, (6.172) becomes

$$\left(\frac{\partial^2}{\partial t^2} - c^2 \frac{\partial^2}{\partial z^2} \right) (\epsilon u_0 + \epsilon^2 u_1 + \epsilon^3 u_2)$$

$$+ \omega_0^2 (\epsilon u_0 + \epsilon^2 u_1 + \epsilon^3 u_2) - \frac{\omega_0^2}{6} \epsilon^3 u_0^3 = 0, \qquad (6.177)$$

and to the same order, if u_0 is given by the first right-hand term in (6.174),

$$\left(\frac{\partial^2}{\partial t^2} - c^2 \frac{\partial^2}{\partial z^2} + \omega_0^2\right) u_0$$

$$= \left\{\epsilon(\omega_0^2 + c^2 k_0^2 - \omega^2(k_0)) + \epsilon\mu\left(2i\omega(k_0)\frac{\partial}{\partial T_1} + 2ic^2 k_0 \frac{\partial}{\partial Z}\right)\right\}$$

$$\times A e^{ik_0 z - i\omega(k_0)t} + (*). \tag{6.178}$$

Substituting (6.178) into (6.177) we obtain at successive orders,

$$\epsilon : \omega_0^2 + c^2 k_0^2 = \omega^2(k_0)$$

$$\epsilon^2 : \frac{\partial^2 u_1}{\partial t^2} - c^2 \frac{\partial^2 u_1}{\partial z^2} + \omega_0^2 u_1 = 0 \tag{6.179}$$

$$\epsilon^3 : \frac{\partial^2 u_2}{\partial t^2} - c^2 \frac{\partial^2 u_2}{\partial z^2} + \omega_0^2 u_2 = \frac{\omega_0^2 A^3}{6} e^{3i(k_0 z - \omega(k_0)t)}$$

$$+ \left(\frac{\omega_0^2}{2} A^2 A^* + 2i\omega(k_0)\frac{\partial A}{\partial T_1} + 2ic^2 k_0 \frac{\partial A}{\partial Z}\right) e^{ik_0 z - \omega(k_0)t}$$

$$+ (*) \tag{6.180}$$

From (6.179), we may take $u_1 = 0$ as the homogeneous solution is already contained in u_0. From (6.180), we look for solutions

$$u_2 = F_3(t)e^{3ik_0 z} + F_1(t)e^{ik_0 z} + (*)$$

and find

$$\frac{d^2 F_3}{dt^2} + (\omega_0^2 + c^2(3k_0)^2)F_3 = \frac{\omega_0^2 A^3}{6} e^{-3i\omega(k_0)t} \tag{6.181}$$

$$\frac{d^2 F_1}{dt^2} + (\omega_0^2 + c^2 k_0^2)F_1 = \left(\frac{\omega_0^2}{2} A^2 A^* + 2i\omega(k_0)\frac{\partial A}{\partial T_1} + 2ic^2 k_0 \frac{\partial A}{\partial Z}\right) e^{-i\omega(k_0)t}. \tag{6.182}$$

Since $3\omega(k_0) \neq \omega(3k_0)$, F_3 is bounded, but F_1 grows like $te^{-i\omega(k_0)t}$ unless we choose

$$\frac{\partial A}{\partial T_1} + \frac{c^2 k_0}{\omega(k_0)}\frac{\partial A}{\partial Z} - \frac{i\omega_0}{4}A^2 A^* = 0, \tag{6.183}$$

whose solution is

$$A(z, T = \epsilon^2 t) = A_0(\zeta = \epsilon^2 z - \omega'(k_0)t) \exp \frac{i\omega_0}{4}|A_0|^2 \epsilon^2 t \qquad (6.184)$$

with A_0 an arbitrary function (determined by the initial state $A(z, 0)$) of the coordinate $\zeta = \epsilon^2(z - \omega'(k_0)t)$ that moves with the group velocity $\omega'(k_0) = c^2 k_0/\omega(k_0)$.

Next let us ask what happens if we take the width of the wavepacket to be of the same order as the amplitude, that is, $\mu = \epsilon$. Then we obtain

$$\epsilon^2 : \frac{\partial^2 u_1}{\partial t^2} - c^2 \frac{\partial^2 u_1}{\partial z^2} + \omega_0^2 u_1 = \left(2i\omega(k_0)\frac{\partial A}{\partial T_1} + 2ic^2 k_0 \frac{\partial A}{\partial Z} \right) e^{ik_0 z - i\omega(k_0)t} + (*)$$

$$(6.185)$$

and

$$\epsilon^3 : \frac{\partial^2 u_2}{\partial t^2} - x^2 \frac{\partial^2 u_2}{\partial z^2} + \omega_0^2 u_2$$

$$= \left(2i\omega(k_0)\frac{\partial A}{\partial T_2} - \frac{\partial^2 A}{\partial T_1^2} + c^2 \frac{\partial^2 A}{\partial Z^2} + \frac{\omega^2}{2} A^2 A^* \right) e^{ik_0 z - i\omega(k_0)t}$$

$$+ \frac{\omega^2}{6} A^3 e^{3ik_0 z - 3i\omega(k_0)t} + (*) - 2\frac{\partial^2 u_1}{\partial t \partial T_1} + 2c^2 \frac{\partial^2 u_1}{\partial z \partial Z}. \qquad (6.186)$$

To avoid secular growths in u_1, u_2 we must choose

$$\frac{\partial A}{\partial T_1} + \omega'(k_0)\frac{\partial A}{\partial Z} = 0 \qquad (6.187)$$

whence we may take $u_1 = 0$, and then from (6.186),

$$2i\omega(k_0)\frac{\partial A}{\partial T_2} - \frac{\partial^2 A}{\partial T_1^2} + c^2 \frac{\partial^2 A}{\partial Z^2} + \frac{\omega^2}{2} A^2 A^* = 0. \qquad (6.188)$$

From (6.176) and (6.187),

$$\frac{\partial A}{\partial T} + \omega'(k_0)\frac{\partial A}{\partial Z} - \frac{i\omega''(k_0)}{2}\epsilon\frac{\partial^2 A}{\partial Z^2} - \frac{i\omega(k_0)\epsilon}{4} A^2 A^* = 0 \qquad (6.189)$$

where $\omega''(k_0) = (\partial^2 \omega/\partial k^2)_{k_0}$, $\omega(k_0) = \sqrt{\omega_0^2 + c^2 k_0}$. We can get rid of the small parameter ϵ in (6.189) by the change of variables

$$\zeta = \epsilon(z - \omega'(k_0)t), \qquad T_2 = \epsilon^2 t = \epsilon T$$

and thus obtain

$$\frac{\partial A}{\partial T_2} - \frac{i\omega''(k_0)}{2}\frac{\partial^2 A}{\partial \zeta^2} - \frac{i\omega(k_0)}{4}A^2 A^* = 0, \tag{6.190}$$

which is the canonical form of the NLS equation. It says that, in a frame of reference moving with the group velocity, the evolution of the envelope $A(\zeta, T_2)$ is determined by a balance of nonlinear modulation and linear dispersion.

Which of the equations (6.183), (6.190) is more relevant? Each is correct. In the first case we began with a wavepacket width μ of order ϵ^2 and much smaller than the width in the second case. In Figure 6.21, we portray the situation schematically. In the first figure the included wavenumbers are each an ϵ^2 distance from k_0 whereas in the second the included wavenumbers are an order ϵ distance from k_0.

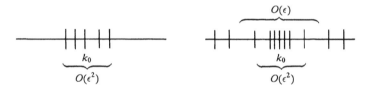

FIGURE 6.21

The point we want to make is that not only is (6.190) better but it is also necessary. The reason is that solutions to (6.190) such as

$$A(T_2) = A_0 \exp \frac{i\omega(k_0)}{4}|A_0|^2 T_2 \tag{6.191}$$

that ignore dispersion can be (and are in this case) unstable (the Benjamin–Feir instability; see Exercises 2.13 and 6.11) to perturbations by wavetrains that are in an order ϵ neighborhood of k_0. This instability is not captured by (6.183). Therefore a lesson to be learned is that in choosing scalings, one must be prepared to consider always "the worst possible scenario."

Remark. The NLS equation is universal in that it obtains in all situations when (1) the medium is strongly dispersive, (2) the wavepacket is weakly nonlinear and almost monochromatic. Its coefficients are determined from the nonlinear dispersion relation for the compounding wavetrain. The only obstruction to its universality is discussed in Section 2h and occurs when the linear dispersion has the property $\omega(0) = 0$. In that case the slow gradient of quadratic nonlinearities can drive slowly varying mean flows which in turn can affect the evolution of A.

EXERCISE 6.12

In an excitable two-level medium that starts in the attenuated state, the equation for the electric field $E = (\partial u/\partial \tau)$ is given by $u_{\tau\tau} = \gamma \sin u$, which is (6.172) with $t - z/c \to 2z, t + z/c \to 2\tau, \gamma = -\omega^2$. Find the NLS equation for the envelope of $A(z, \tau)$ of a wavetrain $e^{-2i\xi\tau + ik(\xi)z}$. Compare with Section 5d.   □

If the leading order approximation contains two wavetrains

$$u_0 = \sum_{j=1}^{2} A_j(Z = \epsilon z, T = \epsilon t)e^{ik_j z - i\omega_j t} + (*), \tag{6.192}$$

$\omega_j = \sqrt{\omega_0^2 + c^2 k_j^2}$, then their evolution is coupled and given by

$$\frac{\partial A_1}{\partial T} + \omega'(k_1)\frac{\partial A_1}{\partial Z} - \frac{i\omega''(k_1)}{2}\epsilon\frac{\partial^2 A_1}{\partial Z^2} - \frac{i\omega_0^2\epsilon}{4\omega(k_1)}(|A_1|^2 + 2|A_2|^2)A_1 = 0$$

$$\frac{\partial A_2}{\partial T} + \omega'(k_2)\frac{\partial A_2}{\partial Z} - \frac{i\omega''(k_2)}{2}\epsilon\frac{\partial^2 A_2}{\partial Z^2} - \frac{i\omega_0^2\epsilon}{4\omega(k_2)}(2|A_1|^2 + |A_2|^2)A_2 = 0.$$

$$\tag{6.193}$$

EXERCISE 6.13

Derive (6.193). Note the connection both with (6.118) and (6.189).   □

Observe that wavepackets of finite width group velocities that differ by more than ϵ separate out on a time scale of order ϵ^{-1} before the effects of nonlinear coupling and dispersion are felt (Figure 6.22a).

In general, then, different wavepackets of finite width do not strongly interact. They interact strongly if the wavepackets

1. are much wider (of width ϵ^{-1} in physical space; Figure 6.22b), in which case initially the dispersion terms $(\partial^2 A_1/\partial Z^2)$ and $(\partial^2 A_2/\partial Z^2)$ can be ignored (see caveat at end of next remark)

2. consist of several lumps each of width ϵ^{-1} spread over a distance ϵ^{-2} (Figure 6.22c), which is the state to which A_1 and A_2 go because of modulational instabilities

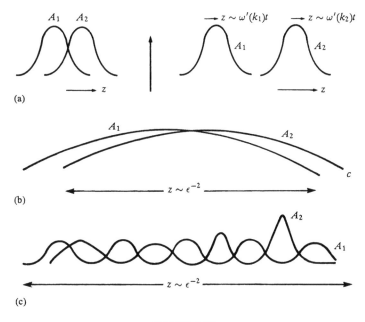

FIGURE 6.22

3. have almost equal group velocities,* which is the case for example if A_1 and A_2 represent different polarization components of the same field (see Section 2i).

We now turn to the interaction of wavepackets whose underlying carrier waves $\exp i(\vec{k}_j \cdot \vec{x} - \omega(\vec{k}_j)t)$ satisfy triad or quartet resonance conditions such as (6.129) and (6.158). In this case,

$$u_0(\vec{x}, t) = \epsilon \sum_{j=1}^{N} A_j(\vec{X} = \epsilon\vec{x}, T = \epsilon t) \exp(i\vec{k}_j \cdot \vec{x} - i\omega(\vec{k}_j)t) + (*). \quad (6.194)$$

EXERCISE 6.14

Show that the equations (6.140) now become

$$\frac{\partial A_1}{\partial T} + \nabla\omega(\vec{k}_1) \bullet \nabla A_1 = \frac{(\vec{k}_2 \times \vec{k}_3)(k_2^2 - k_3^2)}{\alpha^2 + k_1^2} A_2^* A_3^* \quad (6.195)$$

plus two more obtained by cycling 1, 2, 3. Equations (6.195) is relevant in our discussion of Raman and Brillouin scattering of wavepackets. Observe

---

*If the group velocities differ by order one, then removing ϵ from (6.193) is more difficult. Essentially the effect of A_2 on A, is through an average of $|A_2|$ [93].

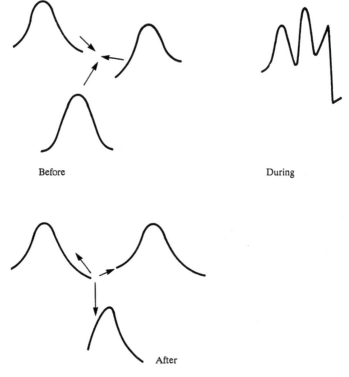

Before   During

After

FIGURE 6.23   Three-wave packet scattering.

that, because of finite velocities, finite width packets only interact for finite times (see Figure 6.23). The equations for the interaction of wavepackets interacting via four-wave mixing processes are found by taking

$$\frac{\partial}{\partial T} \rightarrow \frac{\partial}{\partial T} + \frac{\partial \omega}{\partial k_l} \cdot \frac{\partial}{\partial X_l} - \frac{i}{2}\epsilon \frac{\partial^2 \omega}{\partial k_l \partial k_m} \frac{\partial^2}{\partial X_l \partial X_m}$$

in (6.161), where summation over repeated indices is implied.   □

6i. Bifurcation Theory and the Amplitude Equations for a Laser. [5a]

In this section, we investigate the behavior of driven and damped systems are driven far from some equilibrium state by an external stress [15, 87, 88]. Such systems often undergo a series of symmetry-breaking bifurcation or phase transitions which, as the stress parameter continues to increase, lead to more and more complicated spatial structures and dynamical behavior. Our goal here is to try to understand the nature of the transitions.

For example, consider applying a vertical load to a plane (e.g., a ruler, see Figure 6.24) or cylindrical (see Figure 6.25) elastic bar. At first the bar is simply compressed along its axis AB and the right-left (or axial) symmetry is sustained. As the load is increased, this solution becomes unstable, and bar deflects into one of the two positions ACB or ADB in the plane case, thus breaking the right-left symmetry, or to one of a "circle" of positions ACB in the circular case, thereby breaking the axial symmetry.

A second example that gives rise to more complicated spatial structures is the classical Rayleigh-Bénard problem of a horizontal layer of fluid heated from below (see Figure 6.26). When the temperature difference ΔT achieves a certain value $(\Delta T)_c$, the conduction solution in which heat is carried across

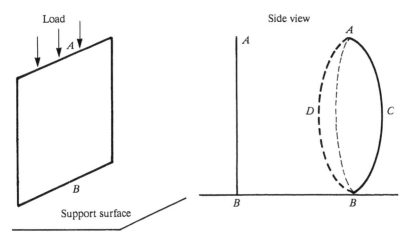

FIGURE 6.24   Deflection of a plane under a vertical load.

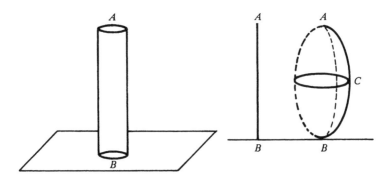

FIGURE 6.25   Deflection of a cylinder under a vertical load.

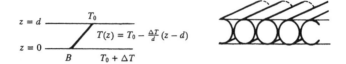

FIGURE 6.26   Convection rolls.

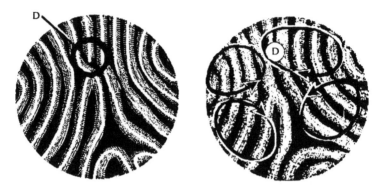

FIGURE 6.27   Natural convection pattern.

from plate B to plate A by molecular diffusion becomes unstable and convection in the form of rolls sets in. The rotational symmetry is broken because, although rolls of all orientations are equally likely, a particular direction is finally chosen by biases in the system too small to measure. As the temperature difference ΔT is increased more, the convection patterns become more and more complicated. Figure 6.27 is a vertical view of the patterns in a cylindrical cell (diameter 42 mm, height 3 mm) containing water at a mean temperature of about $70°$ where $(\Delta T)/(\Delta T)_c$ is about 5. The pattern is fairly complicated spatially (although the basic roll structure is still clearly visible) and its time signal is chaotic.

Nature provides an abundance of examples of such behavior: water turning into ice as the temperature falls below $0°C$, the patterns of highs and lows you see on a weather map (due to the baroclinic instability of the north-south temperature gradient in the earth's atmosphere), the ripples in sand under a shallow layer of moving water (e.g., the oncoming or retreating tide on a flat beach), the ribs on desert cacti, the variety of biological structures and species population patterns, and so on and on and on. What we would like to do is find a mathematical means of analyzing this complicated behavior. As might be expected, of course, a general description is far too much to hope for, and we have to settle for much less. What we describe to you now is a means of handling systems that have just undergone a bifurcation or change from one state to another. To make things as easy as possible, we choose the "before" state to be extremely simple.

The basic ideas and the method will be illustrated using a set of model equations (which we derived in Chapter 4 and discussed further in Chapter 5) suitable for analyzing the onset of lasing action in a cavity containing an active gas of two-level atoms. The main stress parameter is r, the pumping supplied to the system in order to create a population inversion, and is defined as the fractional excess of the number of atoms in the excited state to the number in the ground state. In the Maxwell-Bloch equations (6.196, 6.197, 6.198), $A(x, y, t)$ is the complex valued envelope of the electric field, $P(x, y, t)$, also complex, is the polarization field envelope and $n(x, y, t)$, which is r real, is the instantaneous population inversion. The fast oscillation $\exp(-i\omega_c t)$, where ω_c is the natural frequency of the laser cavity, has been factored out of the electric and polarization fields. The equations are

$$\frac{\partial A}{\partial t} - ia\nabla^2 A = -\sigma A + \sigma P \tag{6.196}$$

$$\frac{\partial P}{\partial t} + (1 + i\Omega)P = (r - n)A \tag{6.197}$$

$$\frac{\partial n}{\partial t} = -bn + \tfrac{1}{2}(AP^* + A^*P) \tag{6.198}$$

The parameters a, σ, Ω, b measure diffraction, cavity losses, cavity detuning (the difference $\omega_{12} - \omega_c$ between the frequency associated with the energy level parameters and the cavity frequency) and homogeneous broadening losses due to spontaneous emissions to other energy levels, all measured using the basic time unit of the homogeneous broadening decay of the polarization field. (In terms of the parameters of Section 5a, $a = \sigma = \kappa/\sigma_{12}, \Omega = (\omega_{12} - \omega_c)/\gamma_{12}b = \gamma_{11}/\gamma_{12}, r = (\omega_c p^2 |N_0|/2\epsilon_0 h\kappa\gamma_{12})$.) We have included diffraction effects to simulate the competition between the various transverse TEM$_{rs}$ modes. We have seen in Section 6f that the frequencies ω_{qrs} of the various harmonic modes can be very close. In fact, $\Delta\omega_{qrs} = c/(nz_0)\Delta(r + s)$ where $z_0 = \tfrac{1}{2}\omega_0^2 k$ is the difffraction distance. If z_0 is sufficiently large, then the frequencies of the transverse modes are so close together that when r exceeds its critical value, a large number of them will be under the gain band. Therefore, in the absence of bias, each transverse mode can compete for the lasing action. In cases where the basic TEM$_{00}$ mode is heavily favored, the diffraction term $-ia\nabla^2 A = -ia((\partial^2/\partial x^2) + (\partial^2/\partial y^2))A$ can be ignored. In that case (6.196–6.198) are called the complex Lorenz equations because when the detuning Ω is zero and A and P are real, equations (6.196–6.198) are precisely the set of equations studied by Lorenz in his seminal 1963 paper on nonperiodic deterministic flow. It is somewhat ironic that the equations are not at all valid for the atmospheric context for which they were originally derived but are exactly valid for the two-level laser. The latter observation is due to Haken.

The nonlasing solution is

$$A = P = n = 0, \tag{6.199}$$

which is stable for values of the population inversion pumping r less than a critical value r_c. Our goal is to analyze what happens near $r = r_c$. The idea of the analysis is simple, and we now outline it in a series of steps.

Step 1. Examine the linear stability of the nonlasing solution and determine the critical value of the parameter r at which various other solutions begin to grow exponentially. We are interested in values of r close to the lowest critical value r_c of this parameter, and we are interested in the shapes and structure of the modes that are either neutrally stable or weakly stable or unstable near $r = r_c$. This set of modes is called the *active* set. Near $r = r_c$, all the other degrees of freedom, which we call the *passive* modes, are slaved to the active modes in the sense that, near $r = r_c$, their amplitudes, which we write schematically as $\mathcal{P}$ (for passive) are given in terms of the amplitudes $\mathcal{A}$ of the active modes by algebraic relations

$$\mathcal{P} = \mathcal{P}(\mathcal{A}). \tag{6.200}$$

The graph $\mathcal{P}(\mathcal{A})$ is called the *Center Manifold.*

Step 2. For r close to r_c, our goal is to write equations for the amplitudes of the active modes $\mathcal{A}$. These amplitudes are called the *order parameters* as they describe the degree of order in the system. We achieve this by expanding the fields $\vec{v} = (A, P, n)$ as an asymptotic series

$$v = \epsilon \vec{v}_0 + \epsilon^2 \vec{v}_1 + \epsilon^3 \vec{v}_2 + \cdots \tag{6.201}$$

where the small parameter ϵ is related to $r - r_c$ and v_0 is a linear combination of all the active modes. The corrections $\vec{v}_1, \vec{v}_2, \ldots$ satisfy linear inhomogeneous equations. The condition that these equations have solutions (the Fredholm alternative theorem, which in our case reduces simply to the condition that the algebraic equation $\Lambda X = F$ has a solution when the matrix Λ is singular) leads to equations for the order parameters, the amplitudes of the active modes.

Step 3. Analyze the equations for the active modes together with the graph (6.200), and this determines the state of the system.

We now carry out this assignment for steps 1 and 2. The complete analysis of the resulting equations will not be given here, but the essential behavior of solutions is discussed in Step 2.

Step 1. Linear Stability Problems of the Nonlasing Solution

Linearize the equations (6.196–6.198) about the solution $A = P = n = 0$ by setting $A = 0 + A'$, $P = 0 + P'$, $n = 0 + n'$ and dropping all terms quadratic in products of A', P', n'. Write

$$\begin{pmatrix} A' \\ P' \\ n' \end{pmatrix} = V_0 e^{\lambda t + i \vec{k} \cdot \vec{x}}, \quad \vec{k} = (k_x, \quad k_y) \tag{6.202}$$

and obtain that V_0 satisfies

$$\Lambda V_0 = \begin{pmatrix} \lambda + \sigma + iak^2 & -\sigma & 0 \\ -r & \lambda + 1 + i\Omega & 0 \\ 0 & 0 & \lambda + b \end{pmatrix} V_0 = 0 \tag{6.203}$$

and the complex growth rate λ is determined by $\det \Lambda = 0$. We find

$$\lambda = -b$$

with eigenvector

$$V_0^{(3)} = \begin{pmatrix} 0 \\ 0 \\ 1 \end{pmatrix}$$

and

$$\lambda^2 + (\sigma + 1 + iak^2 + i\Omega)\lambda + (\sigma + iak^2)(1 + i\Omega) - \sigma r = 0. \tag{6.204}$$

Set $\lambda = \mu - iv$ and find two expressions

$$\mu = \mu(k^2, r) \tag{6.205}$$

and

$$v = v(k^2, r) \tag{6.206}$$

for the growth rate μ and frequency v as functions of $k^2 = k_x^2 + k_y^2$ and the other parameter r. We are interested in the values of r and k^2 for which μ first passes through zero from negative to positive values. Set $\mu = 0$ and obtain easily

$$v = \frac{\sigma \Omega + ak^2}{\sigma + 1} \tag{6.207}$$

$$r = 1 + \frac{1}{(\sigma + 1)^2}(\Omega - ak^2)^2. \tag{6.208}$$

The graph of r versus k, Figure 6.28, is called the *neutral stability curve*. For (r, k) lying above the graph, the nonlasing solution is unstable and the system becomes a laser. We see that when $\Omega < 0$, the $k = 0$ mode is most unstable whereas when $\Omega > 0$ the most unstable modes are $k = \pm\sqrt{\Omega/a}$. The corresponding frequencies are

$$v_< = \frac{\sigma \Omega}{\sigma + 1} \tag{6.209}$$

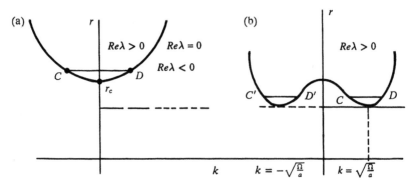

FIGURE 6.28　Neutral stability curve. (a) $\Omega < 0$. (b) $\Omega > 0$.

and

$$v_> = \Omega, \tag{6.210}$$

corresponding to the electric field frequency $\omega_c + v$ of

$$\omega_< = \frac{\sigma\omega_{12} + \omega_c}{\sigma + 1} \tag{6.211}$$

and

$$\omega_> = \omega_{12} \tag{6.212}$$

respectively. The difference is that when $\omega_{12} < \omega_c$, there is no natural frequency ak^2 of the electric field wave that can resonate with the natural frequency of the polarization Ω. The laser is forced to take on a frequency that is a weighted average of the cavity (ω_c) and two-energy level frequency ω_{12}. Indeed when σ is small, called the good cavity limit, $\omega_< \simeq \omega_c$. When $\omega_{12} > \omega_c$, on the other hand, the field can tune (i.e., choose k) its spatial structure so as to resonate with the polarization. We examine each case separately.

We are now in a position to identify the *active* and *passive* modes near $r = r_c$. Because of the transverse (x, y) dependence, there is an infinite number of them. When $r > r_c$, all the $\vec{k}$ modes that lie in a circle or annulus found by rotating the patched bands in Figures 6.29a and 6.29b about the axis are "active." How do we include them all in the description? Before answering this question, we analyze the simpler problem where we ignore diffraction (set $a = 0$). Then the set of equations (6.196–6.198) are finite (five) dimensional (two complex fields A and P and one real field n). This means that we have arranged this geometry of the laser so that the TEM$_{00}$ mode is preferred and dominant. The spatial structure of the active mode is thus fixed once and for all. The task is now to find its time behavior near $r = r_c$.

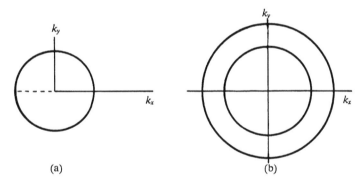

(a)   (b)

FIGURE 6.29

Weakly Nonlinear Behavior of A, P, n Near $r = r_c$

Case 1: $a = 0$ In this case we ignore diffraction and find

$$v = v_c = \frac{\sigma\Omega}{\sigma + 1} \tag{6.213}$$

$$r = r_c = 1 + \frac{\Omega^2}{(\sigma + 1)^2}, \tag{6.214}$$

and there is no distinction between $\Omega \gtrless 0$. The shape of the active mode (the eigenvector associated with the eigenvalue $\lambda = 0 - i\nu_c$ of Λ obtained when $r = r_c$) is

$$\begin{pmatrix} A' \\ P' \\ n' \end{pmatrix} = \begin{pmatrix} 1 \\ x_0 \\ 0 \end{pmatrix} e^{-i\nu_c t} = V_0^{(1)} e^{-i\nu_c t} \tag{6.215}$$

with

$$x_0 = 1 - i \frac{\Omega}{\sigma + 1}. \tag{6.216}$$

We call $V_0^{(2)}$ the eigenvector which goes with the second eigenvalue of (6.204) with negative real part. At $r = r_c$, any vector $\vec{v}$ in complex A, P, n space can be written as a linear combination of $V_1^{(0)}$, $V_2^{(0)}$ and $V_3^{(0)}$,

$$\begin{pmatrix} A \\ P \\ n \end{pmatrix} = B_1 V_0^{(1)} e^{-i\nu_c t} + B_2 V_0^{(2)} c^{\lambda_2 t} + B_3 V_0^{(3)} e^{\lambda_3 t} \tag{6.217}$$

where λ_2 and λ_3 are evaluated at $r = r_c$. Since Real λ_2 and Real $\lambda_3 (= -b)$ are finitely negative at $r = r_c$, any point in A, P, n space is attracted in time toward the vector $V_0^{(1)}$ (see Figure 6.30), because the coefficients $B_2 e^{\lambda_2 t}$ and $B_3 e^{\lambda_3 t}$ of the other two basis vectors $V_0^{(2)}$ and $V_0^{(3)}$ tend to zero. We expect that even when r is slightly greater than r_c, a general starting point (A, P, n) in the phase space is attracted to the neighborhood of $V_0^{(1)}$, which is the initial

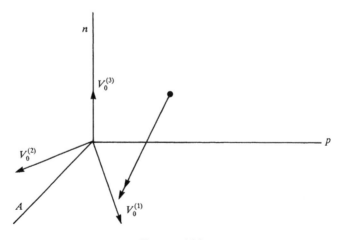

FIGURE 6.30

part of the unstable manifold of the unstable solution $A = P = n = 0$. The basic idea of the subsequent analysis therefore is the recognition that near $r = r_c$, the system becomes simpler to analyze (lowers its dimension) because the point of the phase space that describes the state of the system very quickly relaxes to the neighborhood of the center manifold whose initial direction away from the unstable origin is $V_0^{(1)}$. We now show how to write equations for the slow motion of the phase space point along the center manifold and how to compute the center manifold itself.

Accordingly, we now assume that when

$$r = r_c + R \tag{6.218}$$

for $R \ll 1$, we can seek solutions to the full set of equations (6.196–6.198) as an asymptotic series

$$\vec{v} = \begin{pmatrix} A \\ P \\ n \end{pmatrix} = \epsilon B_1(Rt) V_0^{(1)} e^{-iv_c t} + \epsilon^2 \vec{v}_1 + \epsilon^3 \vec{v}_2 + \cdots \tag{6.219}$$

where, as we shall see, the appropriate choice of ϵ is $\sqrt{R}$. In other words, $V_0^{(1)}$ is the active mode and $V_0^{(2)}$ and $V_0^{(3)}$ are the passive modes. Because $r > r_c$ and because equations (6.196–6.198) are nonlinear, $V_0^{(1)} e^{-iv_c t}$, which satisfies the linearized version of these equations at $r = r_c$, no longer satisfies (6.196–6.198) exactly, and so we allow the amplitude B_1 to be a slowly varying function of time $T = Rt$.

The equation for $\vec{v}_1$ is

$$L\vec{v}_1 = \begin{pmatrix} \frac{\partial}{\partial t} + \sigma & -\sigma & 0 \\ -r_c & \frac{\partial}{\partial t} + 1 + i\Omega & 0 \\ 0 & 0 & \frac{\partial}{\partial t} + b \end{pmatrix} \vec{v}_1 = \begin{pmatrix} 0 \\ 0 \\ \frac{1}{2}(x_0 + x_0^*) \end{pmatrix} B_1 B_1^*, \tag{6.220}$$

which is easy to solve and gives

$$\vec{v}_1 = \begin{pmatrix} 0 \\ 0 \\ \frac{1}{2b}(x_0 + x_0^*) \end{pmatrix} B_1 B_1^* = \begin{pmatrix} 0 \\ 0 \\ \frac{1}{b} \end{pmatrix} B_1 B_1^*. \tag{6.221}$$

The equation for $\vec{v}_2$ at order $\epsilon^3 = R^{3/2}$ is

$$L\vec{v}_2 \equiv -\begin{pmatrix} 1 \\ x_0 \\ 0 \end{pmatrix} \frac{\partial B_1}{\partial T} e^{-iv_c t} + \begin{pmatrix} 0 \\ 1 \\ 0 \end{pmatrix} B_1 e^{-iv_c t} - \frac{1}{b} \begin{pmatrix} 0 \\ 1 \\ 0 \end{pmatrix} B_1^2 B_1^* e^{-iv_c t}. \tag{6.222}$$

Look for solutions

$$\vec{v}_2 = Xe^{-iv_ct} \tag{6.223}$$

and obtain

$$\Lambda_c X = \begin{pmatrix} -iv_c + \sigma & -\sigma & 0 \\ -r_c & -iv_c + 1 + i\Omega & 0 \\ 0 & 0 & -iv_c + b \end{pmatrix} X = F \tag{6.224}$$

with

$$F = -\begin{pmatrix} 1 \\ x_0 \\ 0 \end{pmatrix} \frac{\partial B_1}{\partial T} + \begin{pmatrix} 0 \\ 1 \\ 0 \end{pmatrix} B_1 - \frac{1}{b} \begin{pmatrix} 0 \\ 1 \\ 0 \end{pmatrix} B_1^2 B_1^*. \tag{6.225}$$

Now Λ_c is a singular matrix of rank two, and in order for there to be a solution of (6.224) we need that the vector F lies in the range of Λ_c and be orthogonal to the null space (of dimension one) of the adjoint of Λ_c. This is the Fredholm alternative theorem for the case when the operator Λ_c is a matrix. A very similar kind of condition obtains when $\Lambda_c X = F$ is a differential equation with boundary conditions.

Let U_0^T be the row vector such that

$$U_0^T \Lambda_c = 0. \tag{6.226}$$

$U_0 = \begin{pmatrix} 1 \\ y_0 \\ 0 \end{pmatrix}$, and it is easy to show that $y_0 = (\sigma/r_c)x_0$. Then multiplying (6.224) by U_0^T gives

$$U_0^T F = 0$$

or

$$\frac{\partial B_1}{\partial T}\left(1 + \frac{\sigma}{r_c}x_0^2\right) = \frac{\sigma x_0}{r_c}B_1 - \frac{\sigma x_0}{br_c}B_1^2 B_1^*. \tag{6.227}$$

Then

$$F = \begin{pmatrix} \frac{-1}{r_c} \\ \frac{r_c}{\sigma x_0} \\ 0 \end{pmatrix} \frac{\partial B_1}{\partial T} \tag{6.228}$$

and

$$X = \frac{1}{\sigma} \begin{pmatrix} 0 \\ 1 \\ 0 \end{pmatrix} \frac{\partial B_1}{\partial T}. \tag{6.229}$$

The solution is

$$\begin{pmatrix} A \\ P \\ n \end{pmatrix} = \sqrt{R} \begin{pmatrix} 1 \\ x_0 \\ 0 \end{pmatrix} B_1 e^{-i v_c t} + \frac{1}{b} R \begin{pmatrix} 0 \\ 0 \\ 1 \end{pmatrix} B_1 B_1^*$$

$$+ (\sqrt{R})^3 \frac{1}{\sigma} \begin{pmatrix} 0 \\ 1 \\ 0 \end{pmatrix} \frac{\partial B_1}{\partial T} e^{-i v_c t} + \text{higher order terms}$$

where B_1 satisfies (6.227). Equation (6.230) can be written as a linear combination $b_1 V_0^{(1)} + b_2 V_0^{(2)} + b_3 V_0^{(3)}$ of $V_0^{(1)}$, $V_0^{(2)}$, and $V_0^{(3)}$. Notice that the coefficients b_1, b_2, and b_3 of $V_0^{(1)}$, $V_0^{(2)}$, and $V^{(3)}$ can all be expressed algebraically in terms of B_1 (remember $\partial B_1/\partial T$ is given by (6.227)). These relations,

$$b_1 = b_1(B_1), \quad b_2 = b_2(B_1), \quad b_3 = b_3(B_1) \tag{6.231}$$

define a curve in complex (A, P, n) space called the center manifold and together with (6.227) completely determine the motion of the phase point (A, P, n) except for the very short initial time it takes to relax to (6.231). As time increases, the point $b_1(B_1), b_2(B_1), b_3(B_1)$ moves along the center manifold away from the unstable point $B_1 = 0$ and towards any one of the stable solutions

$$B_1 = \sqrt{b} e^{i\varphi}, \varphi \text{ an arbitrary constant.} \tag{6.232}$$

The laser emits an output with steady power.

If we make Ω, the detuning, equal to zero, then A and P may be chosen real, (6.196–6.198) are the classical Lorenz equations, and $r_c = 1, v_c = 0, x_0 = 1$. After a long time $B_1 \to \pm\sqrt{b}$ corresponding to the lasing solutions $A = \pm\sqrt{b(r-1)}, P = \pm\sqrt{b(r-1)}, n = r - 1$ given in (5.19) and shown in Figure 6.31. Equation (6.227) is called the Landau equation, and it tells us how the order parameter, the amplitude of the active mode, behaves. We talk about the ramification of these results in the laser context in Chapter 5.

We draw the power AA^* of fixed-point solutions as function of r in Figure 6.32.

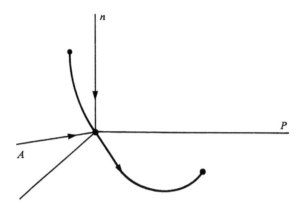

FIGURE 6.31

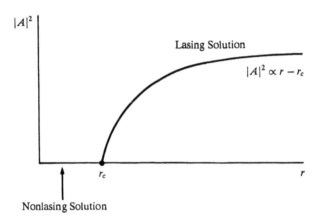

FIGURE 6.32

Case 2: $\Omega < 0, a > 0$ In this case we include in the space of allowable functions all fields $A(x, y, t)$, $P(x, y, t)$, $n(x, y, t)$, that are bounded as $x^2 + y^2 \to \infty$ and which can be represented as a linear combination of Fourier modes $e^{i\vec{k}\cdot\vec{x}}$, $\vec{k} = (k_x, k_y)$. However, when

$$r = r_c + R, \quad 0 < R \ll 1, \tag{6.233}$$

we see from Figure 6.28a, that only these modes $e^{i\vec{k}\cdot\vec{x}}$ whose wavenumber $k = |\vec{k}|$ lies in the band *CD* grow, and the band has a small width of order $\sqrt{R}$. The wavevector $\vec{k}$ lies in a circle b about the origin with diameter *CD*. Therefore instead of representing the active modes as

$$\vec{v}_0 = \sqrt{R} \sum_{\vec{k} \in b} A(\vec{k}, t) e^{i\vec{k}\cdot\vec{x}} e^{-iv_c t}, \tag{6.234}$$

where b is the circular region of Figure 6.29a, then we simply use the wave-packet idea and write

$$\vec{v}_0 = \sqrt{R}B_1(X = \sqrt{R}x, Y = \sqrt{R}y, T = Rt)E^{-iv_ct}V_0^{(1)}. \tag{6.235}$$

Everything proceeds as in case 1, except that now the linear operator L of (6.220) is

$$L = \begin{pmatrix} \frac{\partial}{\partial t} - ia\nabla^2 + \sigma & -\sigma & 0 \\ -r_c & \frac{\partial}{\partial t} + 1 + i\Omega & 0 \\ 0 & 0 & \frac{\partial}{\partial t} + b \end{pmatrix} \tag{6.236}$$

and has the additional part

$$\delta L = \begin{pmatrix} -ia\nabla^2 & 0 & 0 \\ 0 & 0 & 0 \\ 0 & 0 & 0 \end{pmatrix} \tag{6.237}$$

to the L of case 1. But $\delta L V_0^{(1)}$ is small because

$$\nabla^2 B_1 = R\left(\frac{\partial^2 B_1}{\partial X^2} + \frac{\partial^2 B_1}{\partial Y^2}\right), \tag{6.238}$$

and so this extra contribution makes its first appearance at order $R^{3/2}$ in the equation for $\vec{v}_2$. We find

$$\begin{pmatrix} \frac{\partial}{\partial t} + \sigma & -\sigma & 0 \\ -r_c & \frac{\partial}{\partial t} + 1 + i\Omega & 0 \\ 0 & 0 & \frac{\partial}{\partial t} + b \end{pmatrix} \vec{v}_2$$

$$= \vec{F}e^{-iv_ct} + ia\left(\frac{\partial^2 B_1}{\partial X^2} + \frac{\partial^2 B_1}{\partial Y^2}\right)\begin{pmatrix} 1 \\ 0 \\ 0 \end{pmatrix}e^{-iv_ct}. \tag{6.239}$$

The condition for the existence of the solution $\vec{v}_2$ is now

$$\frac{\partial B_1}{\partial T}\left(1 + \frac{\sigma}{r_c}x_0^2\right) - ia\left(\frac{\partial^2 B_1}{\partial X^2} + \frac{\partial^2 B_1}{\partial Y^2}\right) = \frac{\sigma x_0}{r_c}\left(B_1 - \frac{1}{b}B_1^2 B_1^*\right), \tag{6.240}$$

and B_1 satisfies a partial differential equation called the *complex Ginzburg–Landau* (CGL) equation.

The complex Ginzburg–Landau equation

$$B_t + cB_z - \gamma \nabla_1^2 B = \alpha B - \beta B^2 B^*, \quad \alpha, \gamma, \beta \text{ complex, } c \text{ real} \quad (6.241)$$

is universal in a sense similar to what we mean when we say the NLS equation, the conservative limit of (6.241) ($\gamma_r = \alpha_r = \beta_r = 0$),

$$B_t + cB_z - i\gamma_i \nabla_1^2 B = i\alpha_i B - i\beta_i B^2 B^*,$$

is universal. Just as the latter describes the evolution of the envelope of a wavepacket of dispersive waves in a weakly nonlinear conservative system, the CGL equation (6.241) describes the evolution of the envelope of a wavepacket that has been excited by an instability arising when the stress parameter r exceeds its critical value r_c in Figures 6.28a or 1 in Figure 6.28b. Note that in this circumstance, only a small bandwidth (of order $\sqrt{r - r_c}$ or $\sqrt{r - 1}$) of modes is excited, and the wavepacket description is very natural.

While a full discussion of the complete range of solutions of equations (6.240) and (6.241) ((6.240) is (6.241) with $c = 0, \alpha = (1 + (\sigma x_0^2/r_c))^{-1}(\sigma x_0/r_c), \gamma = ia(1 + (\sigma x_0^2/r_c))^{-1}, \beta = (1/b)\alpha$; recall $x_0 = 1 - (i\Omega/\sigma + 1)$) is beyond the scope of this book, two solution classes are important. First, we might ask: in what circumstances is the general solution attracted to the wavetrain centered on the most unstable wavenumber $k = 0$?

EXERCISE 6.15

Show that the space independent solution

$$B = \sqrt{\frac{\alpha_r}{\beta_r}} \exp\left(\left(-i\frac{\beta_i \alpha_r}{\beta_r} + i\alpha_i\right)t + i\phi\right), \quad \phi \text{ constant} \quad (6.242)$$

is neutrally stable if $Re(\beta\gamma^*) = \beta_r \gamma_r + \beta_i \gamma_i > 0$ and unstable if $\beta_r \gamma_r \beta_i \gamma_i < 0$. (For a reference, see [56].) This instability is the analog of the focusing on Benjamin–instability for the CGL equation. As an additional exercise, you might at what positive value of $\beta_r \gamma_r + \beta_i \gamma_i$ (call this value $(\beta_r \gamma_r + \beta_i \gamma_i)(k)$) does the wavetrain solution,

$$B = \sqrt{\frac{\alpha_r - \gamma_r k^2}{\beta_r}}$$

$$\times \exp\left(ik \cdot x - i\left(-\alpha_i + \gamma_i k^2 + \frac{\beta_i}{\beta_r}(\alpha_r - \gamma_r k^2)\right)t + i\phi\right), \quad (6.243)$$

centered on a value of the wavenumber that is not the most unstable, become unstable. This involves some hefty algebra, and the reader should consult reference [90]. Also, Lange and Newell [91] have showen that if the solution grows from *small* fluctuations, and if $\beta_r \gamma_r + \beta_i \gamma_i > 0$, the solution (6.242) is realized. Therefore in this case even though the dimension of the space of active modes is infinite, the solution itself approaches a fixed point of finite dimension as time evolves. □

What happens when the focusing instability sets in? What we find is that a new and important class of solutions, called defects or optical vortices, can be nucleated [89, 92]. The reasons they are important are that (1) they occur easily because they describe the spatial structure of the transition between equally likely finite-amplitude final states; (2) they can be nucleated by instabilities and (3) they can interact strongly with each other and the surrounding field and can give rise to turbulence. We now sketch the main idea of these solutions in the context of (6.240). We have pointed out before that solutions to the space independent version of (6.240), namely (6.227), approach the value $B_1 = \sqrt{b} e^{i\phi}$ where the phase is constant but arbitrary. It is determined by initial conditions. Likewise, space independent solutions to (6.241) with α, β, γ real are given by $B = \sqrt{(\alpha/\beta)} \exp i\phi$. Now suppose the system is spatially extended and that in different parts of the (x, y) plane different phases ϕ are initiated. For example, let us imagine for the moment that B_1 is real and only depends on the X direction. Hence ϕ is constrained to be 0 or π. At some x values, $B \rightarrow +\sqrt{(\alpha/\beta)}$, while at others $B \rightarrow -\sqrt{(\alpha/\beta)}$. The transition between these two states has a hyperbolic tangent structure (Figure 6.33) and is the defect solution of (6.241) with α, β, γ, and $B(X)$ real. The point at which $B = 0$, which you will note is the unstable stationary solution of (6.241), is itself also called a defect. In two space dimensions, with B complex, we can likewise imagine that we draw the curves Re $B(X, Y) = 0$, Im$B(X, Y) = 0$. In general these meet in a point at which $B = 0$ (Figure 6.34). The phase ϕ undergoes a 2π rotation as we circle the defect D. The solution that takes the field from $B = 0$ at D is called the defect solution. The wavenumber k in the far field of the defect is not zero but determined by matching the near and far field solutions [83]. These solutions are observed in many situations (e.g., spiral waves in reaction-diffusion mixtures) where the universal CGL equation obtains. In the present context, they would give rise to spiral patterns in the electric field. They do not arise spontaneously because the space independent solution for (6.240) is stable. The reader should check that Re$(\beta\gamma^*) = -(\sigma a\Omega/br_c(\sigma + 1))|1 + (\sigma x_0^2/r_c)|^2 > 0$ when $\Omega < 0$. However, once nucleated by *finite amplitude* fluctuations, a defect is stable down to

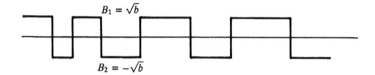

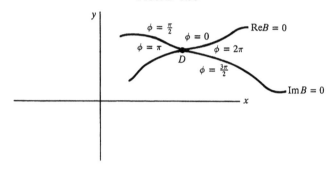

FIGURE 6.33

FIGURE 6.34

that value of $\mathrm{Re}(\beta\gamma^*)$ which corresponds to the value $(\beta_r\gamma_r + \beta_i\gamma_i)(k)$ of Exercise 6.15 where k is the wavenumber in the far field of a defect.

Case 3: $\Omega > 0, a > 0$ In this case, the space of active modes consists of all modes lying in the annulus shown in Figure 6.29b. It is not easy to find a convenient representation for a complex function whose Fourier transform is nonzero in an annulus. We will therefore simplify the problem by restricting ourselves to modes lying near $\vec{k}_0 = (\pm\sqrt{(\Omega/a)}, 0)$. From Figure 6.29b, we see that in this case the appropriate representation of the active modes is

$$\vec{v}_0 = \sqrt{R}B_1(X = \sqrt{R}x, Y = \sqrt[4]{R}y, T_1 = \sqrt{R}t, T_2 = Rt)V_0^{(1)}e^{ik_0x - iv_ct}$$

$$+ \sqrt{R}B_2(X = \sqrt{R}x, Y = \sqrt[4]{R}y, T_1 = \sqrt{R}t, T_2 = Rt)V_0^{(1)}e^{-ik_0x - iv_ct}.$$

We proceed as before, except that this time to leading order

$$L = \begin{pmatrix} -iv_c + ik_0^2 + \sigma & -\sigma & 0 \\ -r_c & -iv_c + 1 + i\Omega & 0 \\ 0 & 0 & b \end{pmatrix}$$

$$= \begin{pmatrix} \sigma & -\sigma & 0 \\ -1 & 1 & 0 \\ 0 & 0 & b \end{pmatrix}$$

and the right and left eigenvectors are $V_0^{(1)} = \begin{pmatrix} 1 \\ 1 \\ 0 \end{pmatrix}$ and $U_0^T = (1, \sigma, 0)$. At

order R, we find B_1 and B_2 satisfy

$$(\sigma + 1)\frac{\partial B_1}{\partial T_1} + 2ak_0\frac{\partial B_1}{\partial X} - ia\frac{\partial^2 B_1}{\partial Y^2} = 0$$

$$(\sigma + 1)\frac{\partial B_2}{\partial T_1} - 2ak_0\frac{\partial B_2}{\partial X} - ia\frac{\partial^2 B_2}{\partial Y^2} = 0$$

$$(6.244)$$

and

$$\vec{v}_1 = \begin{pmatrix} 0 \\ \left(-\frac{\partial B_1}{\partial T_1}e^{ik_0x} - \frac{\partial B_2}{\partial T_1}e^{-ik_0x} \right)e^{-iv_c t} \\ \frac{1}{b}(B_1 B_1^* + B_2 B_2^*) + \frac{1}{b}B_1 B_2^* e^{2ik_0 x} + \frac{1}{b}B_1^* B_2 e^{-2ik_0 x} \end{pmatrix}.$$

At order $R^{3/2}$,

$$(\sigma + 1)\frac{\partial B_1}{\partial T_2} = \sigma\frac{\partial^2 B_1}{\partial T_1^2} + \sigma B_1\left(1 - \frac{1}{b}(B_1 B_1^* + 2B_2 B_2^*)\right) + ia\frac{\partial^2 B_1}{\partial x^2}$$

$$(\sigma + 1)\frac{\partial B}{\partial T_2^2} = \sigma\frac{\partial^2 B_2}{\partial T_1^2} + \sigma B_2\left(1 - \frac{1}{b}(2B_1 B_1^* + B_2 B_2^*)\right) + ia\frac{\partial^2 B_2}{\partial x^2}$$

$$(6.245)$$

Combining (6.244), (6.245) we get

$$(\sigma + 1)\frac{\partial B_1}{\partial t} + 2a\sqrt{\frac{\Omega}{a}}\frac{\partial B_1}{\partial x} - ia\frac{\nabla^2 B}{\nabla^2 B_1} + \frac{\sigma a^2}{(\sigma + 1)^2}\left(2i\sqrt{\frac{\Omega}{a}}\frac{\partial}{\partial x} + \frac{\partial^2}{\partial y^2}\right)^2 B_1$$

$$= \sigma(r - 1)\left(B_1 - \frac{1}{b}(B_1 B_1^* + 2B_2 B_2^*)B_1\right)$$

$$(\sigma + 1)\frac{\partial B_2}{\partial t} + 2a\sqrt{\frac{\Omega}{a}}\frac{\partial B_2}{\partial x} - ia\frac{\nabla^2 B_2}{\nabla^2 B_2} + \frac{\sigma a^2}{(\sigma + 1)^2}\left(-2i\sqrt{\frac{\Omega}{a}}\frac{\partial}{\partial x} + \frac{\partial^2}{\partial y^2}\right)^2$$

$$= \sigma(r - 1)\left(B_2 - \frac{1}{b}(2B_1 B_1^* + B_2 B_2^*)B_2\right) \qquad (6.246)$$

where we have expressed all terms in the original coordinates x, y, and t. To leading order the envelopes B_1 and B_2 move with their respective group velocities $\pm(2\sqrt{a\Omega}/\sigma + 1)$. We will remark on one important solution.

It turns out that the standing-wave solution

$$B_1 B_1^* = B_2 B_2^* = \frac{1}{3}b$$

is unstable while either one of the solutions

$$B_1 B_1^* = b, \quad B_2 = 0 \quad \text{or} \quad B_1 = 0, \quad B_2 B_2^* = b$$

is stable. In other words, the presence of a right-going wave B_1 inhibits the left-going wave B_2. However, there is no initial preference to either, so that we could have left-going waves propagating on one part of the x axis and right-going waves on another. The solutions are joined through the unstable standing wave (see Figure 6.35), which are the defects in this case.

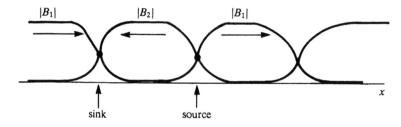

$|B_1|$ $|B_2|$ $|B_1|$

sink   source

FIGURE 6.35   Sources and sinks.

Case 4:   **The connection between cases 2 and 3.** Take $\Omega = \sqrt{R}\Omega_1$ to be small. Again, without loss of generality, we take $a > 0$. Since, to leading order, $\Omega = 0$, the neutral stability curve near its minimum $r = 1, k = 0$ is

$$r = 1 + \frac{a^2 k^4}{(\sigma + 1)^2},$$

namely $r - 1$ is quartic in k. Accordingly the proper scalings are

$$X = R^{\frac{1}{4}}x, Y = R^{\frac{1}{4}}y, T_1 = R^{\frac{1}{4}}y, T_1 = R^{\frac{1}{2}}t, T_2 = Rt$$

where $R = r - 1$. We now take

$$\vec{v} = R^{\frac{1}{2}}B_1 V_0^{(1)} + R\vec{v}_1 + R^{\frac{3}{2}}\vec{v}_2 + \cdots$$

and find at order R

$$L_C \vec{u}_1 = \begin{pmatrix} \sigma & \sigma & 0 \\ -1 & 1 & 0 \\ 0 & 0 & b \end{pmatrix} \vec{u}_1$$

$$= -\frac{\partial B_1}{\partial T_1} \begin{pmatrix} 1 \\ 1 \\ 0 \end{pmatrix} - i\Omega_1 \begin{pmatrix} 0 \\ 1 \\ 0 \end{pmatrix} B_1$$

$$+ ia \left(\frac{\partial^2}{\partial x^2} + \frac{\partial^2}{\partial Y^2} \right) B_1 \begin{pmatrix} 1 \\ 0 \\ 0 \end{pmatrix} + \begin{pmatrix} 0 \\ 0 \\ 1 \end{pmatrix} B_1 B_1^*$$

The solvability condition (note $(1, \sigma, 0) L_C = \vec{0}$) is

$$(\sigma + 1)\frac{\partial B_1}{\partial T_1} = -i\sigma\Omega_1 B_1 + ia \left(\frac{\partial^2}{\partial x^2} + \frac{\partial^2}{\partial Y^2} \right) B_1, \tag{6.247}$$

and then we can solve for $\vec{v}_1$ which is

$$\vec{v}_1 = \frac{1}{b} B_1 B_1^* \begin{pmatrix} 0 \\ 0 \\ 1 \end{pmatrix} - \left(\frac{\partial B_1}{\partial T_1} + i\Omega_1 B_1 \right) \begin{pmatrix} 0 \\ 1 \\ 0 \end{pmatrix}.$$

At order $R^{\frac{3}{2}}$, we obtain

$$L_C \vec{v}_2 = -\frac{\partial B}{\partial T_2} \begin{pmatrix} 1 \\ 1 \\ 0 \end{pmatrix} + \left(\frac{\partial}{\partial T_1} + i\Omega_1 \right)^2 B_1 \begin{pmatrix} 0 \\ 1 \\ 0 \end{pmatrix} + \begin{pmatrix} 0 \\ 1 \\ 0 \end{pmatrix} B_1 - \frac{1}{b} \begin{pmatrix} 0 \\ 1 \\ 0 \end{pmatrix} B_1^2 B_1^*.$$

The solvability condition for $\vec{v}_2$ (i.e. $(1, \sigma, 0) \cdot \text{RHS} = 0$) gives

$$(\sigma + 1)\frac{\partial B_1}{\partial T_2} = \left(\frac{\partial}{\partial T_1} + i\Omega_1 \right)^2 B_1 + \sigma B_1 - \frac{\sigma}{b} B_1^2 B_1^*.$$

But $\left(\frac{\partial}{\partial T_1} + i\Omega_1 \right) B_1 = \frac{ia}{\sigma+1} \left(\frac{\partial^2}{\partial X^2} + \frac{\partial^2}{\partial Y^2} \right) B_1 + i\Omega_1 \left(1 - \frac{\sigma}{\sigma+1} \right) B_1$

and therefore

$$(\sigma + 1)\frac{\partial B_1}{\partial T_2} = -\frac{\sigma}{(\sigma + 1)^2} (a\nabla^2 + \Omega_1)^2 B_1 + \sigma B_1 - \frac{\sigma}{b} B_1^2 B_1^*. \tag{6.248}$$

Now combining (6.247) and (6.248), writing $\sqrt{R}B_1 = \psi$ and then rewriting coordinates gives

$$(\sigma + 1)\frac{\partial \psi}{\partial t} - ia\nabla^2\psi + i\sigma\Omega\psi + \frac{\sigma}{(\sigma + 1)^2}(a\nabla^2 + \Omega)^2\psi$$

$$= \sigma(r - 1)\left(1 - \frac{1}{b}\psi\psi^*\right)\psi. \qquad (6.249)$$

This equation, because of its close resemblance to a complexified version of the real Swift-Hohenberg equation

$$\frac{\partial \omega}{\partial t} + (\nabla^2 + 1)^2\omega + R\omega - \omega^3 = 0$$

is known as the complex Swift-Hohenberg (CSH) equation.

An analysis of its properties has been carried out in [93].

6j. Derivation of the Nonlinear Schrödinger (NLS) Equation in a Waveguide

Consider the geometry shown in Figure 6.36. We wish to consider the evolution of the envelope of a wavepacket confined by refractive index guiding. In Section 6e, we found the electric field $\vec{E}(x, z, t) = \hat{y}E(x, z, t)$ of a wavetrain of frequency ω_0 to have the form

$$\vec{E}(x, z, t) = \hat{y}(e^{ik(\omega_0)z - i\omega_0 t}U(x, \omega_0)A + (*)).$$

Our first task is to develop a representation for a wavepacket that consists of a superposition of wavetrains centered on some frequency ω_0 and wavenumber $k_0 = k(\omega_0)$. We could approximate the transverse shape all the wavetrains in the wavepacket by $U(x, \omega_0)$. However, we shall see that it is wiser to take

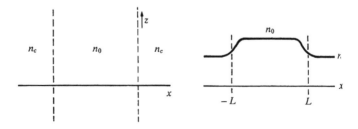

FIGURE 6.36

account of the slight differences in each of the transverse shapes $U(x, \omega)$ as it greatly facilitates subsequent analysis. The correct approximation is

$$\vec{E}(x, z, t) = \hat{y}\left(U(x, \omega_0 + i\epsilon\frac{\partial}{\partial T})A(Z, T)e^{ik_0 z - i\omega_0 t} + (*)\right) \qquad (6.250)$$

where $Z = \epsilon z$, $T = \epsilon t$ and $U(x, \omega_0 + i\epsilon(\partial/\partial T))$ is interpreted by its Taylor expansion around ω_0. To justify the choice of (6.250), let us look at the Fourier representation of $\vec{E}(x, z, t) = \hat{y}E(x, z, t)$,

$$E(x, z, t) = \frac{1}{2\pi}\int_{-\infty}^{\infty} \hat{E}(x, z, \omega)e^{-i\omega t} d\omega,$$

$$\hat{E}(x, z, \omega) = \int_{-\infty}^{\infty} E(x, z, t)e^{i\omega t} dt. \qquad (6.251)$$

If $E(x, z, t)$ is a wavetrain,

$$E(x, z, t) = U(x, \omega_0)Ae^{ik_0 z - i\omega_0 t} + U^*(x, \omega_0)A^* e^{-ik_0 z + i\omega_0 t},$$

then

$$\hat{E}(x, z, \omega) = 2\pi\delta(\omega - \omega_0)AU(x, \omega)e^{ik_0 z}$$

$$+ 2\pi\delta(-\omega - \omega_0)A^* U^*(x, \omega_0)e^{-ik_0 z}.$$

A wavepacket, on the other hand, is represented in Fourier space by

$$\hat{E}(x, z, \omega) = \hat{A}(\omega - \omega_0, Z)U(x, \omega)e^{ik_0 z} + \hat{A}^*(-\omega - \omega_0, Z)U^*(x, -\omega)e^{-ik_0 z} \qquad (6.252)$$

where $\hat{A}(\Omega, Z)$ is localized and is nonzero only in the neighborhood of $\Omega = 0$. The Z dependence is included to take account of both the slightly different propagation characteristics of each wavetrain and nonlinearity. Note that $\hat{E}^*(x, z, -\omega) = \hat{E}(x, z, \omega)$, which ensures that E is real. Now expand $U(x, \omega)$ near $\omega = \omega_0$ and $U^*(x, -\omega)$ near $\omega = -\omega_0$. Then take the inverse transform of $\hat{E}(x, z, \omega)$ and use

$$\left(i\frac{\partial}{\partial t}\right)^n A(Z, t) = \frac{1}{2\pi}\int \Omega^n \hat{A}(Z, \Omega)e^{-i\Omega t} d\Omega, n = 0, 1, \dots . \qquad (6.253)$$

One finds

$$E(x, z, t) = \sum \frac{1}{n!} \frac{d^n U}{d\omega_0^n}$$

$$\times \left(\frac{1}{2\pi} \int (\omega - \omega_0)^n \hat{A}(\omega - \omega_0) e^{-i(\omega-\omega_0)t} d(\omega - \omega_0) \right) e^{ik_0 z - i\omega_0 t}$$

$$+ \sum \frac{(-1)^n}{n!} \frac{d^n U^*}{d\omega_0^n}$$

$$\times \left(\frac{1}{2\pi} \int (\omega + \omega_0)^n \hat{A}^*(-\omega - \omega_0) e^{-i(\omega+\omega_0)t} d(\omega + \omega_0) \right) e^{-ik_0 z + i\omega_0 t},$$

which is precisely (6.250). Since $U(x, \omega)$ is real, we can also write

$$E(x, z, t) = U\left(x, \omega_0 + i\epsilon \frac{\partial}{\partial T} \right) A(Z, T) e^{ik_0 z - i\omega t}$$

$$+ U\left(x, \omega_0 - i\epsilon \frac{\partial}{\partial T} \right) A^*(Z, T) E^{-ik_0 z + i\omega_0 t}. \tag{6.254}$$

The equation satisfied by $E(x, z, t)$ is

$$\frac{\partial^2 E}{\partial x^2} + \frac{\partial^2 E}{\partial z^2} - \frac{1}{c^2} \frac{\partial^2 E}{\partial t^2} - \frac{1}{c^2} \frac{\partial^2}{\partial t^2} \int \chi^{(1)}(x, t - \tau) E(x, z, \tau) d\tau = \frac{1}{c^2} \frac{\partial^2}{\partial t^2}$$

$$\times \int \chi^{(3)}(x, t - \tau_1, t - \tau_2, t - \tau_3) E(x, z, \tau_1) E(x, z, \tau_2) E(x, z, \tau_3) d\tau_1 d\tau_2 d\tau_3. \tag{6.255}$$

We seek solutions

$$E(x, z, t) = \epsilon E_0 + \epsilon^2 E_1 + \epsilon^3 E_2 + \cdots \tag{6.256}$$

where E_0 is given by the wavepacket (6.254). The goal is to choose the long-distance behavior of $A(Z, T)$

$$\frac{\partial A}{\partial Z} = \frac{\partial A}{\partial Z_1} + \epsilon \frac{\partial A}{\partial Z_2} + \cdots, \qquad Z_n = \epsilon^{n-1} Z = \epsilon^n z \tag{6.257}$$

in order to ensure that (6.256) remains a uniformly valid asymptotic expansion in z, t.

We now digress in order to review some useful properties associated the boundary value problem

$$L\left(\frac{d}{dx}, ik, -i\omega \right) U(x, \omega) = \left(\frac{d^2}{dx^2} - k^2 + \frac{n^2 \omega^2}{c^2} \right) U(x, \omega) = 0 \tag{6.258}$$

satisfied by $U(x, \omega)$. We ask that solutions U are smooth and tend to zero as $|x| \to \infty$. We take $n^2 = 1 + \hat{\chi}^{(1)}(x, \omega)$ to be real, although we indicate how to include the affects of absorption (Im $\hat{\chi}^{(1)}(\omega)$) later on. The solution $U(x, \omega)$ can be taken to be real and even in ω. As we have seen in Section 6e, the boundary value problem (6.258) has only smooth solutions that vanish at infinity when the eigenvalue k^2 takes on one of a discrete set of values $\{k_j^2(\omega)\}_1^N$, each of which depends continuously on ω. By choosing the variation $n_0 - n_c$ in $n^2(x, \omega)$ to be sufficiently small, we can arrange that the waveguide parameter $V = (2\pi/\lambda)(2L)\sqrt{n_0^2 - n_c^2}$ is in the range where there is only one eigenvalue $k^2(\omega)$, and its eigenfunction has a Gaussian-like shape shown in the first figure in Figure 6.9. We now discuss two important properties connected with the shape $U(x, \omega)$. Property 1 is worth including although we only need it if we take the less accurate approximation $U(x, \omega_0)A(Z, T)$ instead of $U(\omega_0 + i\epsilon(\partial/\partial T))A(Z, T)$ of the wavepacket. Property 2 is the Fredholm alternative theorem, the condition that the inhomogeneous boundary value problem $L((d/dx), ik_0, -i\omega_0) U = F$ has a solution.

Property 1: For two real functions f, g that are smooth and tend to zero as $|x| \to \infty$, define the inner product $\langle f, g \rangle$ to be $\int_{-\infty}^{\infty} f(x)g(x)dx$. We observe then that the dispersion relation, namely the dependence of k on ω, is given by

$$D([U], k, \omega) = \langle U, LU \rangle = 0. \tag{6.259}$$

We write $D([U], k, \omega)$ because while D only depends on k and ω, its form obviously depends on the choice of U.

EXERCISE 6.16

Show that

$$D([U], k, \omega) = -k^2 \int_{-\infty}^{\infty} U^2 dx + \frac{\omega^2}{c^2} \int n^2 U^2 dx - \int_{-\infty}^{\infty} \left(\frac{dU}{dx}\right)^2 dx = 0. \tag{6.260}$$

(*Hint*: use integration by parts.) $\square$

EXERCISE 6.17

Consider

$$F([U]) = \frac{-\int (dU/dx)^2 + (\omega^2/c^2) \int n^2 U^2 dx}{\int U^2 dx} \tag{6.261}$$

and ask what function $U(x, \omega)$ minimizes F. This is a standard problem in the calculus of variations. Show that the condition that U is minimum is

$$\lim_{\epsilon \to 0} \frac{F([U + \epsilon \delta U]) - F([U])}{\epsilon} = 0$$

$$= \frac{1}{(\int U^2 dx)} \int 2\delta U \left(\frac{d^2 U}{dx^2} + \frac{\omega^2}{c^2} U - FU \right) dx \qquad (6.262)$$

where dU is an arbitrary variation. Therefore the minimum U satisfies (6.258), and the value of F at this solution U is k^2.   □

These results allow us to obtain relations between certain inner products that we would need if we had taken the less accurate representation of the wavepacket. Since (6.259) holds for all ω, differentiate with respect to ω and get

$$\frac{dD}{d\omega} = 0 = \left\langle \frac{dU}{d\omega}, LU \right\rangle + \left\langle U, L \frac{dU}{d\omega} \right\rangle + \left\langle U, \frac{dL}{d\omega} U \right\rangle$$

$$= 2 \left\langle \frac{dU}{d\omega}, LU \right\rangle + \left\langle U, \frac{dL}{d\omega} U \right\rangle. \qquad (6.263)$$

EXERCISE 6.18

Show that by integration by parts $\langle U, LV \rangle = \langle L^A V, U \rangle$ where $L^A = L$. Such an operator L is called self-adjoint.   □

But, on a solution (6.258), $LU = 0$ so that

$$\frac{dD}{d\omega} = 0 = \left\langle U, \frac{dL}{d\omega} U \right\rangle = ik' \langle U, L_2 U \rangle - \langle U, L_3 U \rangle = 0 \qquad (6.264)$$

where $L_2(L_3)$ is the partial derivative of L with respect to the second (third) argument $ik(-i\omega)$. Differentiate (6.264) again with respect to ω,

$$\frac{d^2 D}{d\omega^2} = 0 = \left\langle \frac{dU}{d\omega}, \frac{dL}{d\omega} U \right\rangle + \left\langle U, \frac{dL}{d\omega} \frac{dU}{d\omega} \right\rangle + \left\langle U, \frac{d^2 L}{d\omega^2} U \right\rangle = 0, \qquad (6.265)$$

which gives

$$ik''\langle U, L_2 U\rangle - k'^2\langle U, L_{22}U\rangle + 2k'\langle U, L_{23}U\rangle - \langle U, L_{33}U\rangle$$

$$+ \left\langle \frac{dU}{d\omega}, (ik'L_2 - iL_3)U \right\rangle + \left\langle U, (ik'L_2 - iL_3)\frac{dU}{d\omega} \right\rangle = 0.$$

$$(6.266)$$

Property 2 is the Fredholm alternative theorem. We seek solutions of

$$LX = F \tag{6.267}$$

for which $X \to 0$ as $|x| \to \infty$. Since L is a singular operator (namely has nontrivial homogeneous solutions, U such that $LU = 0$), F needs to satisfy a certain condition—the Fredholm alternative condition—in order that the solution X exists. We can divide the space of functions F into two complementary and orthogonal subspaces, the range of the operator L (the set of functions that can be reached by applying L to a smooth function that tends to zero as $|x| \to \infty$) and the null space of its adjoint operator L^A (the set of functions for which $L^A V = 0$). But $L^A = L$, so that the set of functions V has only one member for each ω, namely the solution $U(x, \omega)$. The condition for a solution X of (6.267) is that the forcing function F has no component in the direction U, namely

$$\langle U, F\rangle = 0. \tag{6.268}$$

We now return to the problem of solving (6.255) for the iterates $E_1, E_2, \ldots$. In performing the analysis we are going to have to calculate

$$\text{(a)} \quad \frac{\partial^2}{\partial t^2} \epsilon \int \chi^{(1)}(x, t - \tau) E_0(x, z, \tau) d\tau \tag{6.269}$$

and

$$\text{(b)} \quad \frac{\partial^2}{\partial t^2} \epsilon^3 \int \chi^{(3)}(x, t - \tau_1, t - \tau_2, t - \tau_3)$$

$$E_0(x, z, \tau_1) E_0(x, z, \tau_2) E(x, z, \tau_3) d\tau_1 d\tau_2 d\tau_3 \tag{6.270}$$

where E_0 is (6.254), and so we will first address this task. The key idea is to take the Fourier transforms and use the fact that $\hat{A}$, the Fourier transform of $A(Z, T)$, is localized at ω_0 and $-\omega_0$.

$$\text{F. T. (a)} = -\epsilon \omega^2 \hat{\chi}^{(1)}(x, \omega) \hat{E}_0(x, z, \omega)$$

$$= -\epsilon \omega^2 \hat{\chi}^{(1)}(x, \omega) U(x, \omega) \hat{A}(\omega - \omega_0) e^{ik(\omega_0)z} \tag{6.271}$$

$$- \epsilon \omega^2 \hat{\chi}^{(1)}(x, \omega) U(x, \omega) \hat{A}^*(-\omega - \omega_0) e^{-ik(\omega_0)z}$$

Now, just as we did before, expand the coefficients multiplying $\hat{A}(\omega - \omega_0)(\hat{A}^* (-\omega - \omega_0))$ in Taylor series about $\omega = \omega_0 (\omega = -\omega_0)$, take the inverse transform, and obtain

$$\text{(a)} = -\epsilon e^{ik(\omega_0)z - i\omega_0 t} \left(\omega_0 + i\frac{\partial}{\partial t} \right)^2 \hat{\chi}^{(1)} \left(x, \omega_0 + i\frac{\partial}{\partial t} \right) U \left(x, \omega_0 + i\frac{\partial}{\partial t} \right)$$

$$- \epsilon e^{-ik(\omega_0)z + i\omega_0 t} \left(\omega_0 - i\frac{\partial}{\partial t} \right)^2 \hat{\chi}^{(1)} \left(x, \omega_0 - i\frac{\partial}{\partial t} \right)$$

$$\times U \left(x, \omega_0 - i\frac{\partial}{\partial t} \right) A^*(Z, T). \tag{6.272}$$

Similarly (for convenience we write $k(\omega_0) = k_0$),

$$\text{F.T.(b)} = -\frac{\epsilon^3}{(2\pi)^2} \int dt d\omega_1 d\omega_2 d\omega_3$$

$$\delta(\omega - \omega_1 - \omega_2 - \omega_3)(\omega_1 + \omega_2 + \omega_3)^2 \hat{\chi}^{(3)}(\omega_1, \omega_2, \omega_3)$$

$$(\hat{A}(\omega_1 - \omega_0) U(x, \omega_1) e^{ik_0 z} + \hat{A}^*(-\omega_1 - \omega_0) U(x, \omega_1) e^{-ik_0 z})$$

$$(\hat{A}(\omega_2 - \omega_0) U(x, \omega_2) e^{ik_0 z} + \hat{A}^*(-\omega_2 - \omega_0) U(x, \omega_2) e^{-ik_0 z})$$

$$(\hat{A}(\omega_3 - \omega_0) U(x, \omega_3) e^{ik_0 z} + \hat{A}^*(-\omega_3 - \omega_0) U(x, \omega_3) e^{-ik_0 z}). \tag{6.273}$$

We are only interested in the $e^{\pm i k_0 z}$ terms in (6.273) as the higher harmonics $e^{\pm 3 i k_0 z}$ do not contribute to the long-distance evolution of $A(Z, T)$ until $O(\epsilon^5)$.

Consider the term

$$-\frac{\epsilon^3}{(2\pi)^2}\int dt d\omega_1 d\omega_2 d\omega_3 \delta(\omega - \omega_1 - \omega_2 - \omega_3)$$

$$\{(\omega_1 + \omega_2 + \omega_3)^2 \hat{\chi}^{(3)}(\omega_1, \omega_2, \omega_3) U(x, \omega_1) U(x, \omega_2) U(x, \omega_3)\}$$

$$\hat{A}^*(-\omega_1 - \omega_0)\hat{A}(\omega_2 - \omega_0)\hat{A}(\omega_3 - \omega_0)e^{ik_0 z} \qquad (6.274)$$

Expand the functions in braces in multivariable Taylor series about $\omega_1 = -\omega_0, \omega_2 = \omega_0, \omega_3 = \omega_0$ and obtain

$$-\frac{\epsilon^3}{(2\pi)^2}\int d\omega_1 d\omega_2 d\omega_3 \delta(\omega - \omega_1 - \omega_2 - \omega_3)$$

$$\times \hat{A}^*(-\omega_1 - \omega_0)\hat{A}(\omega_2 - \omega_0)\hat{A}(\omega_3 - \omega_0)e^{ik_0 z}$$

$$\{(\omega_0^2 + 2\omega_0(\omega_1 + \omega_0) + 2\omega_0(\omega_2 - \omega_0) + 2\omega_0(\omega_3 - \omega_0) + \cdots)$$

$$\times (\chi^{(3)}(-\omega_0, \omega_0, \omega_0) + \chi_1^{(3)}(-\omega_0, \omega_0, \omega_0)(\omega_1 + \omega_0)$$

$$+\chi_2^{(3)}(-\omega_0, \omega_0, \omega_0)(\omega_2 - \omega_0) + \chi_3^{(3)}(-\omega_0, \omega_0, \omega_0)(\omega_3 - \omega_0) + \cdots)\}$$

$$\times \left(U(x, -\omega_0) + (\omega_1 + \omega_0)\frac{\partial U}{\partial \omega}(x, -\omega_0) + \cdots\right)$$

$$\times \left(U(x, \omega_0) + (\omega_2 - \omega_0)\frac{\partial U}{\partial \omega}(x, -\omega_0) + \cdots\right)$$

$$\times \left(U(x, \omega_0) + (\omega_3 - \omega_0)\frac{\partial U}{\partial \omega}(x, -\omega_0) + \cdots\right). \qquad (6.275)$$

In (6.275), $\chi_1^{(3)}$ means $(\partial/\partial\omega_1)\chi^{(3)}(\omega_1, \omega_2, \omega_3)$ estimated at $\omega_1 = -\omega_0, \omega_2 = \omega_0, \omega_3 = \omega_0$.

Take the inverse Fourier transform of (6.275) by multiplying (6.275) by $(1/2\pi)e^{-i\omega t}$ and integrating over ω. It is convenient to write

$$\delta(\omega - \omega_1 - \omega_2 - \omega_3) = \frac{1}{2\pi}\int dt' e^{i(\omega - \omega_1 - \omega_2 - \omega_3)t'}.$$

We find (recall $U(x, \omega_0)$ is even in ω_0)

$$-\frac{\epsilon^3}{(2\pi)^4} \int e^{-i\omega t} d\omega e^{i(\omega - \omega_1 - \omega_2 - \omega_3)t'}$$

$$dt' d\omega_1 d\omega_2 d\omega_3 \hat{A}^*(-\omega_1 - \omega_0)\hat{A}(\omega_2 - \omega_0)\hat{A}(\omega_3 - \omega_0)$$

$$\{\omega_0^2 \chi^{(3)}(-\omega_0, \omega_0, \omega_0)U^3(x, \omega_0) + [\cdots](\omega_1 + \omega_0)$$

$$+ [\cdots](\omega_2 - \omega_0) + [\cdots](\omega_3 - \omega_0) + \cdots\}. \tag{6.276}$$

Where we leave the reader to write down the terms in the brackets. Each of them involves time derivatives of the $A(Z, T)$'s and so is $O(\epsilon)$ lower than the first term. Although we do not calculate these contributions here, they are important, particularly for narrow pulses, and contribute significantly to the non-soliton corrections in the propagation of the wavepacket. Integrate $(1/2\pi) \int e^{-i\omega(t-t')} d\omega = \delta(t - t')$ and then over t'. Write $e^{-i(\omega_1 + \omega_2 + \omega_3)t}$ as $e^{-i\omega_0 t} e^{-i(\omega_1 + \omega_0)t} e^{-i(\omega_2 - \omega_0)t} e^{-i(\omega_3 - \omega_0)t}$ and then do the integration over $\omega_1, \omega_2, \omega_3$, using

$$A(Z, T) = \frac{1}{2\pi} \int \hat{A}(\Omega) e^{-i\Omega T} d\Omega. \tag{6.277}$$

Then (6.276) becomes to leading order

$$-\epsilon^3 \omega_0^2 \hat{\chi}^{(3)}(-\omega_0, \omega_0, \omega_0)A^2(Z, T)A^*(Z, T)U^3(x, \omega_0)e^{ik_0 z - i\omega_0 t}.$$

Recalling that $\hat{\chi}^{(3)}$ is symmetric in its arguments, the total contribution of the terms in (6.273) proportional to $e^{\pm(ik_0 z - i\omega t)}$ is

$$-3\epsilon^3 \omega_0^2 \hat{\chi}^{(3)}(-\omega_0, \omega_0, \omega_0)U^3(x, \omega_0)A^2 A^* e^{ik_0 z - i\omega t}$$

$$-3\epsilon^3 \omega_0^2 \hat{\chi}^{(3)}(-\omega_0, \omega_0, \omega_0)U^3(\chi, \omega_0)A^{*2} A e^{-ik_0 z + i\omega t}. \tag{6.278}$$

We now return to the main analysis. Substitute (6.270) into (6.269). Observe that the $e^{ik_0 z - i\omega t}$ term in

$$\left(\frac{\partial^2}{\partial z^2} + \frac{\partial^2}{\partial x^2} - \frac{1}{c^2}\frac{\partial^2}{\partial t^2} - \frac{1}{c^2}\frac{\partial^2}{\partial t^2} \int \chi^{(1)}(t - \tau)\right)\epsilon E_0 \tag{6.279}$$

is precisely

$$L\left(\frac{d}{dx}, ik_0 + \epsilon\frac{\partial}{\partial Z_1} + \epsilon^2\frac{\partial}{\partial Z_2}, -i\omega_0 + \epsilon\frac{\partial}{\partial T}\right)\epsilon U\left(x, \omega_0 + i\epsilon\frac{\partial}{\partial T}\right)$$

$$\times A(Z_1, Z_2, T). \tag{6.280}$$

Therefore, at order ϵ^2 when we solve for E_1,

$$\frac{\partial^2 E_1}{\partial Z^2} + \frac{\partial^2 E_1}{\partial x^2} - \frac{1}{c^2}\frac{\partial^2}{\partial t^2}\int \chi^{(1)}(t-\tau)E_1 d\tau = F_1 e^{ik_0z - i\omega_0 t} + (*). \tag{6.281}$$

F_1 consists exactly of the order ϵ^2 component of (6.280), which is the negative of

$$\left(L_2 U\frac{\partial A}{\partial Z_1} + L_3 U\frac{\partial A}{\partial T}\right) + iL\frac{dU}{d\omega}\frac{\partial A}{\partial T}, \tag{6.282}$$

where

$$L_2 = (\partial L/\partial(ik))((d/dx), ik_0, -i\omega_0),$$

$$L_3 = (\partial L/\partial(-i\omega))((d/dx), ik_0, -i\omega_0)$$

as defined in (6.264). Look for solutions

$$E_1 = U_1 e^{ik_0z - i\omega t} + U_1 e^{-ik_0z + i\omega t} \tag{6.283}$$

and obtain

$$LU_1 = F_1 = -L_2 U\frac{\partial A}{\partial Z_1} - L_3 U\frac{\partial A}{\partial T} - iL\frac{dU}{d\omega}\frac{\partial A}{\partial T}. \tag{6.284}$$

But differentiating $LU = 0$ with respect to ω gives

$$L\frac{dU}{d\omega} = -ik_0' L_2 U + iL_3 U$$

and so

$$F_1 = -L_2 U\left(\frac{\partial A}{\partial Z} + k_0'\frac{\partial A}{\partial T}\right).$$

But the Fredholm alternative condition (6.268) for (6.280) demands that

$$\langle U, L_2 U\rangle\left(\frac{\partial A}{\partial Z_1} + k_0'\frac{\partial A}{\partial T}\right) = 0,$$

which can only be true if

$$\frac{\partial A}{\partial Z_1} + k_0' \frac{\partial A}{\partial T} = 0 \tag{6.285}$$

because

$$\langle U, L_2 U \rangle = 2i k_0 \langle U, U \rangle \neq 0. \tag{6.286}$$

Since $F_1 = 0$, we can also take $U \equiv 0$.

At order ϵ^3, we set

$$E_2(x, z, t) = U_2 e^{ik_0 z - i\omega t} + U_2^* e^{-ik_0 z + i\omega t} + V_2 e^{3ik_0 z - 3i\omega t} + V_2^* e^{-3ik_0 z + 3i\omega t}$$

and find

$$L\left(\frac{d}{dx}, ik_0, -i\omega_0\right) U_2 = F_2 \tag{6.287}$$

where

$$F_2 = -L_2 U \frac{\partial A}{\partial Z_2} - \frac{1}{2} L_{22} U \frac{\partial^2}{\partial Z_1^2} A - L_{23} U \frac{\partial^2 A}{\partial Z_1 \partial T}$$

$$- \frac{1}{2} L_{33} U \frac{\partial^2 A}{\partial T^2} - i L_2 \frac{dU}{d\omega} \frac{\partial^2 A}{\partial Z_1 \partial T} - i L_3 \frac{dU}{d\omega} \frac{\partial^2 A}{\partial T^2} \tag{6.288}$$

$$+ \frac{1}{2} L \frac{d^2 U}{d\omega^2} \frac{\partial^2 A}{\partial T^2} - \frac{3\omega_0^2}{c^2} \hat{\chi}^{(3)}(-\omega_0, \omega_0, \omega_0) U^3(x, \omega_0) A^2 A^*,$$

the sum of the negative of the $O(\epsilon^3)$ component in (6.280) and the last term. But

$$L\frac{d^2 U}{d\omega^2} = -2i k_0' L_2 \frac{dU}{d\omega} + 2i L_3 \frac{dU}{d\omega} - i k_0'' L_2 U + k_0'^2 L_{22} U \tag{6.289}$$

$$- 2k_0' L_{23} U + L_{33} U \tag{6.290}$$

and

$$\frac{\partial A}{\partial Z_1} = -k' \frac{\partial A}{\partial T}$$

so that

$$F_2 = -L_2 U \left(\frac{\partial A}{\partial Z_2} + i \frac{k_0''}{2} \frac{\partial^2 A}{\partial T^2} \right) - \frac{3\omega_0^2}{c^2} \hat{\chi}^{(3)}(-\omega_0, \omega_0, \omega_0) U^3 A^2 A^*.$$

The solvability condition (6.268) is

$$\frac{\partial A}{\partial Z_2} + i \frac{k_0''}{2} \frac{\partial^2 A}{\partial T^2} - i \frac{3\omega_0^2}{2c^2 k_0} \frac{\int \chi^{(3)} U^4 dx}{\int U^2 dx} A^2 A^* = 0. \tag{6.291}$$

We write $k_0 = \beta(\omega_0)\omega_0/c$ and define

$$\frac{3 \int \chi^{(3)} U^4 dx}{\int U^2 dx} = 2\beta n_2 \tag{6.292}$$

so that

$$\frac{\partial A}{\partial Z_2} + i \frac{k_0''}{2} \frac{\partial^2 A}{\partial T^2} - i \frac{n_2 \omega_0}{c} A^2 A^* = 0. \tag{6.293}$$

Combining (6.285) and (6.293) and introducing the coordinates

$$\tau = \epsilon(t - k_0' z)$$

$$\xi = \epsilon^2 z$$

in order to remove the small parameter ϵ gives us the canonical form of the NLS equation,

$$\frac{\partial A}{\partial \xi} + i \frac{k_0''}{2} \frac{\partial^2 A}{\partial \tau^2} - i \frac{n_2 \omega_0}{c} A^2 A^* = 0. \tag{6.294}$$

We conclude this section with two remarks.

- Perturbations to (6.293) due to the effects of absorption ($\text{Im}\,\hat{\chi}^{(1)}(\omega) \neq 0$), higher-order dispersion and the effects of delay in the nonlinear term are discussed when we consider the propagation of wavepacket pulses in optical fibers in Section 3c.

- If we had taken the wavepacket to be

$$E_0 = U(x, \omega_0) A(Z, T) e^{ik_0 z - i\omega t} + (*) \tag{6.295}$$

then, when we are calculating E_1, we would need to use (6.264) when we apply the solvability condition (6.281). We would find F_1 to be the negative of $L_2 U(\partial A/\partial Z_1) + L_3 U(\partial A/\partial T)$, the solvability condition is $(\partial A/\partial Z_1) + k_0'(\partial A/\partial T) = 0$ so that $F_1 = (k_0' L_2 U - L_3 U)(\partial A/\partial T)$. Then U_1 would be nonzero and equal to $i(dU/d\omega)(\partial A/\partial T)$. Thus

$$E_0 + \epsilon E_1 = \left(U(x, \omega_0) + i\epsilon \frac{\partial U}{\partial \omega}(x, \omega_0)\frac{\partial}{\partial T} + \cdots \right)$$

$$\times A(Z, T)e^{ik_0 z - i\omega t} + (*),$$

(6.296)

which is the first two terms in (6.254). Likewise, when we apply the solvability condition at $O(\epsilon^3)$, we need to use the relations (6.266) in order to reduce the solvability condition to the form (6.291). After applying the solvability condition we would find

$$E_2 = -\tfrac{1}{2}\frac{\partial^2 U}{\partial \omega^2}(x, \omega_0)\frac{\partial^2 A}{\partial T^2}e^{ik_0 z - i\omega_0 t} + (*)$$

$$+ \text{ third order harmonics } e^{\pm(3ik_0 z - 3i\omega t)}$$

(6.297)

We stress this point, therefore, that the correct representation (6.254) of the wavepacket E_0, while not absolutely necessary, greatly facilities computation.

6k.   Solitons and the Inverse Scattering Transform. [3b,c]

In the late 1960s and early 1970s, mathematicians discovered that classes of nonlinear partial differential equations (p.d.e.'s) of evolution type containing equations such as

$$q_t + 6qq_x + q_{xxx} = 0 \quad \text{(Korteweg–de Vries or KdV)} \qquad (6.298)$$

$$q_t - iq_{xx} \mp 2iq^2 q^* = 0 \quad \text{(Nonlinear Schrödinger or NLS)} \qquad (6.299)$$

$$u_{tt} - c^2 u_{xx} + \omega_0^2 \sin u = 0 \quad \text{(Sine–Gordon or SG)} \qquad (6.300)$$

$$A_{jt} + c_j A_{jx} = \alpha_j A_k^* A_l^* \quad \text{Three-wave interaction or TWI} \qquad (6.301)$$

with j, k, l cycled over 1, 2, 3

were exactly integrable in the sense that there was a transformation which would convert the p.d.e.'s into an uncoupled set of ordinary differential equations (o.d.e.'s) for the amplitudes and phases of the normal modes (see references [9–12, 94–96]). Let us illustrate what we mean by an example,

$$q_t = iq_{xx}, \tag{6.302}$$

the linearized form of the NLS equation. Suppose we are given the complex valued function

$$q(x, 0), \quad -\infty < x < \infty \tag{6.303}$$

whose real and imaginary parts are smooth and decaying at $x = \pm\infty$ such that its Fourier transform

$$\hat{q}(k, 0) = \frac{1}{2\pi} \int_{-\infty}^{\infty} q(x, 0)e^{-ikx} dx \tag{6.304}$$

exists. Actually, all we require is that

$$\int_{-\infty}^{\infty} |q(x, 0)| dx < \infty. \tag{6.305}$$

Now, equation (6.302) is a highly coupled system of equations in which the evolution of q at one value of x depends on its values at neighboring sites (think of the finite difference approximation to (6.302)). How do we separate all the components of the equation? The answer is the Fourier transform

$$q(x, t) = \int_{-\infty}^{\infty} \hat{q}(k, t)e^{ikx} dk, \tag{6.306}$$

which converts (6.302) to a set of easily solved uncoupled equations

$$\hat{q}_t = -ik^2 \hat{q} \tag{6.307}$$

for $\hat{q}(k, t)$, one for each k. So the solution algorithm is:

$$q(x, 0) \xrightarrow{\text{direct transform}} \hat{q}(k, 0)$$

$$\downarrow \text{ evolve with (6.307)} \tag{6.308}$$

$$q(x, t) \xleftarrow{\text{inverse transform}} \hat{q}(k, t) \tag{6.309}$$

Readers familiar with Hamiltonian mechanics can also see this as a canonical transformation to action angle variables [70]. We can write (6.302) as

$$q_t = \frac{\delta H}{\delta q^*}, \quad q_t^* = -\frac{\delta H}{\delta q}, \tag{6.310}$$

where

$$H = -i \int_{-\infty}^{\infty} q_x q_x^* dx$$

and $\delta H / \delta q$ is the Frechet or variational derivative, familiar from the calculus of variations, defined as

$$\lim_{\epsilon \to 0} \frac{H[q + \epsilon \eta, q^*] - H[q, q^*]}{\epsilon} = \int_{-\infty}^{\infty} \frac{\delta H}{\delta q} \cdot \eta \, dx. \tag{6.311}$$

In the Hamiltonian formulation, then, $(q^*, q) = (P, Q)$ where

$$P_t = -\frac{\delta H}{\delta Q}, \quad Q_t = \frac{\delta H}{\delta P}. \tag{6.312}$$

The equations (6.312) solve easily if we can go to a new set of coordinates $\overline{P}, \overline{Q}$ that depend on P and Q in such a way that the *form* of the equation (6.312) are preserved and that the Hamiltonian H when written in the new variables only depends on $\overline{P}$. If we can do this, then

$$\overline{P} = \overline{P}_0, \quad \overline{Q} = \left(\frac{\delta H}{\delta \overline{P}} \right)_{\overline{P}_0} t + \overline{Q}_0 \tag{6.313}$$

and everything is solved. In Fourier coordinates,

$$H = -2\pi i \int k^2 \hat{q}(k, t) \hat{q}^*(k, t) dk. \tag{6.314}$$

Now choose

$$\overline{P}(k, t) = 2\pi \hat{q}(k, t) \hat{q}^*(k, t), \quad \overline{Q}(k, t) = i \operatorname{Arg} \hat{q}(k, t) = \frac{i}{2} \tan^{-1} \frac{\hat{q}}{\hat{q}^*}. \tag{6.315}$$

Then

$$H = -i \int k^2 \overline{P}(k, t) dk \tag{6.316}$$

$$\overline{P}_t = 0, \quad \overline{Q}_t = i \frac{d}{dt} \operatorname{Arg} \hat{q}(k, t) = -ik^2. \tag{6.317}$$

The way to assure ourselves that (6.312) is indeed a canonical transformation (that the coefficients are correct so that the form of (6.312) is exactly preserved) is to show that the two-form or symplectic form $\int \delta P \wedge \delta Q dx$ is preserved by proving that

$$\int \delta P \wedge \delta Q dx = \int \delta \overline{P} \wedge \delta \overline{Q} dk \qquad (6.318)$$

The expression $\delta P \wedge \delta Q$ simply means $\delta_1 P \delta_2 Q - \delta_1 Q \delta_2 P$, where δ_1 and δ_2 are independent variations. Note $\delta P \wedge \delta P = 0$ and $\delta P \wedge \delta Q = -\delta Q \wedge \delta P$.

EXERCISE 6.19

Let $Q = \int_{-\infty}^{\infty} \hat{q}(k, t) e^{ikx} dk$, $P = \int_{-\infty}^{\infty} \hat{q}^*(k, t) e^{-ikx} dk$, and $\overline{P}, \overline{Q}$ be given by (6.315). Prove (6.318).

$$\int \delta P \wedge \delta Q dx = \int \int \int \delta \hat{q}^*(k) e^{-ikx} dk \wedge \delta \hat{q}(k') e^{ik'x} dk' dx$$

$$= \int \int 2\pi \delta(k - k') \delta \hat{q}^*(k) \wedge \delta \hat{q}(k') dk dk'$$

$$= 2\pi \int_{-\infty}^{\infty} \delta \hat{q}^*(k) \wedge \delta \hat{q}(k) \, dk.$$

Now, from (6.315),

$$\hat{q}(k) = \frac{1}{\sqrt{2\pi}} \sqrt{\overline{P}(k)} e^{\overline{Q}(k)}, \quad \hat{q}^*(k) = \frac{1}{\sqrt{2\pi}} \sqrt{\overline{P}(k)} e^{-\overline{Q}(k)}.$$

Therefore

$$2\pi \delta \hat{q}^* \wedge \delta \hat{q} = \left(\frac{\delta \overline{P}}{2\sqrt{\overline{P}}} - \sqrt{\overline{P}} \delta \overline{Q} \right) \wedge \left(\frac{\delta \overline{P}}{2\sqrt{\overline{P}}} + \sqrt{\overline{P}} \delta \overline{Q} \right)$$

$$= \delta \overline{P} \wedge \delta \overline{Q}.$$

Hence

$$\int \delta P \wedge \delta Q dx = \int \delta \overline{P} \wedge \delta \overline{Q} dk. \qquad \square$$

So we see that the Fourier transform is, in the language of Hamiltonian mechanics, simply a canonical transformation that takes us from a highly coupled linear system (6.302) to a separable one, (6.307) or (6.317). The inverse scattering transform or IST plays an exactly analogous role for the nonlinear p.d.e.'s (6.298), (6.299), (6.300), (6.301); it is a canonical transformation that when applied to these equations separates each into an uncoupled set of equations for the action $(\overline{P})$ and angle $(\overline{Q})$ variables of the natural normal modes of the nonlinear system. Let us emphasize however: very few nonlinear p.d.e.'s have the property that they are equivalent to an uncoupled set of o.d.e.'s. *When they do we will say they are integrable.*

Now let us introduce the IST for (6.299). Consider $V = \begin{pmatrix} v_1(x,t) \\ v_2(x,t) \end{pmatrix}$, where

$$V_x = \begin{pmatrix} -i\zeta q(x,t) \\ r(x,t) \quad i\zeta \end{pmatrix} V \tag{6.319}$$

$$V_t = \begin{pmatrix} -2i\zeta^2 - iqr & 2\zeta q + iq_x \\ 2\zeta r - ir_x & 2i\zeta^2 + iqr \end{pmatrix} V + cV, \tag{6.320}$$

and the parameters ζ and c are arbitrary. The latter is in some sense trivial as it can be removed by setting $V \rightarrow Ve^{ct}$, but it is useful in order to normalize certain solutions of (6.319) which we will shortly define. The first parameter ζ is not trivial. It cannot be removed and it plays a pivotal role in the theory. For the moment the reader can think of it as the Fourier transform coordinate k. Now here is the remarkable thing: The compatibility of (6.319) and (6.320) gives, for arbitrary ζ,

$$q_t = i(q_{xx} - 2q^2 r)$$

$$r_t = -i(r_{xx} - 2qr^2). \tag{6.321}$$

EXERCISE 6.20

By adding $a \begin{pmatrix} -i\zeta & q \\ r & i\zeta \end{pmatrix} V$ to (6.320), show that (6.321) is changed by adding aq_x and ar_x to the left-hand side of (6.321).   □

The solution of (6.321) proceeds in three steps, each analogous to the steps drawn schematically in (6.309):

$$q(x,0), r(x,0) \xrightarrow{\text{direct transform}} S(t=0)$$

$$\downarrow \text{time evolution}$$

$$q(x,t), r(x,t) \xleftarrow{\text{inverse transform}} S(t=t). \tag{6.322}$$

First, given $q(x, 0)$, $r(x, 0)$, we must define and characterize (understand the properties of) the scattering data $S(t = 0)$. Second, we determine its time evolution. Third, we show how to reconstruct $q(x, t)$, $r(x, t)$ given $S(t = t)$. We follow reference [66] closely.

Step 1: The Direct Transform

We define solutions to (6.319) called Jost functions, which have a specified behavior as $x \to \pm\infty$. Remember $q, r \to 0$ as $x \to \pm\infty$ so that (6.319) is easily solved there. Define solutions $\phi(x, t; \zeta)$, $\overline{\phi}(x, t; \zeta)\psi(x, t; \zeta)$, $\overline{\psi}(x, t; \zeta)$ with the asymptotic properties

$$\phi \to \begin{pmatrix} 1 \\ 0 \end{pmatrix} e^{-i\zeta x}, \quad x \to -\infty$$

$$\overline{\phi} \to \begin{pmatrix} 0 \\ -1 \end{pmatrix} e^{i\zeta x}, \quad x \to -\infty$$

$$\psi \to \begin{pmatrix} 0 \\ 1 \end{pmatrix} e^{i\zeta x}, \quad x \to +\infty$$

$$\overline{\psi} \to \begin{pmatrix} 1 \\ 0 \end{pmatrix} e^{-i\zeta x}, \quad x \to +\infty \quad (6.323)$$

Clearly $\phi, \overline{\phi}$ are linearly independent, and their Wronskian $W(\phi, \overline{\phi}) = \phi_1\overline{\phi}_2 - \overline{\phi}_1\phi_2$ is constant and, by (6.323), equal to -1. Similarly $W(\psi, \overline{\psi}) = -1$.

EXERCISE 6.21

Prove that if $\phi = \begin{pmatrix} \phi_1 \\ \phi_2 \end{pmatrix}$ and $\overline{\phi} \begin{pmatrix} \overline{\phi}_1 \\ \overline{\phi}_2 \end{pmatrix}$ are two solutions of (6.319), then $d/dx(\phi_1\overline{\phi}_2 - \overline{\phi}_1\phi_2) = 0$. □

Because (6.319) is a second-order system, there are only two linearly independent solutions, and so the following relations must hold:

$$\phi(x, t; \zeta) = a(\zeta, t)\overline{\psi}(x, t; \zeta) + b(\zeta, t)\psi(x, t; \zeta)$$

$$\overline{\phi}(x, t; \zeta) = \overline{b}(\zeta, t)\overline{\psi}(x, t; \zeta) - \overline{a}(\zeta, t)\psi(x, t; \zeta) \quad (6.324)$$

or

$$\Phi = (\phi, -\overline{\phi}) = \Psi A = (\overline{\psi}, \psi) \begin{pmatrix} a & -\overline{b} \\ b & \overline{a} \end{pmatrix}. \quad (6.325)$$

We call A the "scattering" matrix (although there is no "scattering time" in the problem) because it tells us how the fundamental solution matrix, with asymptotic behavior

$$\begin{pmatrix} e^{-i\zeta x} & 0 \\ 0 & e^{i\zeta x} \end{pmatrix}$$

at $x = -\infty$, looks like at $x = +\infty$. Because $W(\phi, \bar{\phi}) = W(\psi, \bar{\psi}) = -1$, we have

$$a\bar{a} + b\bar{b} = 1. \tag{6.326}$$

Roughly speaking, (6.326) expresses a "conservation of energy."

EXERCISE 6.22

Show that as $x \to +\infty$,

$$\phi \to \begin{pmatrix} ae^{-i\zeta x} \\ be^{i\zeta x} \end{pmatrix}, \quad \bar{\phi} \to \begin{pmatrix} \bar{b}e^{-i\zeta x} \\ -\bar{a}e^{i\zeta x} \end{pmatrix}$$

and as $x \to -\infty$,

$$\psi \to \begin{pmatrix} \bar{b}e^{-i\zeta x} \\ ae^{i\zeta x} \end{pmatrix}, \quad \bar{\psi} \to \begin{pmatrix} \bar{a}e^{-i\zeta x} \\ -be^{i\zeta x} \end{pmatrix}. \tag{6.327}$$

□

EXERCISE 6.23

Show that $a(\zeta) = W(\phi, \psi), \bar{a}(\zeta) = W(\bar{\phi}, \bar{\psi})$.

Having defined the scattering matrix A for real values of ζ, let us look at its properties. First, let us construct the solution $\phi(x, t; \zeta)$. From (6.319)

$$\phi_1 e^{i\zeta x} = 1 + \int_{-\infty}^{x} q(z)\phi_2 e^{i\zeta z} dz \tag{6.328}$$

$$\phi_2 e^{-i\zeta x} = \int_{-\infty}^{x} r(y)\phi_1 e^{-i\zeta y} dy \tag{6.329}$$

Substitute (6.329) in (6.328) and obtain

$$\phi_1 e^{i\zeta x} = 1 + \int_{-\infty}^{x} q(z) dz \int_{-\infty}^{z} r(y)(\phi_1 e^{i\zeta y})e^{2i\zeta(z-y)} \, dy,$$

which by changing order of integration (see Figure 6.37)

$$= 1 + \int_{-\infty}^{x} M(x, y; \zeta)(\phi_1 e^{i\zeta y})dy \tag{6.330}$$

where

$$M(x, y; \zeta) = r(y) \int_{y}^{x} q(z)e^{2i\zeta(z-y)}dz. \tag{6.331}$$

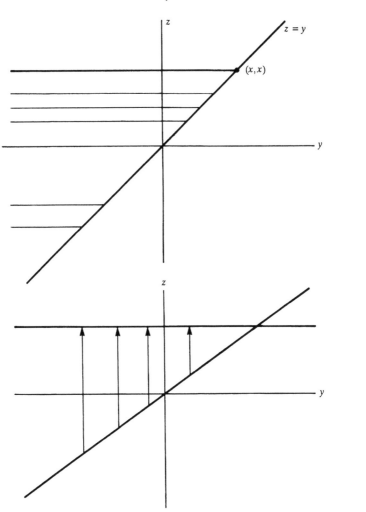

FIGURE 6.37 The domain of integration. In the top figure the domain is shaded so as to indicate we integrate first from $y = -\infty$ to $y = z$ and then from $z = -\infty$ to $z = x$. In the lower figure, we interchange the order of integration by integrating first in z from $z = y$ to $z = x$ and then in y from $y = -\infty$ to $y = x$.

We construct the solution by iterating,

$$\phi_1 e^{i\zeta x} = 1 + \int_{-\infty}^{x} M(x, y; \zeta)dy \left(1 + \int_{-\infty}^{y} M(x, z; \zeta)(1 + \cdots)\right). \quad (6.332)$$

The solution $\phi_2 e^{i\zeta x}$ is constructed in a similar fashion. Now we are not going to go through the details of the proof that $\phi_1 e^{i\zeta x}$ exists for all x for ζ real. We refer the reader instead to reference [66]. However, we will list the properties.   □

1. Given $\int_{-\infty}^{\infty} |x^n||q(x)|dx, \int_{-\infty}^{\infty} |x^n||r(x)|dx$ are finite, then $\phi e^{i\zeta x}$ and its first n derivatives with respect to ζ exist for all real ζ. Similarly $\psi e^{-i\zeta x}, \overline{\phi} e^{-i\zeta x}$, and $\overline{\psi} e^{i\zeta x}$ and their first n derivatives exist for all real ζ.

2. Given $\int_{-\infty}^{\infty} |q(x)|dx, \int_{-\infty}^{\infty} |r(x)|dx$ are finite, $\phi e^{i\zeta x}, \psi e^{-i\zeta x}$ and $a(\zeta)$ $[= W(\phi, \psi)]$ are analytic functions of ζ for Im $\zeta > 0$ and $\overline{\phi} e^{-i\zeta x}$; $\overline{\psi} e^{i\zeta x}$ and $\overline{a}(\zeta)$ are analytic functions of ζ for Im $\zeta < 0$. To see why this is so, just look at the expression (6.331) for $M(x, y, \zeta)$ and observe that, in the range of integration, $z - y > 0$, and therefore if $\zeta = \xi + i\eta, \eta > 0$, the exponentially decaying factor $e^{-2\eta(z-y)}$ is introduced into the integrand. Even when one differentiates $\phi_1 e^{i\zeta x}$ as given by (6.332) with respect to ζ, the exponentially decaying factor dominates any algebraic growth like $(z - y)$ that appears when we differentiate $M(x, y; \zeta)$. Taking ζ into the upper half-plane enhances convergence. Therefore we can analytically continue $\phi e^{i\zeta x}, \psi e^{-i\zeta x}$, and $a(\zeta) = W(\phi, \psi)(\overline{\phi} e^{-i\zeta x}, \overline{\psi} e^{i\zeta x}$ and $\overline{a}(\zeta))$ into the upper (lower) half complex ζ-plane.

3. Furthermore, the zeros $\zeta_k, k = 1, \ldots N (\overline{\zeta}_k, k = 1, \ldots \overline{N})$ of $a(\zeta)$ $(\overline{a}(\zeta))$ in the upper (lower) half ζ-planes are the bound state eigenvalues of (6.319). Why? If $a(\zeta_k) = 0$, then the Wronskian $W(\phi(x, \zeta_k), \psi(x, \zeta_k))$ is zero, which means that $\phi(x, \zeta_k)$ and $\psi(x, \zeta_k)$ are linearly dependent; that is,

$$\phi(x, \zeta_k) = b_k \psi(x, \zeta_k). \quad (6.333)$$

But if Im $\zeta_k > 0$, then $\phi(x, \zeta_k)$ (and $\psi(x, \zeta_k)$) decays exponentially both at $x = -\infty$ and $x = +\infty$. Therefore it is a bound state eigenfunction. Likewise at $\overline{\zeta}_k, \text{Im}\overline{\zeta}_k < 0$, where $\overline{a}(\overline{\zeta}_k) = 0$,

$$\overline{\phi}(x, \zeta_k) = \overline{b}_k \overline{\psi}(x, \overline{\zeta}_k) \quad (6.334)$$

and $\overline{\phi}(x, \zeta_k)$ (and $\overline{\psi}(x, \overline{\zeta}_k)$) decay as $x \to +\infty$ and as $x \to -\infty$.

4. It is also useful to look at the "linear" limit obtained by ignoring all quadratic and higher products of q and r whenever they occur. Then

$$\phi_1 e^{i\zeta x} = 1$$

and

$$\phi_2 e^{-i\zeta x} = \int_{-\infty}^{x} r(y) e^{-2i\zeta y} dy.$$

In the limit $x \to +\infty$, from (6.327), $\phi_2 e^{-i\zeta x} \to b(\zeta)$ so that

$$b(\zeta) = \int_{-\infty}^{\infty} r(y) e^{-2i\zeta y} dy$$

$$= 2\pi \hat{r}(2\zeta)$$

where $\hat{r}(k)$ is the Fourier transform of $r(x)$. This limit is not strictly correct, because we will come across situations where q and r are small but their areas $\int_{-\infty}^{\infty} q dx$, $\int_{-\infty}^{\infty} r dx$ are sufficiently large so as to negate this step. Nevertheless, the reader can see that the coefficients $b(\zeta)$, $\bar{b}(\zeta)$ play the role of nonlinear analogs to the Fourier coefficients.

5. The asymptotic behavior of $\phi e^{i\zeta x}$, $\psi e^{-i\zeta x}$, and $a(\zeta)$ as a $\zeta \to \infty$, $\text{Im}\zeta > 0$ can be read off from (6.328), (6.329), and the analogous expressions for $\psi e^{-i\zeta x}$. We note in particular that

$$\phi_2 e^{-i\zeta x} \sim \int_{-\infty}^{x} r(y) dy e^{-2i\zeta y} = \frac{i}{2\zeta} r(x) e^{-2i\zeta x} + O\left(\frac{1}{\zeta^2}\right). \quad (6.335)$$

(recall $\phi_1 e^{i\zeta x} \to 1 + O(\frac{1}{\zeta})$). Therefore,

$$\lim_{\zeta \to \infty} -2i\zeta \phi_2 e^{i\zeta x} = r(x). \quad (6.336)$$

Similarly one can show

$$\lim_{\zeta \to \infty} 2i\zeta \psi_1 e^{-i\zeta x} = q(x). \quad (6.337)$$

and that as $\zeta \to \infty$, $\text{Im}\zeta \geq 0$,

$$a(\zeta) \to 1. \quad (6.338)$$

Both the expressions (6.336) and (6.337) are useful when we reconstruct $q(x)$ and $r(x)$ from the scattering data S. From (6.338), we see that $a(\zeta)$ has its zeros in the finite part of the upper half ζ-plane. Therefore the number N of zeros of $a(\zeta)$ is finite as otherwise there would have to be a point of accumulation of zeros of $a(\zeta)$ and then since it is analytic, it would be identically zero.

6. In summary, then, we define the scattering data to be the set

$$S\{a(\zeta), b(\zeta), \bar{a}(\zeta), \bar{b}(\zeta), \quad \text{real}; \quad (\zeta_k, b_k)_1^N, (\bar{\zeta}_k, \bar{b}_k)_1^{\bar{N}}\}. \quad (6.339)$$

Because of the analytic properties of $a(\zeta)$ and $\bar{a}(\zeta)$ and the fact that (6.326) holds for real ζ, not all these quantities are independent, but we will not worry about this point for now.

7. When r and q are related, then there also exist relations in the scattering data. In particular, if $r = \alpha q^*$, α real, then

$$\bar{\phi}(x, \zeta) = \begin{pmatrix} -\frac{1}{\alpha}\phi_2^*(x, \zeta^*) \\ -\phi_1^*(x, \zeta^*), \end{pmatrix}, \quad \bar{\psi}(x, \zeta) = \begin{pmatrix} \psi_2^*(x, \zeta^*) \\ \alpha\psi_1^*(x, \zeta^*), \end{pmatrix}$$

$$\bar{a}(\zeta^*) = a^*(\zeta), \quad \bar{\zeta}_k = \zeta_k^*, \quad \bar{N} = N, \quad \bar{b}_k = -\frac{1}{\alpha}b_k^* \quad (6.340)$$

$$\bar{b}(\zeta) = -\frac{1}{\alpha}b^*(\zeta), \quad \zeta \quad \text{real}$$

and if $r = \alpha q$, then

$$\bar{\phi}(x, \zeta) = \begin{pmatrix} -\frac{1}{\alpha}\phi_2(x, -\zeta) \\ -\phi_1(x, -\zeta) \end{pmatrix}, \quad \bar{\psi}(x, \zeta) = \begin{pmatrix} \psi_2(x, -\zeta) \\ \alpha\psi_1(x, -\zeta) \end{pmatrix},$$

$$\bar{a}(-\zeta) = a(\zeta), \quad \bar{\zeta}_k = -\zeta_k, \quad \bar{N} = N, \quad \bar{b}_k = -\frac{1}{\alpha}b_k,$$

$$\bar{b}(\zeta) = -\frac{1}{\alpha}b(-\zeta).$$

To prove these, note that if $r = \alpha q^*$, then if $\phi_1(x, \zeta), \phi_2(x, \zeta)$ solves (6.319) so does $-(1/\alpha)\phi_2^*(x, \zeta^*), -\phi_1^*(x, \zeta^*)$. Similarly if $\psi_1(x, \zeta), \psi_2(x, \zeta)$ solves (6.319) so does $(\psi_2^*(x, \zeta^*), \alpha\psi_1^*(x, \zeta^*))$, and it has the same boundary conditions as $\bar{\psi}_1(x, \zeta), \bar{\psi}_2(x, \zeta)$ as $x \to +\infty$.

In the case of the self-focusing NLS equation, $r = -q^*$, and the real and imaginary parts of the bound state eigenvalues $\zeta_k = \xi_k + i\eta_k$ give the speed and amplitude of the kth soliton. For the defocusing NLS equation $r = +q^*$, we find that (6.319) is self-adjoint and has only real eigenvalues ζ. Therefore $a(\zeta)$ has no zeros, $\text{Im}\,\zeta > 0$, and there are no solitons in this case. There are solitons in the defocusing case, however (dark solitons), when $q \to q_0 \neq 0$ as $x \to \pm\infty$, but they are not seen in this analysis.

In the case of the sine–Gordon equation, which we meet when we study propagation of coherent pulses through resonant media, $r = -q$, and in that case the zeros of $a(\zeta)$ either lie on the imaginary axis ($\zeta_k = i\eta_k$) or occur in conjugate pairs ($\pm\xi_k + i\eta_k$). The former give rise to kinklike solutions for which the time integral of the electric field $\int_{-\infty}^{\infty} E\,dt$ is 2π or a multiple thereof. The latter give rise to paired pulses of zero area called breathers.

8. For the self-focusing ($r = -q^*$) and defocusing (($r = +q^*$) NLS equation with zero boundary conditions at $x = \pm\infty$, the direct scattering problem involves calculating the map

$$q(x, 0) \to S(a(\zeta), b(\zeta), \zeta\,\text{real}; (\zeta_k, b_k)_1^N). \qquad (6.341)$$

The bound state eigenvalues ζ_k, which we will see are time independent, give the speed and amplitude of the solitons. The coefficients b_k give their positions. The other parts of the scattering data and in particular $b(\zeta)$ give rise to oscillatory-like solutions that disperse and decay with time, like the solutions we calculated for $q_t = iq_{xx}$ in (6.300). Therefore the asymptotic state of the solution consists of solitons. If $b(\zeta)$ is identically zero, then the solution $q(x, t)$ consists of pure solitons for all time. For the most part, in this book, we are concerned with this case, called the reflectionless potential case.

We will now work out a few examples of (6.341).

EXERCISE 6.24

Given

$$-r^*(x, 0) = q(x, 0) = \begin{cases} 0, & x < 0, x > L \\ Q, & 0 < x < L. \end{cases}$$

Show, using continuity of ϕ at $x = 0$ and $x = L$, that

$$a(\zeta)e^{-i\zeta L} = \cos\sqrt{\zeta^2 + Q^2}L - \frac{i\zeta}{\sqrt{\zeta^2 + Q^2}}\sin\sqrt{\zeta^2 + Q^2}L \qquad (6.342)$$

$$b(\zeta)e^{i\zeta L} = -\frac{Q}{\sqrt{Q^2 + \zeta^2}}\sin\sqrt{Q^2 + \zeta^2}L \qquad (6.343)$$

Remark 1. Observe that $a(\zeta)$ is an entire function, as is $b(\zeta)$. This is because $q(x, 0)$ was on compact support and all functions can be analytically extended to the whole complex ζ-plane. In general $b(\zeta)$ is only defined on the real axis.

Remark 2. The ordinary Fourier transform of $r(x, 0)$ is $\hat{r}(k) = (1/2\pi)\int(-Q)e^{-ikx}dx = (-1/\pi k)Qe^{(-ikL/2)}\sin kL/2$. Observe that $2\pi\hat{r}(2\zeta) = (-Q/\zeta)e^{-i\zeta L}\sin\zeta L$, which for $Q^2 \ll \zeta^2$ is the limit of $b(\zeta)$. However, also notice that it is a nonuniform limit and does not hold near $\zeta = 0$.

Remark 3. The solitons. It is not difficult to show that all zeros of $a(\zeta)$ in the upper half-plane lie on the imaginary axis between 0 and Q. Let $\zeta = iQ\cos\theta$ and find

$$a(\zeta)e^{-i\zeta L} = \frac{\sin(QL\sin\theta + \theta)}{\sin\theta} \qquad (6.344)$$

whose zeros are given by the intersection of $y = QL\sin\theta$ and the set of curves $y_n = n\pi - \theta$, $n = 1, 2, \ldots$ in the interval $0 < \theta < \pi/2$ (see Figure 6.38). Note that for $QL < \pi/2$, no solitons exist. $\qquad\square$

EXERCISE 6.25

We examine what happens to the scattering data and in particular the soliton spectrum $\zeta_j, a(\zeta_j) = 0$, when the square wave potential of Exercise 6.24 is multiplied by a "wavenumber chirp" $\exp i\Delta x^2$. In optics where we follow the evolution of the initial ($z = 0$) pulse $q(t, 0)$ in z, the chirp is a frequency chirp $\exp i\Delta t^2$; that is, the frequency changes linearly across the pulse. Here we just quote the results. The reader may consult reference [97] for details. The nature of the result changes depending on whether $\sqrt{\Delta}L$ is much less (slow chirp) or greater (fast chirp) than unity.

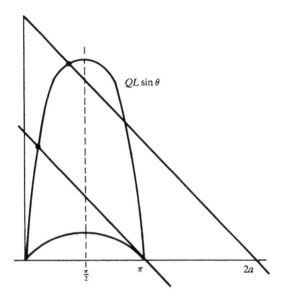

FIGURE 6.38

The equations $a(\zeta) = 0$ in each case are

$$\sqrt{\Delta}L \ll 1, \quad \tan\sqrt{\zeta^2 + Q^2}\,L = -i\frac{\sqrt{\zeta^2 + Q^2}}{\zeta} + \frac{\Delta^2 Q^4 L^3}{24(\zeta^2 + Q^2)^{3/2}\zeta^2}$$

$$\sqrt{\Delta}L \gg 1, \quad \tan\left\{\frac{1}{\Delta}\left(\zeta + \tfrac{1}{2}\Delta L\right)S_x(L) - \frac{1}{\Delta}\left(\zeta - \tfrac{1}{2}\Delta L\right)S_x(0)\right.$$

$$\left. +Q^2 \ln\frac{\zeta + \tfrac{1}{2}\Delta L + S_x(L)}{\zeta - \tfrac{1}{2}\Delta L + S_x(0)}\right\} = -i\frac{S_x(L) + S_x(0) + \Delta L}{S_x(L) - S_x(0) + 2\zeta}$$

where

$$S_x(L) = \sqrt{(\zeta + \Delta L)^2 + Q^2}$$

$$S_x(0) = \sqrt{\zeta^2 + Q^2}.$$

Note that when $\Delta = 0$, this gives what we would obtain from (6.342) with $a = 0$. The transcendental equation is complicated, but a fundamental property of its solutions is that fast chirping enchances soliton production. Namely, at the value $QL = 3\pi/2$, a second soliton for the unchirped problem is about to appear (the second eigenvalue ζ_2 is at the origin) whereas in the chirped case, it has already been born (i.e., $\zeta_2 = i\eta_2$, $\eta_2 > 0$). $\qquad\square$

EXERCISE 6.26

$q(x, 0) = A\text{sech}x$. The reader should consult reference [88] for details. The results are

$$a(\zeta) = \frac{\Gamma^2\left(\frac{1}{2} - i\zeta\right)}{\Gamma\left(-i\zeta + A + \frac{1}{2}\right)\Gamma\left(-i\zeta - A + \frac{1}{2}\right)}$$

$$b(\zeta) = i\text{sech}\,\pi\zeta\,\sin\pi A,$$

when $\Gamma(z)$ is the gamma function whose poles occur at the negative integers $z = -n$. Hence the zeros of $a(\zeta)$ in the upper half-plane occur at

$$\zeta_n = i\left(A - n - \frac{1}{2}\right), \quad n = 0, 1, \ldots N, \quad \text{for} A - \frac{1}{2} > N.$$

The number N of solitons is equal to the integer lying between $A - (1/2)$ and $A + (1/2)$. Observe that if $A = m$ and m is an integer, $b(\zeta) = 0$, there is no radiation and $N = m$. ☐

EXERCISE 6.27

The case of two humps.

$$q(x, 0) = \text{sech}(x - x_0) + e^{i\alpha}\text{sech}(x + x_0).$$

The reference for details is [98, 99]. The upshot is that if the potential sech x has an eigenvalue ζ on the imaginary axis $\zeta = i/2$, the potential $q(x, 0) = \text{sech}(x - x_0) + e^{i\alpha}\text{sech}(x + x_0)$ for large x_0 two eigenvalues, which to leading order are given by

$$\zeta_\pm = i\left(\frac{1}{2} + \left(\frac{2x_0}{\sinh 2x_0}\right)\cos\alpha \pm \frac{1}{2}\left(1 + \frac{2x_0}{\sinh 2x_0}\right)\text{sech}\,x_0\cos\frac{\alpha}{2}\right)$$

$$+ \left(\frac{2x\coth 2x_0 - 1}{\sinh 2x_0}\right)\sin\alpha \pm \frac{1}{2}\left(1 - \frac{2x_0}{\sinh 2x_0}\right)\text{cosech}\,x_0\sin\frac{\alpha}{2}\right)$$

Observe that for $\alpha = 0$, ζ_+ and ζ_- are both pure imaginary whereas $\alpha = \pi$, they develop nonzero real values. ☐

Step 2: The Time Evolution of The Scattering Data

This is a very easy step. Observe that ϕ and $\overline{\psi}$ satisfy (6.320) with $c = +2i\zeta^2$ and $\overline{\phi}$ and ψ satisfy (6.320) with $c = -2i\zeta^2$ in order that their respect asymptotic behaviors at $x = -\infty$ and $x = +\infty$ are as given by (6.323). To see this, substitute $\phi = \begin{pmatrix} 1 \\ 0 \end{pmatrix} e^{-i\zeta x}$ into (6.320) and get $c = 2i\zeta^2$. Now let

$x \to +\infty$ and substitute $\phi = \begin{pmatrix} ae^{-i\zeta x} \\ be^{i\zeta x} \end{pmatrix}$ into (6.320) and find

$$a_t = 0$$
$$b_t = 4i\zeta^2 b.$$

(6.345)

Since $a(\zeta, t)$ is time independent, so are its zeros $\zeta_k, k = 1, \ldots N$ in the upper half ζ-plane. To find the time dependence of $b_k(t)$, use its definition $\phi(x, \zeta_k) = b_k \psi(x, \zeta_k)$ and differentiate with respect to time. We find

$$\phi_t(x, \zeta_k) = \begin{pmatrix} -iqr & 2\zeta_k q + iq_x \\ 2\zeta_k r - ir_x & 4i\zeta_k^2 + iqr \end{pmatrix} \phi(x, \zeta_k)$$

$$= b_{kt} \psi(x, \zeta_k) + \begin{pmatrix} -4i\zeta_k^2 - iqr & 2\zeta_k q + iq_x \\ 2\zeta_k r - ir_x & iqr \end{pmatrix} b_k \psi(x, \zeta_k),$$

which implies

$$b_{kt} = 4i\zeta_k^2 b_k.$$

(6.346)

Step 3: The Inverse Transform

The goal here is to reconstruct $q(x, t)$ and $r(x, t)$ given

$$S(a(\zeta), b(\zeta), \zeta \text{ real}; (\zeta_k, b_k)_1^N).$$

(6.347)

We do this by first constructing the solutions $\phi e^{-i\zeta x}$ and then use (6.337) to get $q(x, t)$. Consider $\bullet(\phi(x, \zeta) e^{i\zeta x}/a(\zeta))$, a function that is meromorphic (analytic except at poles $z = \zeta_k$, which we assume are simple) in the upper half-plane. From (6.324), we see

$$\frac{\phi e^{i\zeta x}}{a} = \overline{\psi} e^{i\zeta x} + \frac{b}{a} \psi e^{i\zeta x}$$

(6.348)

on the real ζ axis. What we want to do is construct functions $\phi e^{i\zeta x}/a$, meromorphic with a finite number of poles at given locations ζ_k in the upper half ζ-plane and tending to $\begin{pmatrix} 1 \\ 0 \end{pmatrix}$ as $\zeta \to \infty$, and $\overline{\psi} e^{i\zeta x}$, analytic in the lower half-plane and tending to $\begin{pmatrix} 0 \\ 1 \end{pmatrix}$ as $\zeta \to \infty$, Im $\zeta \le 0$, whose difference on the real axis separating the domains of meromorphic and analytic behavior is given by $\bullet(b(\zeta,t)/a(\zeta)\psi(x,\zeta))e^{i\zeta x}$. This is a version of what is called the Riemann-Hilbert problem and it is solved like this. Evaluate

● Q2

$$\int_{-\infty}^{\infty} \frac{\phi(x,\zeta')e^{i\zeta'x}}{a(\zeta')(\zeta'-\zeta^*)}d\zeta' \tag{6.349}$$

for Im $\zeta > 0$ by moving to the contour $C_1, \zeta' = Re^{i\theta}, 0 < \theta < \pi$ (Figure 6.39), and we get $(a'_k = (da/d\zeta)_{\zeta_k})$,

$$-i\pi \begin{pmatrix} 1 \\ 0 \end{pmatrix} + 2\pi i \sum_{k=1}^{N} \frac{b_k}{a'_k} \frac{\psi(x,\zeta_k)e^{i\zeta_k x}}{\zeta_k - \zeta^*}. \tag{6.350}$$

Now (6.349) can also be evaluated by using (6.348) and evaluating the first term in

$$\int_{-\infty}^{\infty} \frac{\overline{\psi}(x,\zeta')e^{i\zeta'x}}{\zeta'-\zeta^*}d\zeta' + \int_{-\infty}^{\infty} \frac{b(\zeta,t)}{a(\zeta')}\frac{\psi(x,\zeta')e^{i\zeta'x}}{\zeta'-\zeta^*}d\zeta'$$

by going to the contour $C_2, \zeta' = Re^{i\theta}, -\pi < \theta < 0$, whereupon we get

$$i\pi \begin{pmatrix} 1 \\ 0 \end{pmatrix} - 2\pi i \overline{\psi}(x,\zeta^*)e^{+i\zeta^*x} + \int_{-\infty}^{\infty} \frac{b}{a}\frac{\psi(x,\zeta')e^{i\zeta'x}}{\zeta'-\zeta^*}d\zeta'$$

$$= i\pi \begin{pmatrix} 1 \\ 0 \end{pmatrix} - 2\pi i \begin{pmatrix} \psi_2^*(x,\zeta) \\ -\psi_1^*(x,\zeta) \end{pmatrix} e^{+i\zeta^*x} + \int_{-\infty}^{\infty} \frac{b}{a}\frac{\psi(x,\zeta')e^{i\zeta'x}}{\zeta'-\zeta^*}d\zeta' \tag{6.351}$$

from (6.340). Equating (6.350) and (6.351) gives an integral equation $\psi(x,\zeta)$, which can be reduced to an integral equation of Gelfand–Levi–Marcenko type

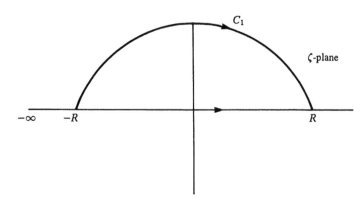

FIGURE 6.39

by taking a Fourier transform. We refer the reader to reference [66] for details. Here we deal only with the reflectionless potential and set $b \equiv 0$, whereupon, for Im $\zeta > 0$, and $\gamma_k = (b_k)/(a'_k)$,

$$\begin{pmatrix} \psi_2^*(x, \zeta) \\ -\psi_1^*(x, \zeta) \end{pmatrix} e^{i\zeta^* x} = \begin{pmatrix} 1 \\ 0 \end{pmatrix} - \sum_1^N \frac{\gamma_k \psi(x, \zeta_k) e^{i\zeta_k x}}{\zeta_k - \zeta^*} \tag{6.352}$$

From the second of these equations and (6.337),

$$q(x, t) = -2i \sum_1^N \gamma_k^* \psi_2^*(x, \zeta_k) e^{-i\zeta_k x}. \tag{6.353}$$

Therefore our goal is to find $\psi_1(x, \zeta_k), \psi_2(x, \zeta_k), k = 1 \ldots N$, and this can done by setting $\zeta = \zeta_j$ in (6.352), $j = 1 \ldots N$, and obtaining $2N$ equation for the $2N$ unknowns:

$$u_{lk} = \sqrt{\gamma_k} \psi_l(x, \zeta_k), \quad l = 1, 2; \quad k = 1, \ldots N. \tag{6.354}$$

It is convenient to set

$$\sqrt{\gamma_k} e^{i\zeta_k x} = \lambda_k, \quad k = 1 \ldots N \tag{6.355}$$

and then (6.352) can be rewritten

$$u_1 = -Bu_2^*$$

$$(I + B^*B)u_2^* = \lambda^* \tag{6.356}$$

$$B = \left(\frac{\lambda_j \lambda_k^*}{\zeta_j - \zeta_k^*} \right), \tag{6.357}$$

where u_1 and u_2 are the column vectors $(u_{11}, \ldots u_{1N})$ and $(u_{21}, \ldots u_{2N})$ respectively. From (6.353) and (6.356)

$$q(x, t) = -2i \sum_{k=1}^N \lambda_k^* u_{2k}^*. \tag{6.358}$$

The one-soliton solution is given by

$$\zeta_1 = \xi_1 + i\eta_1, \quad a = \frac{\zeta - \zeta_1}{\zeta - \zeta_1^*}, \quad \gamma_1 = \sqrt{2\eta_1}\, e^{2\bar{\theta}_1 + 2i\bar{\sigma}_1}, \quad b_1 = a_1'\gamma_1,$$

$$a_1' = \frac{da}{a\zeta}(\zeta_1) \quad \lambda_1 = \sqrt{2\eta_1}\, e^{\bar{\theta}_1 + i\sigma_1}, \quad \theta_1 = -\eta_1 x + \bar{\theta}_1, \quad \sigma_1 = \xi_1 x + \bar{\sigma}$$

$$B = -ie^{2\theta_1}, \quad u_2^* = \frac{1}{1 + e^{4\theta_1}}\lambda^*,$$

with $\bar{\theta}_{1t} = -4\xi_1\eta_1$ and $\bar{\sigma}_{1t} = 2(\xi_1^2 - \eta_1^2)$ from (6.346).

$$q(x, t) = 2\eta \operatorname{sech} 2\eta_1(x + 4\xi_1 t - x_0)\exp(-2i\xi_1 x - 4i(\xi_1^2 - \eta_1^2)t - 2i\sigma_0).$$
$$(6.359)$$

We now show some pictures of solitons. Figure 6.40a shows a single soliton (6.359) as functions of x. Figure 6.40b shows a double soliton solution with $\xi_1 = \xi_2 = 0$ and $\eta_1 = \eta_2 = \eta$. It starts out looking like two separated solitons, which then come together in the manner shown after a time $t = t_0$ and then return to the original state at $t = 2t_0$. Figure 6.40c shows the interaction of two solitons. Figure 6.40d shows the dark soliton $|q| = \eta \tanh^2 \eta x$, which also solves (6.299) in the defocusing case.

Other Important Properties

Now that we have introduced the inverse scattering transform, we turn to two important properties that are useful in applications, the generation of the constants of the motion and the behavior of the soliton parameters when the NLS equation (6.299) is perturbed by other influences such as damping, periodic forcing, nonlinear delay effects, and higher-order dispersion.

The Conserved Quantities [79]

We have seen that the function $a(\zeta)$ is independent of time for all ζ. It contains information about all the conserved quantities. We will first express it as a functional of the fields q and r and then in terms of the scattering data. Equating these two expressions then gives us relations between the conserved

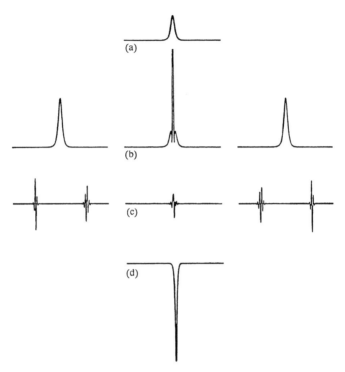

FIGURE 6.40   Pictures of solitons of the NLS equation $q_z - iq_{tt} - 2isq^2q^* = 0$ in the focusing ($s = 1$, (a), (b), (c)) and defocusing ($s = -1$, (d)) cases: (a) bright soliton, (b) periodic evolution of initial condition $q(0, t) = 2\mathrm{sech}(t)$, (c) interaction of two solitons, (d) dark soliton.

quantities

$$C_1 = \int_{-\infty}^{\infty} qr\,dx, \quad C_2 = \tfrac{1}{2}\int_{-\infty}^{\infty}(qr_x - q_x r)\,dx,$$

$$C_3 = \int_{-\infty}^{\infty}(-q_x r_x - q^2 r^2)\,dx, \tag{6.360}$$

and the soliton parameters.

To write $a(\zeta)$ as a functional of q and r, we set

$$\phi_1 e^{i\zeta x} = e^{\Phi} \tag{6.361}$$

and eliminate ϕ_2 from (6.319) to obtain ($D = d/dx$)

$$(D - i\zeta)\frac{1}{q}(D + i\zeta)\phi_1 = r\phi_1. \tag{6.362}$$

A little calculation gives

$$2i\zeta\,\Phi_x = -qr - \frac{q_x}{q}\Phi_x + \Phi_{xx} + \Phi_x^2. \tag{6.363}$$

Solving iteratively (thinking of $\zeta \gg 1$), we obtain

$$(\Phi_x)^{(1)} = \frac{-1}{2i\zeta}qr$$

$$(\Phi_x)^{(n)} = \frac{1}{2i\zeta}\left(-\frac{q_x}{q}(\Phi_x)^{(n-1)} + (\Phi_x)_x^{(n-1)} + \sum_{j+k=n-1}(\Phi_x)^j(\Phi_x)^k\right).$$
$$\tag{6.364}$$

Now as $x \to -\infty$, $\Phi \to O(\phi e^{i\zeta x} \to 1)$ and as $x \to +\infty$, $\Phi \to \ln a$. Therefore integrating Φ_x from $-\infty$ to $+\infty$ we obtain

$$\ln a \simeq -\sum_1^\infty \frac{1}{(2i\zeta)^n}C_n, \tag{6.365}$$

where the first three C_n are listed in (6.360).

To write $a(\zeta)$ as a functional of the scattering data we observe that the function

$$\ln f(\zeta) = \ln\left(a(\zeta)\prod_{j=N}^N \frac{\zeta - \zeta_j^*}{\zeta - \zeta_j}\right) \tag{6.366}$$

is analytic in the upper half ζ-plane and tends to zero as $\zeta \to \infty$ because the zeros of $a(\zeta)$ are canceled. Therefore, by Cauchy's integral theorem, for Im $\zeta > 0$,

$$\ln f(\zeta) = \frac{1}{2\pi i}\int_{-\infty}^\infty \frac{\ln f(\xi)}{\xi - \zeta}d\xi. \tag{6.367}$$

Similarly,

$$\ln \overline{f}(\zeta) = \ln\left(\overline{a}(\zeta)\prod_{j=1}^N \frac{\zeta - \overline{\zeta}_j^*}{\zeta - \overline{\zeta}_j}\right)$$

is analytic in the lower half-plane. We now assume $r = -q^*$ so that $\overline{\zeta}_j = \zeta_j^*$ and that on the real axis $\overline{a}(\zeta) = a^*(\zeta)$. Then, we have that, for Im $\zeta > 0$

(evaluate by going to the semicircle contour in the lower half-plane),

$$0 = \frac{1}{2\pi i} \int \frac{\ln \overline{f}(\xi)}{\xi - \zeta} d\xi = \frac{1}{2\pi i} \int \frac{\ln f^*(\xi)}{\xi - \zeta} d\xi. \tag{6.368}$$

Adding, we obtain

$$\ln a(\zeta) = \sum_{j=1}^{N} \ln(\zeta - \zeta_j) - \sum_{j=1}^{N} \ln(\zeta - \zeta_j^*) + \frac{1}{2\pi i} \int_{-\infty}^{\infty} \frac{\ln a a^*}{\xi - \zeta} d\xi,$$

which when expanded about $\zeta = \infty$ gives

$$\ln a(\zeta) = -\sum_{1}^{\infty} \frac{1}{(2i\zeta)^m} C_m$$

where the C_m of (6.365) are now expressed in terms of the scattering data

$$\frac{C_m}{(2i)^m} = \sum_{k=1}^{N} \frac{\zeta_k^m - \zeta_k^{*m}}{m} + \frac{1}{2\pi i} \int_{-\infty}^{\infty} \xi^{m-1} \ln a a^* d\xi$$

$$= \sum_{k=1}^{N} \frac{\zeta_k^m - \zeta_k^{*m}}{m} - \frac{1}{2\pi i} \int \xi^{m-1} \ln\left(1 + \frac{bb^*}{aa^*}\right) d\xi \tag{6.369}$$

because $aa^* + bb^* = 1$ for real ξ. Observe that if there are no bound states and $|b/a|$ is "small,"

$$\frac{1}{(2i)^m} C_m = -\frac{1}{2\pi i} \int \xi^{m-1} \left|\frac{b}{a}\right|^2 d\xi, \tag{6.370}$$

which in the "linear" limit obtained by only retaining quadratic products in the C_m are Parseval's relations. The formulae (6.369) are called the *Trace Formulae*. They are very useful when we consider perturbations of (6.299).

They also allow us to see IST as a canonical transformation through which (6.299) is converted to action-angle variables. First, observe that

$$q_t = i(q_{xx} + 2q^2 q^*) = \frac{\delta}{\delta q^*} H_3, \quad q_t^* = -\frac{\delta H_3}{\delta q}$$

where $H_3 = -iC_3$. In terms of the scattering data

$$H_3 = 8 \sum_{1}^{N} \frac{\zeta_k^3 - \zeta_k^{*3}}{3} + \frac{4}{\pi i} \int_{-\infty}^{\infty} \xi^2 \ln a a^* d\xi. \tag{6.371}$$

The new momenta variables, the action variables, are

$$2i\zeta_k, 2i\zeta_k^*, \quad \text{and} \quad \frac{1}{\pi} \ln aa^*,$$

and the new angle variables are

$$\ln b_k, \quad \ln b_k^*, \quad \text{and} \quad \ln b(\xi). \tag{6.372}$$

Perturbation equations

If $r = -q^*$ and

$$q_t - iq_{xx} - 2iq^2q^* = F, \tag{6.373}$$

then we can show (but will not show here; see reference [86]), that the time evolution of the scattering data is

$$\zeta_{kt} = -\gamma_k \int_{-\infty}^{\infty} \left(F\psi_1^2(x, \zeta_k) + F^*\psi_2^2(x, \zeta_k) \right) dx \tag{6.374}$$

$$\beta_{kt} + \frac{a_k''}{a_k'} \zeta_{kt} \beta_k = -4i\zeta_k^2 \beta_k \tag{6.375}$$

$$- \frac{1}{a_k'^2} \int_{-\infty}^{\infty} dx \left(F \frac{\partial}{\partial \zeta} \psi_1^2(x, \zeta) + F^* \frac{\partial}{\partial \zeta} \psi_2^2(x, \zeta) \right) \tag{6.376}$$

$$\left(\frac{b^*}{a} \right)_t = 4i\xi^2 \frac{b^*}{a} - \frac{1}{a^2} \int_{-\infty}^{\infty} \left(F\psi_1^2(x, \xi) + F^*\psi_2^2(x, \xi) \right) dx \tag{6.377}$$

where $\gamma_k = 1/\beta_k a_k'^2$. We observe that the presence of the perturbation terms serves to couple the normal modes together. However, if the perturbation F is in some sense small, of order $\epsilon, 0 < \epsilon \ll 1$, then we can attempt to iterate the equations (6.374–6.377) by first calculating ψ_1^2 and ψ_2^2 based on some initial solution, say a pure soliton solution. However, such a perturbation series ordinarily exhibits the small-denominator or resonant behavior we encountered already in Sections 6g and 6h when we studied weakly coupled oscillators. In principle, then, the analysis of the perturbed NLS equation is very difficult and requires a much more detailed analysis than we are able to go into here. What we are going to do instead is to assume that the perturbation F belongs to a class that does not excite new normal modes but merely serves to modulate slowly the parameters (the amplitudes and speeds) of an initial pure soliton state. Therefore we only use (6.374) and ignore the perturbation terms in (6.376) (which modify the soliton's position) and (6.376) (which describes the evolution of the radiation component).

The easiest way to derive the equations for $\{\zeta_k\}$ is to use the conservation laws directly. For example, it is easy to show that

$$\frac{d}{dt} \int_{-\infty}^{\infty} qq^* dx = \int_{-\infty}^{\infty} (Fq^* + F^*q)dx, \qquad (6.378)$$

and similar relations can be derived for the other $\{C_m\}$, $m > 1$. Now suppose the perturbation F was due to damping, possibly time dependent:

$$F = -\alpha(t)q. \qquad (6.379)$$

Then (6.378) gives

$$\frac{d}{dt} \int_{-\infty}^{\infty} qq^* dx = -2\alpha(t) \int_{-\infty}^{\infty} qq^* dx. \qquad (6.380)$$

Now, suppose $q(x,t)$ is given approximately by a single soliton solution

$$q(x,t) = 2\eta \operatorname{sech} 2\eta(x - \bar{x})e^{-2i\xi x - 2i\sigma}, \qquad (6.381)$$

corresponding to the single eigenvalue $\zeta_1 = \xi + i\eta$ of (6.319). From the Trace Formula with $m = 1$,

$$\int_{-\infty}^{\infty} qq^* dx = 4\eta,$$

and (6.380) gives

$$\frac{d\eta}{dt} = -2\alpha(t)\eta, \qquad (6.382)$$

or

$$\eta(t) = \eta(0) \exp\left(-2 \int_0^t \alpha(\tau)d\tau\right); \qquad (6.383)$$

the soliton amplitude decreases the width increases at an exponential rate. Observe that the amplitude decay is twice what it would have been had the problem been linear $q_t = -\alpha q$. The reason for this is that as the soliton loses amplitude it must also reshape, that is, get wider, and this leads to the additional loss. From the second conservation law found by evaluating

$$\frac{d}{dt} \int (iqq_x^* - iq^*q_x)dx, \qquad (6.384)$$

we find for $F = -\alpha q$ that

$$\frac{d}{dt}\xi\eta = -2\alpha\xi\eta \qquad (6.385)$$

so that $d\xi/dt$ is zero. Therefore damping (or linear amplification if $\alpha(t) < 0$) does not affect either the speed (-4ξ) or the wavenumber (-2ξ) of the soliton. The time dependence of the position $\bar{x}$ and phase of the soliton is given by

$$\bar{x}_t = -4\xi + O(\alpha^2)$$

$$\bar{\sigma}_t = 2(\xi^2 - \eta^2) + O(\alpha^2) \qquad (6.386)$$

and it can be shown, although we do not do it here, that the perturbation terms remain of order α. Thus $\bar{x} = -4\xi t + \bar{x}_0$ and $\bar{\sigma} = 2(\xi^2 t - \int \eta^2 dt) + \bar{\sigma}_0$ to leading order. It can also be shown that the amount of radiation created by the perturbation also remains small.

Next we study a class of perturbations that affect the speed and wavenumber, but not the amplitude and width, of the soliton. Consider

$$q_t - iq_{xx} - 2iq^2q^* = -iV(x)q \qquad (6.387)$$

in which the space dependent potential $V(x)$ can be due to variations in the linear refractive index of the material. This kind of perturbation was extremely important when we considered the effect on a confined beam that propagates near weak discontinuities in the linear and nonlinear refractive indices. From the first conservation law, we find

$$\frac{d}{dt}\int_{-\infty}^{\infty} qq^*dx = 0, \qquad (6.388)$$

which means that the beam power

$$P = \int_{-\infty}^{\infty} qq^*dx \qquad (6.389)$$

remains constant. If we take q to be the single soliton given by (6.381), then $P = 4\eta$. Let us define the mean position of the pulse to be

$$\bar{x} = \frac{\int_{-\infty}^{\infty} xqq^*dx}{\int_{-\infty}^{\infty} qq^*dx}, \qquad (6.390)$$

which is just the $\bar{x}$ in the formula for the soliton (6.381) if we substitute q from (6.381) into (6.390). As yet we have not made this assumption. Next, verify that

$$\bar{x}_t = v = \frac{i \int_{-\infty}^{\infty} x(qq_x^* - q^*q_x)}{\int_{-\infty}^{\infty} qq^* dx}. \qquad (6.391)$$

Again if we substitute (6.381) for q, we get $v = -4\xi$. After a little calculation we can show that

$$\frac{dv}{dt} = -2 \frac{\int_{-\infty}^{\infty} V_x qq^* dx}{\int_{-\infty}^{\infty} qq^* dx} \qquad (6.392)$$

There are two cases of immediate interest. First, we might consider that the origin of the perturbation potential $V(x)$ is a variation in the refractive index that we take to be given. Second, let us imagine that the perturbation $-iV(x)q$ arises because of sudden changes in both the linear and nonlinear refractive indices that occur at interfaces between two dielectrics. For example, in Section 3g, we look at the case where $V(x) = 0, x < 0$, and $V(x) = -1 - 2(\alpha^{-1} - 1)|q|^2, x > 0$. In each case, using an approximation that must be justified a posteriori, we can express the right-hand side of (6.392) as the gradient of a potential. The reader will find the second case discussed in great detail in Section 3g. Here we will carry out the calculation for the first case.

Suppose we assume that the amplitude of q (i.e., qq^*) moves with as a single unit and is a function of $x - \bar{x}$. Then integration by parts gives us that

$$\int_{-\infty}^{\infty} \frac{dV}{dx} qq^* dx = -\int_{-\infty}^{\infty} V(x) \frac{d}{dx} qq^* dx$$

$$= \int_{-\infty}^{\infty} V(x) \frac{d}{d\bar{x}} qq^* dx = \frac{d}{d\bar{x}} \int_{-\infty}^{\infty} V(x) qq^* dx.$$

Therefore

$$\frac{dv}{dt} = -\frac{\partial U(\bar{x})}{\partial \bar{x}} \qquad (6.393)$$

where the equivalent potential

$$U(\bar{x}) = 2 \int_{-\infty}^{\infty} \left(\frac{qq^*(x - \bar{x})}{\int qq^* dx} \right) V(x) dx \qquad (6.394)$$

is found by taking the weighted average of the original potential $V(x)$ by twice the normalized amplitude. Therefore if, to a good first approximation, the pulse or wavepacket $q(x, t)$ stays intact, it behaves like a particle in the potential $U(\bar{x})$.

If q has the shape of the soliton (6.381), then

$$v = -4\xi, \quad \bar{x} = \bar{x}, \quad P = 4\eta,$$

$$\text{and} \quad U(\bar{x}) = \int_{-\infty}^{\infty} V(x)\text{sech}^2 2\eta(x - \bar{x})d(2\eta x). \tag{6.395}$$

Another perturbation of particular importance in short pulses (tens of femtoseconds) arises from the delay in the nonlinear response of the system and leads to an F,

$$F = -i\beta q \frac{\partial}{\partial x}|q|^2. \tag{6.396}$$

Using the first two conservation laws gives

$$\frac{d\eta}{dt} = 0$$

$$\frac{d\xi}{dt} = -\frac{128}{15}\beta\eta^4. \tag{6.397}$$

The wavenumber (-2ξ) and velocity (-4ξ) of the soliton increases with time. In the fiber context, where we shall be studying the effect, t is z, distance along the fiber, x is $t - k'z$, the retarded time, and the total frequency of the electric field is $\omega + 4\xi$, which we see decreases with distance traveled along the fiber. This frequency downshift can be very important in short pulses. Because it bears similarity to and is connected with the excitation of neighboring frequencies by Raman scattering, it is called the Raman downshift.

In the next example, we consider higher-order dispersion as a small perturbation. In this case the equation of interest is

$$q_t - iq_{xx} - 2i|q|^2 q = \beta q_{xxx}, \quad |\beta| \ll 1. \tag{6.398}$$

Following the procedure of the previous example, one obtains for the first two conserved quantities that

$$\frac{d}{dt}\int_{-\infty}^{\infty}|q|^2 dx = 0$$

and

$$\frac{d}{dt}\left(i\int_{-\infty}^{\infty}(qq_x^* - q^*q_x)dx\right) = 0,$$

which upon the assumption that q takes the one-soliton solution gives

$$\eta = \eta_0 + 0(\beta^2)$$
$$\xi = \xi_0 + 0(\beta^2). \tag{6.399}$$

One, however, has on $O(\beta)$ correction in the time evolution for $\bar{x}$:

$$\frac{d\bar{x}}{dt} = \left[i \int_{-\infty}^{\infty} (qq_x^* - q^*q_x)dx + 3\beta \int_{-\infty}^{\infty} |q_x|^2 dx \right] \Big/ \int_{-\infty}^{\infty} |q|^2 dx,$$

which after substitution of the one-soliton solution gives

$$\frac{d\bar{x}}{dt} = -4\xi + 4\beta(\eta^2 + 3\xi^2) \tag{6.400}$$

or

$$\bar{x} = [-4\xi_0 + 4\beta(\eta_0^2 + 3\xi_0^2)]t + \bar{x}_0. \tag{6.401}$$

In other words, the velocity of the solution has an $O(\beta)$ correction from that depends on η_0. A consequence of this is that a bound N-soliton solution of the unperturbed problem characterized by N eigenvalues $\xi_0 + i\eta_j, 1, \ldots N$ asymptotically splits into N individual solitons, each of amplitude η_j and velocity $-4\xi_0 + 4\beta(\eta_j^2 + 3\xi_0^2)$.

The final example we consider is the behavior of the soliton in the complex Ginzburg–Landau (CGL) equation when the real coefficients multiplying $\partial^2 q/\partial x^2$ and $|q|^2 q$ are considered to be small as well as the loss (or gain) term. Thus we write the CGL equation as the perturbed NLS equation,

$$q_t - iq_{xx} - 2i|q|^2 q = \alpha q + \gamma q_{xx} - \beta |q|^2 q, \tag{6.402}$$

where $|\alpha|, |\beta|, |\gamma| \ll 1$.

The first two conservation laws give the following evolution equations

$$\frac{d}{dt} \int_{-\infty}^{\infty} |q|^2 dx = 2\alpha \int_{-\infty}^{\infty} |q|^2 dx - 2\gamma \int_{-\infty}^{\infty} |q_x|^2 dx - 2\beta \int_{-\infty}^{\infty} |q|^4 dx$$

$$\frac{d}{dx} i \int_{-\infty}^{\infty} (qq_x^* - q^*q_x)dx = 2\alpha \int_{-\infty}^{\infty} (qq_x^* - q^*q_x)dx - 2\gamma \int_{-\infty}^{\infty} (q_x q_{xx}^* - q_x^* q_{xx})$$

$$- 2\beta \int_{-\infty}^{\infty} (qq_x^* - q^*q_x)|q|^2 dx \tag{6.403}$$

which after substituting the one-soliton solutions of NLS gives a system for the soliton parameters η and ξ:

$$\frac{d\eta}{dt} = 2\alpha\eta - \frac{8}{3}(\gamma + 2\beta)\eta^3 - 8\gamma\eta\xi^2$$

$$\frac{d\xi}{dt} = -\frac{16}{3}\gamma\xi\eta^2. \tag{6.404}$$

A phase plane analysis of this system shows that the line $\eta = 0$ is a set of "trivial" critical points as well as the point $\eta_c = \sqrt{(3\alpha/4(\gamma + 2\beta))}, \xi_c = 0$ corresponding to a soliton of zero velocity whose amplitude depends on the coefficients of the perturbation terms. Clearly, the condition $(\alpha/\gamma + 2\beta) > 0$ is necessary for the critical point to exist.

We want to make a very important point in connection with (6.404). Note that the fixed point is stable provided $\gamma + 2\beta > 0$, which means that as long as diffusion is large enough it can overcome neutral ($\beta = 0$) or even destabilizing ($\beta < 0$) nonlinearity. Trajectories of solutions to (6.404) are shown in Figure 6.41 which corresponds to the supercritical case where α and $\gamma + 2\beta$ are positive.

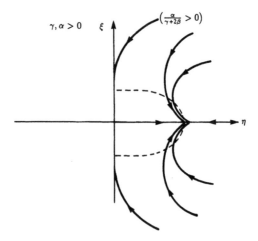

FIGURE 6.41   Solitary waves of CGL.

6l.   Chaos and Turbulence

A detailed survey of the behavior of dynamical systems is way beyond the present scope of this book, but it would be remiss of us not to include a brief mention of this topic (see also references [15, 16, 100]). The study of nonlinear optics is still in its infancy, and so often special efforts are made to suppress and inhibit the full range of dynamical behavior of which optical systems are capable in order to concentrate on gaining a complete understanding of "simple" processes. However, as it was made clear in Chapters 4 and 5, lasers and feedback cavities are dynamical systems with many degrees of freedom, and unless special arrangements are made to suppress most of them, each of these degrees of freedom can compete on a more or less equal basis for the available energy supplied by external sources. As a result, the observed output can often appear highly complicated, and as the subject of nonlinear optics develops to maturity, it will become more important and relevant to study the full range of dynamical behavior and gain a more complete understanding of chaotic processes.

What do we mean by chaos? The word itself originally derives from the Greek "$\chi\alpha os$" and in ancient times meant the primordial infinite expanse of space. The Romans interpreted chaos as being the shapeless mass on which order and harmony were superimposed. Modern day usage connotes a state of disorder such as we might find in the kitchen the Sunday morning after the local rugby team has celebrated its Saturday night rituals.

In dynamics, the term *deterministic chaos* refers to the seemingly unpredictable output of a dynamical system whose continuous or discrete time evolution is defined by a set of ordinary or partial differential equations or maps, to which equations no external noise is added and in which equations none of the coefficients are random variables. You might ask: how can one possibly get chaotic output from a deterministic system?

The simplest example of deterministic chaos is the nonlinear map

$$x_{n+1} = 2x_n (\mathrm{mod}\ 2). \tag{6.405}$$

where $0 < x_n < 1, n = 0, 1, \ldots$. If we forgot about the (mod 2) part, the map in linear and each application simply moves the "decimal" point of a number written in binary one step to the right: $x_0 = 0.10110xxxx$ (where $xxxx$ refers to the unknown digits beyond the fifth significant figure) becomes $x_1 = 1.0110xxxx$, which then becomes $x_2 = 10.110xxx$ and so on. Note that the error remains at one part in 2^5. However, if we put back in the (mod 2), $x_0 = 0.10110xxxx, x_1 = 0.0110xxxx, x_3 = 0.110xxxx, x_4 = 0.10xxxx, x_5 = 0.0xxxx, x_6 = 0.xxxx$. After five applications of the map, the output is a random variable which, assuming that each of the unknown digits is as likely to be a zero as it is a one, is distributed uniformly on the unit interval. What the nonlinear map (6.405) has done is uncover the uncertainty in our knowledge of the initial conditions. The randomness of the output comes from our ignorance of the initial state. The dynamics of the map were such that this uncertainty was magnified at each step. Note that after five steps, two points, $x_0^{(1)} = 0.101101$ and $x^{(2)} = 0.101100$, initially very close together, are half the unit interval apart.

We say a dynamical system has the property of *sensitive dependence on initial conditions* if points near a given orbit move on trajectories that on the average diverge from the given orbit at an exponential rate. This average rate of divergence is measured by what is called the Lyapunov exponents of the dynamical system. If any of the Lyapunov exponents are positive, then the dynamical system exhibits sensitive and chaotic behavior. Many of the equations and maps we encounter in mathematical physics—the Navier–Stokes equations of fluid dynamics, the reaction–diffusion equations of chemical kinetics, the Maxwell–Bloch equations of optics, the Ikeda map defining the feedback loop in an optical ring cavity—have this property for certain ranges of their parameters. This property, when coupled with the always present uncertainty about the initial state, results in a long-time behavior of solutions that, on the surface, is difficult to distinguish from the output produced by the influence of random noise.

There are all sorts of measures of the nature of a chaotic system, including Lyapunov exponents and the dimension (Hausdorff, correlation, etc.) of the attractor (the set of points in the phase space of a dissipative system to which the solution tends as $t \to \infty$) if there is one. A discussion of these is beyond the scope of this book, but the reader will find the references helpful. However, we will discuss one measure of random behavior and that is the power spectrum $P(\omega)$, which is the Fourier transform

$$P(\omega) = \int R(\tau)e^{i\omega\tau}\, d\tau \qquad (6.406)$$

of the autocorrelation function

$$R(\tau) = \langle x(t)x(t+\tau)\rangle = \lim_{N\to\infty} \frac{1}{N} \sum_{i=1}^{N} x(t_i)x(t_i+\tau) \qquad (6.407)$$

where $x(t)$ is the measured output of some variable with a zero mean value; that is, $\langle x(t)\rangle = \lim(1/N)\sum_{i=1}^{N} x(t_i) = 0$. (If $\langle x(t)\rangle \neq 0$, simply replace $x(t)$ by $y(t) = x(t) - \langle x(t)\rangle$.) A periodic solution $x(t)$ of period T would give rise to a power spectrum $P(\omega)$ concentrated (like a Dirac delta function) at $\omega = 2\pi/T$ and integer multiples thereof. A quasiperiodic solution $x(t)$ with several distinct frequencies $\omega_1, \omega_2, \ldots \omega_r$ would again give rise to a spectrum like Dirac delta functions concentrated on all the ω_i's and their sums and differences. Chaotic behavior, due to the deterministic chaos we have described or due to external noise, is revealed by a broad-band power spectrum in which the spikes display a finite width so that the energy is distributed over a continuous band of frequencies.

In spatially extended systems whose field variables $u(\vec{x}, t)$ evolve according to partial differential equations, we can look at both the "spatial" and "temporal" power spectra

$$P(\vec{k}, \omega) = \frac{1}{(2\pi)^3} \int \langle u(\vec{x}, t) u(\vec{x} + \vec{r}, t + \tau) \rangle e^{-i\vec{k}\cdot\vec{r} + i\omega\tau} d\vec{x} d\tau \qquad (6.408)$$

where $\langle u(\vec{x}, t) \rangle = 0$. We often say that a system whose output is broad-band in time but ordered in space (has several distinct wavevectors) is weakly turbulent (wimpy turbulence). On the other hand, we say that a system that exhibits broad-band power spectra in both frequency and wavevector space is fully or strongly turbulent (macho turbulence!). For example, the sea surface immediately below a hurricane would be strongly turbulent whereas the short surface waves one obtains by shaking a coffee cup at high frequencies but at very small amplitudes might very well exhibit regular spatial structures but a chaotic temporal behavior. Much of the chaotic behavior observed in connection with optical systems has been of the weak type, but studies on the transverse behavior of optical beams whose dynamics are described by partial differential equations of nonlinear Schrödinger type are revealing fully developed turbulent patterns.

Chaotic behavior does not require many degrees of freedom. Indeed, systems of ordinary differential equations of order three, maps of dimension two, or noninvertible maps of one dimension can exhibit chaotic behavior. Three well-known illustrative examples are

$$\frac{dA}{dt} = -\sigma A + \sigma P$$

$$\frac{dP}{dt} = rA - P - An \qquad (6.409)$$

$$\frac{dn}{dt} = -bn + AP \qquad (6.410)$$

the Lorenz equations already discussed in Section i of this chapter,

$$g_{n+1} = a + Re^{i\phi + iN(|g_n|^2)L} g_n \qquad (6.411)$$

the Ikeda map discussed in Chapter 5, and 6.405 introduced earlier. A better known example of the noninvertible one-dimensional map is the famous logistic map

$$x_{n+1} = \mu x_n (1 - x_n) \tag{6.412}$$

from which model Feigenbaum extracted many wonderful universal properties. The graphs of (6.405) and (6.412), shown in Figure 6.42, are seen to be noninvertible since one output value has two possible input values.

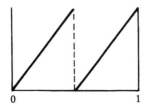 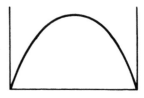

FIGURE 6.42

The Lorenz equations (6.409) are valid for a perfectly tuned two-level laser. The real variables A, P and n are the amplitudes of the electric, polarization, and population inversion fields, which have the well-defined and nonturbulent spatial structure of the fundamental TEM$_{00}$ mode. The equations are dissipative in the sense that if we started off with a cube of initial states in A, P, n space with volume $V(0)$, that volume is reduced at an exponential rate $V(t) = V(0)e^{-(\sigma+b+1)t}$. (In contrast, for a conservative system such as a pendulum, the volume of a set of initial status is conserved in time.) As we have already seen, for $r < 1$ (recall r is the amount of population inversion induced by external pumping), the nonlasing solution ($A = P = n = 0$) is stable. At $r = 1$, this solution becomes unstable to the lasing solution $A = P = \pm\sqrt{b(r-1)}, n = r - 1$ and the trajectories through all initial conditions (except the point $A = P = n = 0$) end up at either one of the two fixed points. However, for r large enough, and in certain ranges of the parameters σ and b (the bad cavity limit, $\sigma > b + 1$, discussed in Chapter 5), both the lasing and nonlasing solutions are unstable and the motion of the point $A(t)$, $P(t)$, $n(t)$ is very complicated, remains time dependent and after a long time appears to lie on a fractal set of dimension greater than two but strictly less than three, as shown in Figure 5.3. This fractal set is called the Lorenz attractor.

Note that not only is r very large, but σ is also. It is unlikely that a two-level laser would be operated at such parameter values. However, it should not be too surprising to learn that in more realistic models of lasing, which involve more energy levels or which involve more than one of the cavity modes, the output exhibits this kind of chaotic behavior for values of the parameters much closer to the critical values at which lasing action first begins.

We have just seen that dissipation by itself does not necessarily inhibit chaotic behavior. While the dissipation causes phase space volumes to contract, it cannot always control their distortion. Indeed, a cube of initial states becomes spread more or less evenly over this very complicated fractal set known as the Lorenz attractor. This distortion and spreading out of the box of initial states is closely tied to the existence of objects called homoclinic orbits, which join the unstable fixed point $A = P = n = 0$ to itself. In other words, we have already seen in Section 6i that an orbit starting near $A = P = n = 0$ will leave on the curve $A = P, n = 0$. Where does this orbit go? For a certain range of r values it goes to one or other of the lasing solutions. For r just greater than one it goes to the lasing solution closest to the initial direction. For example, if $A = P = +\epsilon, n = 0$, it goes to $A = P = \sqrt{b(r-1)}, n = r - 1$. But as r gets bigger, the attraction towards this solution can get weaker, and at a certain value of r, the unstable manifold leaving $A = P = n = 0$ eventually joins with the stable manifold as shown in Figure 6.43 so as to complete a loop in which the trajectory leaving $A = P = n = 0$ returns along the $A = P = 0$ vertical axis. Now one can begin to see one of the causes of sensitivity and the distortion of the box of initial conditions. Imagine the orbits in the neighborhood of the homoclinic orbit $0BC0$. Trajectories on either side of $A = P = 0$ diverge as they approach the unstable fixed point at the origin $A = P = n = 0$. When the fixed points $\pm\sqrt{b(r-1)}, \pm\sqrt{b(r-1)}, r - 1$ corresponding to the lasing solutions eventually become unstable, the orbits continue to wind about these unstable solutions and trace out the complicated Lorenz attractor.

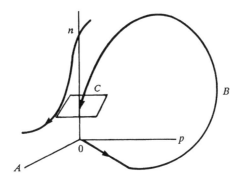

FIGURE 6.43

In conservative systems, homoclinic orbits (often called separatrices) play a similar role. Consider the phase plane (Figure 6.44) of the pendulum. Now suppose one were to apply a small periodic perturbation,

$$\theta_{tt} + \sin\theta = \epsilon \cos \omega t, \quad 0 < \epsilon \ll 1.$$

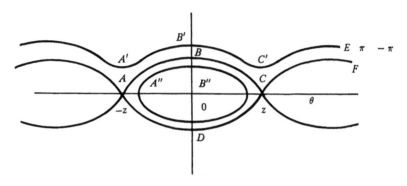

FIGURE 6.44

Imagine what happens to an orbit in the neighborhood of the separatrix ABC (the homoclinic orbit, $\theta(t) = 2\sin^{-1}\tanh t$, connecting the identical points $\theta = -\pi$ and $\theta = \pi$). As the pendulum comes close to the unstable equilibrium C, teetering precariously in the unstable position, its future course will depend on which direction it gets pushed by the external force $\epsilon\cos\omega t$. Since the separatrix orbit ABC has an infinite period, a nearby orbit will have a very long period, and the phase point on the orbit will spend most of that period in the neighborhood of the unstable point C. The direction of the force will change sign many times during that interval, and a prediction as to whether the pendulum reverses its direction and falls back along CDA or continues forward along CEF requires impossibly precise information on the previous history of the trajectory. Two different orbits which initially differ only slightly over ABC can end up traveling along divergent paths CDA or CEF. In fact, it is a theorem that if one assigns a head (tail) to the event that the trajectory of an orbit crosses the stable position with θ increasing (decreasing), then the sequence of heads and tails one would obtain is as random as one would obtain by tossing a coin. If one plots the projections of the trajectory in θ_t, θ, t space onto the θ_t, θ phase plane at times $t = n(2\pi/\omega)$, $n = 1, 2, 3, \ldots$ (this construction is called a Poincaré map), one would construct the graph shown in Figure 6.45.

In conservative systems, therefore, the chaotic behavior, exhibited by systems that are perturbations of systems which have simple dynamical behavior, like the periodic (the orbit $A''B''C''$ of Figure 6.44) and rotating (the orbit $A'B'C'$ of Figure 6.44) solutions of the pendulum, or quasiperiodic behavior, like the periodic solutions of uncoupled nonlinear oscillators (whose periods may depend on their respective amplitudes), occurs either near separatrices or near certain orbits in the unperturbed phase plane that correspond to resonances such as these discussed in Section 6g. The motion of orbits that project onto other parts of the original unperturbed phase plane stays fairly regular

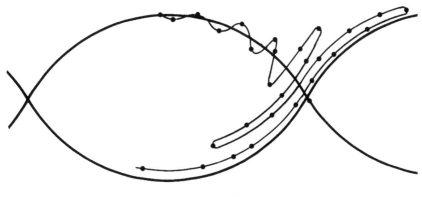

FIGURE 6.45

although, when one looks closely enough, every region of phase space exhibits both regular and irregular behavior.

The upshot is that the phase space of chaotic conservative systems is very complicated, consisting of some regions in which the motion is fairly regular and others in which it is very irregular. Moreover, as long as the dimension of the system is sufficiently large (Hamiltonian systems with one constant of the motion and at least two degrees of freedom), the irregular orbits can migrate from one area of chaos to another.

The astute reader may then very well question the validity of all the "averaging" and "multiple scale" methods for obtaining the equations that describe the long-time behavior of the amplitudes of weakly coupled oscillators. Over the long run, many of these systems exhibit chaotic behavior. The reader would be quite right to ask this question. However, the point is that in using the "averaged" equations for slowly varying amplitudes, we are only interested in finite, albeit long (of order ϵ^{-n}), times, and over these times the chaotic regions in phase space are well confined to the neighborhoods of the "generalized" separatrices. Over longer times, one would indeed have to account for the effects of the random fluctuations that arise from the small-amplitude, fast oscillating nonresonant terms that we ignored in the so-called rotating wave approximation.

6m. Computational Methods

The computer plays a central role in all areas of nonlinear science, and nonlinear optics is no exception. Much of our understanding of the interaction of intense laser light fields with materials comes from computer-generated solutions of the appropriate coupled field–matter equations. Particularly note-worthy in the mid-1960s to mid-1970s were the computer study of critical self-focusing of laser beams in bulk media, stimulated scattering (Raman and Brillouin), and the coherent interaction of intense optical pulses with absorbing (SIT) and amplifying media. The rapid progress in our recent understanding of the phenomena of pulse propagation in fibers (single- and multi-mode), of self-trapped light channel propagation in nonlinear waveguiding media, of laser dynamics, and so forth could not be envisaged without computer-generated solutions and, in particular, sophisticated graphical tools. It is appropriate therefore that we should provide a brief section outlining some of the more useful computational techniques and strategies that are applicable to the class of nonlinear optics problem discussed in this monograph. These techniques tend to be buried in the technical research literature even though they can be implemented with relative ease on many of the multitude of available computer workstations. The methods that we choose form a small subset of the available computational tools (see also reference [91]).

Solvers and Phase Portrait Analysis

Reliable ordinary differential equation solvers are available in all scientific subroutine packages (IMSL, NAG, ODE,...) and it is more economical to use one of these directly rather than write one from scratch. A good robust scheme for solving the types of o.d.e.'s encountered in this book is the variable step fourth-order Runge–Kutta method. In the event that a black-box scheme is unavailable, the authors strongly recommend the one provided in the book "Numerical Recipes" [101], which has the source code on accompanying diskettes.

The o.d.e. solvers can be used to generate time series of the appropriate solution components, but a much more powerful geometric approach involves phase portrait analysis. This allows one to anticipate much of the complex dynamic behavior ahead of time and avoid apparent discontinuous jumps in behavior as a physical parameter is varied in a conventional time series analysis. The idea is to track the principal unstable fixed points in phase space (these are hyperbolic in all our optics examples), construct the global stable and unstable manifolds of these fixed points, and track their change in topology under parameter variation. The special types of orbits that we wish to isolate are homoclinic and heteroclinic orbits that organize much of the dynamical systems global bifurcation behavior; these are orbits of infinite period that terminate on one or more hyperbolic fixed points respectively [63].

The basic computational strategy is to locate the relevant fixed point in phase space, linearize about it to determine its stability and if unstable, determine the local stable and unstable manifold directions from the eigenvectors of the stable and unstable eigenvalues. The local unstable manifold W_{loc}^u lies along the direction of the exponential growth away from the hyperbolic fixed point, while the local stable manifold W_{loc}^s lies along the exponentially contracting directions. These manifolds may be simple curves or hypersurfaces in phase space. The following simple algorithm constructs these manifolds for a map.

1. Place a circle of points of small radius around the unstable fixed point.

2. Iterate the map forward in time using this circle of points as initial values.

3. Repeat (2) by applying the inverse of the map (if it exists!) to the initial circle.

If a flow, solve the system of o.d.e.'s forward and backwardin time using this set of initial conditions.

Auto Bifurcation Package

The AUTO bifurcation package was developed by Eusebius Doedel while at Caltech and Concordia University (Montreal). This is one of a number of such packages that are currently available from their authors at a nominal charge. AUTO performs many sophisticated functions in addition to bifurcation analysis, but the latter application is the only one of direct relevance here. The full package has been written in FORTRAN and has a very useful graphics utility for plotting the types of bifurcation diagrams shown in Figure 5.6. The graphics package is written so as to be essentially device independent with specific graphics hardware device options grouped in a single module for easy access or modification. Anybody interested in obtaining the package from the author should keep in mind that the files are rather large so disk space may pose a problem. A very useful User manual with many test examples accompanies the package, and the user can quickly get to grips with the various types of problem-solving options by following these examples. AUTO provides important bifurcation and graphical information in a set of default data files that are written on FORTRAN units 7, 8, and 9. Unit 6, which is usually the screen, writes an abbreviated bifurcation analysis that identifies all of the special points encountered in a particular run. For example, under a single parameter variation, bifurcation points from equilibrium solutions (pitchfork,

Hopf), limit points, and endpoints are tagged and labeled. Branches of periodic solutions can be tracked from Hopf bifurcation points by restarting the package at a tagged point. Period doubling or bifurcations to a torus may be detected on the periodic solution branches. Orbits of fixed period can be tracked under two-parameter variation. This is a particularly powerful feature for tracking curves of homoclinic or heteroclinic orbits (a large period orbit is a good numerical approximation to such infinite period orbits).

We now outline, in point form, the general procedure for applying the AUTO bifurcation package to, say the Lorenz equations (6.409) discussed in the text. The first step is to generate the equilibrium branches of solutions starting from a given location that is not a bifurcation point. The obvious choice of principal bifurcation parameter is the "pump" parameter r.

Step 1.   First find at least one equilibrium solution to the problem at a fixed set of parameter values. The trivial solution is most easily computed in many cases. Evaluate the Jacobian matrix for the right-hand side of the equations, analytically if possible. If not, AUTO will compute it numerically. Decide which is to be your principal bifurcation parameter from the available set and evaluate the derivative of the Jacobian with respect to that parameter. It is a good idea at this stage to decide on a secondary bifurcation parameter for a later two-parameter variation and provide the derivative of the Jacobian matrix with respect to this parameter also.

Step 2.   The equilibrium solution at the chosen set of parameter values is specified in a user-supplied subroutine called "STPNT." The right-hand side of your equations are given together with the Jacobian matrix in a user-supplied subroutine called "FUNC." The type of problem is defined by overriding the default set of problem parameters initialized in the AUTO library. This is done through the user-supplied subroutine "INIT." In our case we seek to map out branches of equilibrium solutions starting from the steady-state solution ($A = P = n = 0$) given in Section 5a.

Step 3.   Run the preprocessor stage of AUTO. This feature is included to incorporate more than one class of problem in the library. For example, to specify an algebraic eigenvalue problem we set a parameter $IP = 0$ in the subroutine INIT. The accompanying AUTO manual provides all of the relevant parameter settings. The role of the preprocessor is to generate a short main

FORTRAN program that calls the relevant subroutines from the library package. A preprocessor command file is supplied for a Vax, although this can be trivially modified to a Unix script on a workstation.

Step 4. Run AUTO. A Vax command file is supplied to compile, link, and execute the code. This may be changed easily to a Unix script, or the appropriate commands can be executed explicitly. The abbreviated output of AUTO which is written to FORTRAN unit 6 is listed in Table 6.1 for the Lorenz problem. The accompanying picture in Figure 6.46 is generated by running the "PLAUT" plot package for a Textronix Plot-10 device. The labels "EP" in the listing denote endpoints of the branch and specify the range over which we instructed AUTO to vary the bifurcation parameter r. This latter parameter is labeled "PAR(1)" in the listing. The point "BP" denotes a pitchfork bifurcation corresponding to the first laser threshold and is shown as an open square in the diagram. Here the new branches of equilibrium solutions labeled "Branch 2" in the listing correspond to a constant amplitude output of the laser. The symmetry of the Lorenz system ensures that there is a pair of such branches corresponding to positive and negative field amplitudes. An intensity bifurcation diagram would show just a single branch. The closed squares on the lasing equilibrium branches represent Hopf bifurcations (HB in the listing) at the second lasing threshold. Solid curves represent stable equilibrium solutions and dashed curves unstable equilibrium solutions.

Important detailed subsidiary information on eigenvalues of the Jacobian matrix along a branch, adaptive step size used, and warnings on singular point detection are all printed on FORTRAN unit 9. These should always be examined carefully even when AUTO appears to work smoothly. Of course, AUTO can run into difficulty relatively easily for reasons that are too numerous list.

Step 5. Generate branches of periodic orbits.

Computation of branches of periodic solutions constitutes a different problem to the algebraic problem above, and so we must specify that the problem is different by setting $IP = 1$ in the INIT subroutine. The information already accumulated on FORTRAN units 7, 8, and 9 is also needed to restart AUTO. To specify that we are restarting AUTO on an existing problem and indicate from which Hopf bifurcation point that the branch of periodic solutions is to be generated (here we have a choice of two) we set the variable $IRS = 4$, which corresponds to the label that tags that point in the previous run. The information on FORTRAN unit 8 must be copied to unit 3, and we are ready to go.

BR	PT	TY	LAB	PAR(1)	L2-NORM	U(1)	U(2)	U(3)	PERIOD
1	1	EP	1	0.000000E+00	0.000000E+00	0.000000E+00	0.000000E+00	0.000000E+00	
1	5	BP	2	9.999999E-01	0.000000E+00	0.000000E+00	0.000000E+00	0.000000E+00	
1	45	EP	3	4.080000E+01	0.000000E+00	0.000000E+00	0.000000E+00	0.000000E+00	
2	52	HB	4	2.473683E+01	2.626849E+01	7.956018E+00	7.956018E+00	2.373683E+01	
2	74	EP	5	4.012741E+01	4.170903E+01	1.021470E+01	1.021470E+01	3.912752E+01	
2	52	HB	6	2.473683E+01	2.626849E+01	−7.956018E+00	−7.956018E+00	2.373683E+01	
2	74	EP	7	4.012741E+01	4.170903E+01	−1.021470E+01	−1.021470E+01	3.912752E+01	
3	25		8	1.653788E+01	1.534623E+01	1.202755E+01	1.439731E+01	2.458927E+01	1.212038E+00
3	45		9	1.392656E+01	2.154120E+00	1.147046E+01	1.362098E+01	2.166324E+01	4.999997E+01
3	50		10	1.392656E+01	7.190631E−01	1.146573E+01	1.363560E+01	2.164414E+01	4.487205E+02
3	68	MX	11	1.392656E+01	1.887328E−02	1.148152E+01	1.365964E+01	2.166873E+01	7.858649E+05

TABLE 6.1 Listing of the output from AUTO to Unit 6. Each branch of equilibrium solutions is labeled with the numbers 1–2 in the first column and the branch of periodic solutions is labeled with 3. PAR(1) refers to the principal bifurcation parameter (or r for the classical Lorenz equation). The L2-NORM and individual solution components are printed for each equilibrium branch. The periodic branch shows the maximum of each periodic solution component, and the period of the solution is shown in the final column.

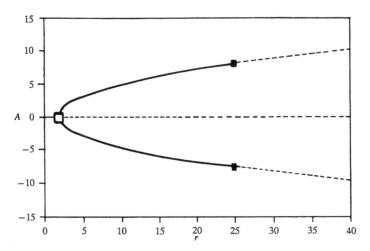

FIGURE 6.46 This figure shows the equilibrium solutions and local bifurcation points for the Lorenz system with $\sigma = 10$ and $b = 8/3$. This plot was generated using the "PLAUT" code. See the discussion in the text for an explanation of this diagram.

Note: this new run of AUTO will overwrite the information in units 7, 8, 9, and it is necessary to save these as separate files beforehand if we are building up a bifurcation diagram piecemeal as in the present case.

The part of the listing denoted by branch 3 shows the output of AUTO on unit 6 from this run, and it looks rather uninteresting. Appending the data in FORTRAN units 7, 8, 9 to the saved data from the previous run allows us to use the plot package PLAUT to generate the more complete bifurcation diagram in Figure 6.47. We see that the Hopf bifurcation is subcritical (a result known analytically [15]) and the branch of unstable periodic orbits (open circles) moves to the left and terminates. Inspection of the listing on unit 6 given here or the detailed listing on unit 7 shows that the period of the unstable orbits that is printed in the last column is increasing rather rapidly towards the end of the branch. The indication is that the branch ends at a point labeled "MX," which means that AUTO could not proceed further subject to the constraints imposed on it. The increasing period of the unstable periodic orbits indicates approach to a homoclinic orbit in phase space (an infinite period orbit). We can get closer by refining the mesh required to compute the periodic orbits, but we do not do so here. Instead we show a picture of this terminal orbit in Figure 6.48 to illustrate that it is a good numerical approximation to the actual homoclinic orbit of the Lorenz attractor. This is one of an infinite sequence of homoclinic orbits called the principal homoclinic orbit as it lies furthest to the left in the pump parameter value. FORTRAN unit 9 now has accumulated information on the Floquet exponents at each point on the branch. Caution should be used when interpreting the graphical output of branches of periodic

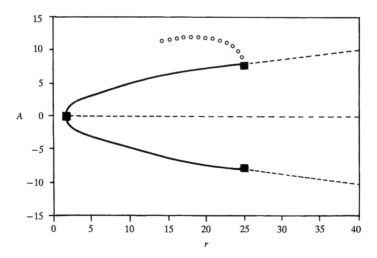

FIGURE 6.47 Complete bifurcation diagram with the branch of periodic solutions appended. The open circles indicate unstable periodic solutions and the subcritical bifurcation leads to a branch of periodic solutions terminating on a homoclinic orbit at the extreme left.

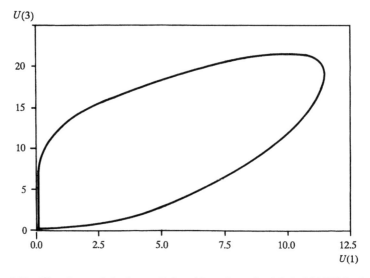

FIGURE 6.48 The shape of the homoclinic orbit at the point labeled "MX" in the unit 6 listing above. The orbit is homoclinic to the trivial solution.

orbits as the computer does not recognize 0.999998 or 1.00001 as unity to the available accuracy. A periodic orbit has one Floquet exponent equal to unity always to represent the neutral direction. If all others are less than unity in magnitude, the orbit is stable. If one crosses through unity, this signifies a bifurcation from the branch of periodic solutions. The problem is that AUTO often gets confused when the Floquet exponents are close to unity and gives incorrect stability information if the orbits are not accurately computed. It is important to inspect unit 9 carefully in these circumstances.

The AUTO package is a sophisticated software library that can be an extremely powerful tool in the hands of the experienced user. This section is meant to provide the reader with a brief glimpse of its power and sophistication without getting swamped in technicalities. The AUTO manual is an excellent users guide to the package.

A Nonlinear P.D.E. Solver: Split-Step FFT

Many of the p.d.e.'s appearing in this monograph are of the NLS type where we need to solve a specific initial value problem. We now describe a numerical algorithm that provides accurate stable solutions to this class of p.d.e.'s and can be run with relative ease on a variety of computer workstations with floating-point chips. The split-step method, also called the beam propagation method in optics applications, is a well-tested robust computational scheme that lends itself naturally to solving the initial value problems of interest. This numerical method can be used to study the development of the Benjamin–Feir or modulational instability in space or time, the evolution of transverse ring structures in a bistable cavity and pulse propagation in nonlinear fibers. Consider the general initial value problem described by the NLS-type equation

$$iq_t = Lq + N(t)q \qquad (6.413)$$

where $q = q(x, t)$ and L is a linear operator (i.e., $i\partial^2/\partial x^2$) and $N(t)$ is a nonlinear operator that depends on the solution q. Let us formally integrate this equation in time over a small interval Δt to give

$$q(x, \Delta t) = \exp -i \left(L\Delta t + \int_0^{\Delta t} dt' N(t') \right) q(x, 0). \qquad (6.414)$$

This formal operator expression allows us to propagate the solution by a single time step. As the operators L and N do not commute, in general, the evaluation of this expression is complicated. Let us approximate the exponential operator as follows:

$$e^{-i(A+B)} \approx e^{-i/2A} e^{-iB} e^{-i/2A} \qquad (6.415)$$

where we let $A = L\Delta t$ and $B = \int_0^{\Delta t} N(t') dt'$. In effect, we have split the composite operator up into a set of simple exponential operations that can be carried out on the function $q(x, 0)$ in a sequential fashion. The error in approximating the original exponential operator by this split operator is $O((\Delta t)^3)$, as can be seen by Taylor expanding the terms on both sides of the equality.

EXERCISE 6.28

Show that the split operator expression in (6.415) approximates the composite operator expression up to $O((\Delta t)^3)$ by Taylor series expanding the exponentials on both sides of the equality and equating terms on either side. Take care to keep the order of the operator products in your calculation.   □

Notice that the operator B involves an integral over the time interval t in general and this integral is implicit as we do not know its value ahead of time. However, in the special case where $N = N(|q|^2)$, which we encounter in relevant exercises, N is a constant of the motion, and we can write $B = N(|q(0, x|^2)\Delta t$. This makes the entire propagation sequence explicit.

The power of the split-step method in solving NLS-type equations is that the linear operator $L = i\partial^2/\partial x^2$, and we already know how to solve this problem using the Fourier transform. In other words, in Fourier space, the effect of this exponential operator is a simple scalar multiplication by $\exp(-iK^2\Delta t/2)$. The basic algorithm for a single propagation step can be summarized schematically by the sequence of operations:

$$q(\Delta t, x) =$$

$$FFT^{-1}[e^{-iK^2\Delta t/2}FFT[e^{iN(|q(0,x|^2)\Delta t}FFT^{-1}[e^{-iK^2\Delta t/2}FFT[q(0, x)]]]]$$

where $FFT(FFT^{-1})$ refer to the forward (inverse) Fast Fourier Transform. This propagation algorithm is very natural in optics if we replace the t variable by z, the direction of propagation. It involves a sequence of steps that includes a free space propagation by half a step (solve the linear paraxial ray equation), a nonlinear correction in the middle of the step, and a final-free space propagation over the final half step. Once we have $q(\Delta t, x)$, we can use the identical sequence of operations to generate $q(2\Delta t, x)$ and so on. A schematic of the propagation of a laser beam through a sequence of thin slabs of thickness Δz is shown in Figure 6.49. It is easy to see that the two back-to-back free-space half steps can be combined into a single free-space step over Δz.

A simple algorithm for propagating a laser beam over an interval L follows. All that is needed is access to an *FFT* routine in any software package.

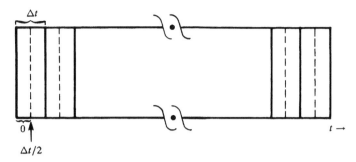

FIGURE 6.49 Propagation scheme using the split-step *FFT*. The first ($\Delta t/2$) step involves free-space propagation. The nonlinear correction is then applied at the dashed vertical line. After this a full free-space propagation step (Δt) occurs between nonlinear correction steps (dashed vertical lines are separated by Δt) until the final step.

Step 1. Define the initial data (either a Gaussian beam or an apertured plane wave).

Step 2. Free-space propagate half a step $\Delta z/2$ (i.e., Fourier transform the data, multiply by the quadratic phase factor $e^{-(i/2)K^2\Delta z}$, and invert the transform).

Step 3. Multiply by the nonlinear exponential term.

Step 4. Free-space propagate a full step Δz (i.e., Fourier transform the data, multiply by the quadratic phase factor $e^{-iK^2\Delta z}$, and invert the transform).

Step 5. Repeat step 3 until the point $L - (\Delta z)/2$ is reached, and then branch to step 6.

Step 6. Free-space propagate half a step $\Delta z/2$ (i.e., Fourier transform the data, multiply by the quadratic phase factor $e^{(-i/2)K^2\Delta z}$, and invert the transform).

This simple algorithm is straightforward to implement, yielding satisfactory results in most cases with either 128 or 256 transverse points (2^n points are needed in the *FFT*; n an integer). Note that the pulse propagation problem is identical except that we replace x by t in the second derivatives. The scheme can be easily generalized to coupled NLS-type problems and, with more work, to counter-propagating beams.

We end with a few words about boundary conditions. In investigating the Benjamin–Feir instability we add a small perturbation $\epsilon \cos(K_{max}x)$ to an infinite plane wave. The equivalent problem on a computer requires us to assume a finite transverse aperture that periodically repeats itself outside this interval. This is precisely how the *FFT* operates in a periodic window, so no special precautions need to be taken at the boundary. When propagating a finite width beam, however, we must ensure that our window is wide enough and a proper windowing function is applied to avoid leakage or aliasing, where frequencies that appear outside the finite window in transform space are folded back into the current window as a consequence of periodicity. It is good practice to monitor any conserved quantities that are at hand as the computation proceeds.

A FORTRAN program to solve the one-dimensional NLS equation starting from one of three initial conditions follows. The user can choose to study the time evolution of a single soliton, a double ($N = 2$) soliton, or two distinct ($N = 1$) solitons. This split-step *FFT* algorithm can be adapted to use arbitrary initial conditions including periodic boundary conditions.

Program NLSN

```
cccccccccccccccccccccccccccccccccccccccccccccccccccccccccccccccccccc
c      solve the nls equation using a split step method.
c
c      The equation to be solved:
c
c                      2 *
c      q  = i q   + 2iq q
c       t      xx
c
c      with periodic boundary conditions.
c
c      Starting from one of three initial conditions depending on the value
c      of IC.
c
c      1 soliton          : IC=1
c                                            2
c                                 _   4i eta t
c      q  = 2 eta sech 2 eta   (x - x) e
c       0
c
c      double 1 soliton   : IC=2
c                                            2
c                                 _   4i eta t
c      q  = 4 eta sech 2 eta   (x - x) e
c       0
c
c      2 distinct 1 solitons : IC=3, q = q  + q
c                                     0   1    2
c
c                                        2    2
c                                 _   -4i(xi  - eta )t
c      q  = 4 eta sech 2 eta (x - x ) e      i      i
c       i       i            i     i
c
c
c      In order for the algorithm to approximate the real solution, the
c      amount of phase shift in either the linear or nonlinear steps should
c      be small.  The code will warn you if it is too large, and suggest a
```

```
c      factor by which to increase nsteps. Refer to INILIN and CHECK
c      subroutines for the criteria used. Note that the step size decreases
c      rapidly with an increasing number of transverse points (NUMPTS) or
c      amplitude (ETA).
c
c      Some suggested reasonable values
c
c      xmin=-10, xmax=10, nsteps=8000, tmax=1
c
c      IC=1 : eta=1, xbar=0
c      IC=2 : eta=1, xbar=0
c      IC=3 : eta1=1.5, xbar1 = 1, xi1 = 0.1
c             eta2=1.0, xbar2 =-1, xi1 =-0.1
c
ccccccccccccccccccccccccccccccccccccccccccccccccccccccccccccccc
c
c      If your compiler supports a double precision complex data type
c      you may want to use double precision throughout in these
c      calculations.
c
c      N is the parameter controlling the number of points in the
c      discretization.  Some of the FFT routines require this number
c      to be a power of two, while others allow it to be a product of
c      powers of small primes
c
c      PARAMETER(N=256)
c
c      The size of the workspace needed for the FFT will depend on the FFT
c      routine being used.  For CFFT from NCAR (Swarztrauber)\index{Swarztrauber}
c      PARAMETER(NWORK=4*N+15)
c      for FFTCC from imsl v9    use 6*N+150. (Note you may be able to use
c      less, consult IMSL manual)
c      PARAMETER(NWORK=6*N+150)
c      for F2TCF(B) from imsl v10 use 6*N+15
c      PARAMETER(NWORK=6*N+15)
c      for C06FCF   from NAG use  use N
c      PARAMETER(NWORK=N)
c
c      Q is the array holding the complex field values being calculated
c
c      COMPLEX Q(0:N-1)
c
c      AMUL is an array holding the multiplier that is used in the spectral
c      portion of the spectral portion of the split step.  This array is
c      independent of time, so it is calculated once at the start by inilin
c      and then used in linear.
c
c      COMPLEX AMUL(0:N-1),AMULH(0:N-1)
c
c      Work space needed by FFT
c
c      COMMON /WORKSP/WORK(NWORK)
c
c      Parameters used in the initial condition, set in GETICP, and used
c      in INITQ
c
c      COMMON /INITS/IC,ETA1,ETA2,XBAR1,XBAR2,XI1,XI2
c
c      Constants set in SETUP
c
c      COMMON /CONSTS/ PI,DT,DX,XMIN
c
c      GETICP gets the parameters for the initial conditions
c
c      CALL GETICP
c
c      SETUP puts in values for the common block CONSTANTS, and calls
c      the initialization program INILIN
c
c      CALL SETUP(TMAX,NSTEPS,AMUL,AMULH)
c
c      INITQ puts in the initial value for Q, and sets T to 0
c
c      CALL INITQ(Q,T)
c
c      Check to see if the initial values of Q are reasonable for
```

```
c     the choice of timestep etc.
c
      CALL CHECK(Q)
c
c     Call LINEAR with AMULH to do the initial half step
c
      CALL LINEAR(Q,AMULH)
c
c     Alternate full steps of LINEAR and NLINEA
c
      DO 100 ISTEP=1,NSTEPS-1
         CALL NLINEA(Q)
         CALL LINEAR(Q,AMUL)
  100 CONTINUE
c
c     Finish off a last NLINEA and another LINEAR with the final half step
c
      CALL NLINEA(Q)
      CALL LINEAR(Q,AMULH)
c
c     Check to see if the final values of Q are reasonable for
c     the choice of timestep etc.
c
      CALL CHECK(Q)
c
c     Plot the result
c
      CALL PLOT(Q,TMAX)
      STOP
      END
c
      SUBROUTINE SETUP(TMAX,NSTEPS,AMUL,AMULH)
ccccccccccccccccccccccccccccccccccccccccccccccccccccccccccccccccccccccc
c     Get the length of time for the run (TMAX) and the number of steps to
c     use (NSTEPS), calculate constants and call INILIN
ccccccccccccccccccccccccccccccccccccccccccccccccccccccccccccccccccccccc
      PARAMETER(N=256)
      COMMON /CONSTS/ PI,DT,DX,XMIN
      COMPLEX AMUL(0:N-1),AMULH(0:N-1)
      WRITE(6,*)'TMAX,NSTEPS,XMIN,XMAX?'
      READ(5,*)TMAX,NSTEPS,XMIN,XMAX
c
c     Calculate PI as accurately as possible for this machine
c
      PI=4.0D0*ATAN(1.0D0)
      DT=TMAX/(NSTEPS-1)
      DX = (XMAX-XMIN)/N
      CALL INILIN(AMUL,AMULH)
      RETURN
      END
c
      SUBROUTINE GETICP
ccccccccccccccccccccccccccccccccccccccccccccccccccccccccccccccccccccccc
c
c     GETICP gets the parameters for the initial conditions
c
c     1 soliton        : IC=1
c                                                2
c                               _        4i eta t
c     q  = 2 eta sech 2 eta  (x - x) e
c      0
c
c     double 1 soliton : IC=2
c                                                2
c                               _        4i eta t
c     q  = 4 eta sech 2 eta  (x - x) e
c      0
c
c     2 distinct 1 solitons : IC=3, q = q  + q
c                                        0    1    2
c
c                                            2    2
c                               _     -4i(xi - eta )t
c     q  = 4 eta sech 2 eta (x - x ) e       i      i
c      i       i               i
c
```

```
cccccccccccccccccccccccccccccccccccccccccccccccccccccccccccccccccccc
c
      COMMON /INITS/IC,ETA1,ETA2,XBAR1,XBAR2,XI1,XI2
      WRITE(6,*)'IC? (1: 1-SOLITON, 2: 2-SOLITON ,',
     $          '3: 2 DIFFERENT 1-SOLITONS)'
      READ(5,*)IC
      IF(IC.LT.3.AND.IC.GT.0)THEN
         WRITE(6,*)'ETA XBAR?'
         READ(5,*)ETA1,XBAR1
      ELSE IF(IC.EQ.3)THEN
         WRITE(6,*)'ETA1 XBAR1,XI1?'
         READ(5,*)ETA1,XBAR1,XI1
         WRITE(6,*)'ETA2 XBAR2,XI2?'
         READ(5,*)ETA2,XBAR2,XI2
      ELSE
         WRITE(6,*)'Bad value for IC: ',IC
         STOP
      ENDIF
      RETURN
      END
c
      SUBROUTINE INITQ(Q,T)
cccccccccccccccccccccccccccccccccccccccccccccccccccccccccccccccccccc
c
c     INITQ puts in the initial value for Q, and sets T to 0
c
cccccccccccccccccccccccccccccccccccccccccccccccccccccccccccccccccccc
c
      PARAMETER(N=256)
      DIMENSION X(0:N-1)
      COMPLEX Q(0:N-1)
      COMMON /INITS/IC,ETA1,ETA2,XBAR1,XBAR2,XI1,XI2
      COMMON /CONSTS/ PI,DT,DX,XMIN
c
      T=0.0E0
      DO 10 I=0,N-1
         X(I)=XMIN + DX*I
 10   CONTINUE
c
      IF (IC.EQ.1)THEN
C************************************
c     1 soliton        : IC=1
c                                        2
c                          _    4i eta t
c     q  = 2 eta sech 2 eta (x - x) e
c      0
c
C************************************

         DO 100 I=0,N-1
            Q(I)=2.0E0*ETA1/COSH(2.0*ETA1*(X(I)-XBAR1))
 100     CONTINUE
      ELSEIF (IC.EQ.2) THEN
C************************************
c     double 1 soliton : IC=2
c                                        2
c                          _    4i eta t
c     q  = 4 eta sech 2 eta  (x- x) e
c      0
c
C************************************
         DO 200 I=0,N-1
            Q(I)=4.0E0*ETA1/COSH(2.0*ETA1*(X(I)-XBAR1))
 200     CONTINUE
      ELSEIF (IC.EQ.3) THEN
C************************************************
c     2 distinct 1 solitons : IC=3, q = q  + q
c                                    0   1    2
c
c                                       2     2
c                          _    -4i(xi - eta )t
c     q  = 4 eta sech 2 eta (x - x ) e     i     i
c      i       i          i    i
c
C************************************************
         DO 300 I=0,N-1
```

```
                 ARG1 = -2.0E0*XI1*X(I)
                 ARG2 = -2.0E0*XI2*X(I)
                 Q(I)=EXP(CMPLX(0.0E0,ARG1))
       $              *2.0E0*ETA1/COSH(2.0E0*ETA1*(X(I)-XBAR1))
       $              +EXP(CMPLX(0.0E0, ARG2))
       $              *2.0E0*ETA2/COSH(2.0E0*ETA2*(X(I)-XBAR2))
    300   CONTINUE
        ELSE
           WRITE(0,*)'Bad value for IC: ',IC
           STOP
        ENDIF
        RETURN
        END
c
        SUBROUTINE INILIN (AMUL, AMULH)
        PARAMETER(N=256)
cccccccccccccccccccccccccccccccccccccccccccccccccccccccccccccccccccccc
c
c       INILIN calculates the values of AMUL and AMULH, and initializes the
c       FFT routine if necessary.  AMUL is used in LINEAR to integrate
c       the linear half of the problem for one time step.  The value of
c       AMUL is
c                      2
c                   -ik dt
c                  e
c                  ----------
c                      N
c
c       Note that k is the index value minus 1 times 2 pi divided by the width
c       of the domain for indices less than N/2, and
c       the index value minus 1 minus N times 2 pi divided by
c       the width of the domain for indices greater than N/2.
c
c       AMULH is used similarly, but for a time step of half the size
c       Where k is the frequency. The factor of 1/N is needed to normalize
c       the FFT.
c
cccccccccccccccccccccccccccccccccccccccccccccccccccccccccccccccccccccc
c
c       The size of the WORKSP needed for the FFT will depend on the FFT
c       routine being used.  For CFFT from NCAR (Swartztrauber)
c
        PARAMETER(NWORK=4*N+15)
c       for FFTCC from imsl v9     use 6*N+150
c       PARAMETER(NWORK=6*N+150)
c       for F2TCF(B) from imsl v10 use 6*N+15
c       PARAMETER(NWORK=6*N+15)
c       for C06FCF  from NAG use  use N
c       PARAMETER(NWORK=N)
c
        PARAMETER (AMAX=0.50E0)
        COMPLEX AMUL(0:N-1),AMULH(0:N-1)
        COMMON /WORKSP/WORK(NWORK)
        COMMON /CONSTS/ PI,DT,DX,XMIN
        FACTOR=-DT/2.0E0*(2.0E0*PI/(N*DX))**2
        FACTN = 1.0/FLOAT(N)
        DO 10 I=0,N/2
           ARG=FACTOR*FLOAT(I)**2
           AMULH(I)=CMPLX(COS(ARG),SIN(ARG))*FACTN
           AMUL(I)=CMPLX(COS(2.0E0*ARG),SIN(2.0E0*ARG))*FACTN
     10 CONTINUE
        DO 20 I=N/2+1,N-1
           K=I-N
           ARG=FACTOR*FLOAT(K)**2
           AMULH(I)=CMPLX(COS(ARG),SIN(ARG))*FACTN
           AMUL(I)=CMPLX(COS(2.0E0*ARG),SIN(2.0E0*ARG))*FACTN
     20 CONTINUE
        ARGMAX = FACTOR*(N/2)*(N/2)
        IF (ABS(ARGMAX) .GT. AMAX) THEN
           WRITE(6,1000)ABS(ARGMAX)/AMAX
        ENDIF
        CALL CFFTI(N, WORK)
        RETURN
   1000 FORMAT('WARNING:  STEP SIZE MAY BE TOO LARGE FOR THE CHOSEN VALUE',
       $    /,'           OF DX, INCREASE NSTEPS BY A FACTOR OF AT LEAST',
       $    G12.4)
```

```
      END
c
      SUBROUTINE LINEAR (Q, AMUL)
cccccccccccccccccccccccccccccccccccccccccccccccccccccccccccc
c
c     Linear propagation step.  Fourier\index{Fourier} transform, multiply and
c     inverse transform
c
cccccccccccccccccccccccccccccccccccccccccccccccccccccccccccc
c
      PARAMETER(N=256)
c
c     The size of the WORKSP needed for the FFT will depend on the FFT
c     routine being used.  For CFFTF from NCAR (Swarztrauber)\index{Swarztrauber}
c
      PARAMETER(NWORK=4*N+15)
c     for FFTCC from imsl v9      use 6*N+150
c     PARAMETER(NWORK=6*N+150)
c     for F2TCF(B) from imsl v10 use 6*N+15
c     PARAMETER(NWORK=6*N+15)
c     for C06FCF    from NAG use  use N
c     PARAMETER(NWORK=N)
c
      COMPLEX AMUL(0:N-1),Q(0:N-1)
      COMMON /WORKSP/WORK(NWORK)
      COMMON /CONSTS/ PI,DT,DX,XMIN
      CALL CFFTF(N,Q,WORK)
      DO 10 I=0,N-1
         Q(I)=Q(I)*AMUL(I)
 10   CONTINUE
      CALL CFFTB(N,Q,WORK)
      RETURN
      END
c
      SUBROUTINE NLINEA(Q)
cccccccccccccccccccccccccccccccccccccccccccccccccccccccccccc
c
c     Propagate the field under the influence of the
c     nonlinearity only for a step of DT
c
cccccccccccccccccccccccccccccccccccccccccccccccccccccccccccc
c
      PARAMETER(N=256)
      COMPLEX Q(0:N-1)
      COMMON /CONSTS/ PI,DT,DX,XMIN
      TDT=2.0E0*DT
      DO 10 I=0,N-1
         ARG=TDT*(REAL(Q(I))**2+AIMAG(Q(I))**2)
         Q(I)=Q(I)*CMPLX(COS(ARG),SIN(ARG))
 10   CONTINUE
      RETURN
      END
c
      SUBROUTINE CHECK(Q)
cccccccccccccccccccccccccccccccccccccccccccccccccccccccccccc
c
c     Check to see if the magnitude of Q is too large compared
c     to the size of the timestep.
c
cccccccccccccccccccccccccccccccccccccccccccccccccccccccccccc
c
      PARAMETER(N=256)
      PARAMETER(AMAX=0.50E0)
      COMPLEX Q(0:N-1)
      COMMON /CONSTS/ PI,DT,DX,XMIN
      TDT=2.0E0*DT
      ARGMAX=TDT*(REAL(Q(0))**2+AIMAG(Q(0))**2)
      DO 10 I=1,N-1
         ARG=TDT*(REAL(Q(I))**2+AIMAG(Q(I))**2)
         ARGMAX=MAX(ARG,ARGMAX)
 10   CONTINUE
      IF (ABS(ARGMAX) .GT. AMAX) THEN
         WRITE(6,1000)ABS(ARGMAX)/AMAX
      ENDIF
      RETURN
 1000 FORMAT('WARNING: MAGNITUDE MAY BE TOO LARGE FOR THE CHOSEN VALUE',
```

```
      $    /,'          OF NSTEPS, INCREASE NSTEPS BY A FACTOR OF AT',
      $    /,'          LEAST ',G12.4)
      END
c
      SUBROUTINE PLOT(Q,T)
ccccccccccccccccccccccccccccccccccccccccccccccccccccccccccccccccccccccc
c
c     Plot plots the amplitude of the field using NCAR graphics calls.
c     It also anotates the plot with the date and time and with the values
c     of the input parameters.  This routine is not portable because of the
c     use of NCAR graphics calls and of the call to FDATE and LNBLNK.
c     FDATE returns a 24 character string with the date and time.  LNBLNK
c     returns the index of the last non-blank character in a string.
ccccccccccccccccccccccccccccccccccccccccccccccccccccccccccccccccccccccc
c
      PARAMETER(N=256)
      COMPLEX Q(0:N-1)
      REAL Y(0:N-1),X(0:N-1)
      CHARACTER*60 LABEL(3)
      CHARACTER*24 DATE
      COMMON /INITS/IC,ETA1,ETA2,XBAR1,XBAR2,XI1,XI2
      COMMON /CONSTS/ PI,DT,DX,XMIN
c
c     Initialize NCAR graphics
c
      CALL OPNGKS
c
c     Get the date and time
c
      CALL FDATE(DATE)
c
c     convert date string to all uppercase
c
      DO 10 I=1,24
         IF(DATE(I:I).LE.'z'.AND.DATE(I:I).GE.'a')
     $        DATE(I:I)=CHAR(ICHAR(DATE(I:I))-ICHAR('a')+ICHAR('A'))
 10   CONTINUE
c
c     Calculate the amplitude and x coordinate vector
c
      DO 150 I=0,N-1
         X(I)=XMIN+I*DX
         Y(I)=ABS(Q(I))
 150  CONTINUE
c
c     prepare the title for the plot
c
      WRITE(LABEL(1),1000)T
 1000 FORMAT('TIME=',F5.2,'$')
c
c     prepare the anotation
c
      IF(IC.LT.3)THEN
         WRITE(LABEL(2),2000)IC,ETA1,XBAR1,XI1
 2000    FORMAT('IC=',I1,' ETA=',1PE9.2,' XBAR=',1PE9.2,' XI=',1PE9.2)
         LABEL(3)=' '
      ELSE
         WRITE(LABEL(2),2500)IC,ETA1,XBAR1,XI1
         WRITE(LABEL(3),3000)ETA2,XBAR2,XI2
 2500    FORMAT('IC=',I1,' ETA1=',1PE9.2,
     $        ' XBAR1=',1PE9.2,' XI1=',1PE9.2)
 3000    FORMAT('ETA2=',1PE9.2,' XBAR2=',1PE9.2,' XI2=',1PE9.2)
      ENDIF
      LNB=LNBLNK(LABEL(3))
      LABEL(3)(LNB+2:LNB+26)=DATE(1:24)
c
c     Put the parameter values and the date in the bottom corner of the plot
c
      CALL PWRITX(0.95,0.05,LABEL(2),LNBLNK(LABEL(2)),8,0,1)
      CALL PWRITX(0.95,0.03,LABEL(3),LNBLNK(LABEL(3)),8,0,1)
c
c     Draw the plot of amplitude vs. x
c
      CALL EZXY(X,Y,N.,LABEL(1))
c
c     Close down the graphics
```

```
c
      CALL CLSGKS
      RETURN
      END
c
c.................................................................
c
c     Now come the routines for calling the various libraries' FFTs.
c     These routines translate the CFFTI, CFFTF and CFFTB calls to the
c     NCAR P. N. Swarztrauber\index{Swarztrauber} (November 1986 version) routines
c     to match the requirements of the available libraries.  At
c     most one of these sets of routines should be uncommented.
c.................................................................
c
c     IMSL version 9: If another library is available, it will probably be
c     more efficient to use it. Note the use of the work array twice, and
c     the reversal of forward and backward as mentioned in the IMSL 9 manual.
c     See the IMSL version 9 manual entry for FFTCC for details.
c
c     SUBROUTINE CFFTI(N,WORK)
c     RETURN
c     END
c     SUBROUTINE CFFTF(N,Q,WORK)
c     COMPLEX Q(N)
c     REAL WORK(*)
c     DO 100 I=1,N
c         Q(I)=CONJG(Q(I))
c 100 CONTINUE
c     CALL FFTCC(Q,N,WORK,WORK)
c     DO 200 I=1,N
c         Q(I)=CONJG(Q(I))
c 200 CONTINUE
c     RETURN
c     END
c     SUBROUTINE CFFTB(N,Q,WORK)
c     COMPLEX Q(N)
c     REAL WORK(*)
c     CALL FFTCC(Q,N,WORK,WORK)
c     RETURN
c     END
c
c.................................................................
c
c     IMSL version 10: Note the use of the work array in two places.
c
c     SUBROUTINE CFFTI(N,WORK)
c     REAL WORK(*)
c     CALL FFTCI(N,WORK)
c     RETURN
c     END
c     SUBROUTINE CFFTF(N,Q,WORK)
c     COMPLEX Q(N)
c     REAL WORK(*)
c     CALL F2TCF(N,Q,Q,WORK,WORK(4*N+15))
c     RETURN
c     END
c     SUBROUTINE CFFTB(N,Q,WORK)
c     COMPLEX Q(N)
c     REAL WORK(*)
c     CALL F2TCB(N,Q,Q,WORK,WORK(4*N+15))
c     RETURN
c     END
c.................................................................
c
c     NAG:  Since the NAG routines work with the real and imaginary parts
c     in separate arrays, if you use NAG, it will be more efficient to
c     recode everything to make use of the field Q in that form. That will
c     avoid the overhead of copying back and forth, at the expense of making
c     the code somewhat more difficult to read.
c
c     SUBROUTINE CFFTI(N,WORK)
c     RETURN
c     END
c     SUBROUTINE CFFTF(NP,Q,WORK)
c     PARAMETER(N=256)
c     COMPLEX Q(NP)
```

```
c       REAL QR(N),QI(N)
c       REAL WORK(*)
c       DO 100 I=1,N
c          QR(I)=REAL(Q(I))
c          QI(I)=AIMAG(Q(I))
c 100   CONTINUE
c       CALL C06FCF(QR,QI,N,WORK,IFAIL)
c       DO 200 I=1,N
c          Q(I)=CMPLX(QR(I),QI(I))
c 200   CONTINUE
c       RETURN
c       END
c       SUBROUTINE CFFTB(NP,Q,WORK)
c       PARAMETER(N=256)
c       COMPLEX Q(NP)
c       REAL QR(N),QI(N)
c       REAL WORK(*)
c       DO 100 I=1,N
c          QR(I)=REAL(Q(I))
c          QI(I)=-AIMAG(Q(I))
c 100   CONTINUE
c       CALL C06FCF(QR,QI,N,WORK,IFAIL)
c       DO 200 I=1,N
c          Q(I)=CMPLX(QR(I),-QI(I))*N
c 200   CONTINUE
c       RETURN
c       END
c
c.........................................................................
```

REFERENCES

[1]   M. Born and E. Wolf, *Principles of Optics: Electromagnetic Theory of Propagation Interference and Diffraction of Light, 6th edition* (Pergamon Press, New York) (1989).

[2]   H. Haken, *Light* (North-Holland, New York) (1981).

[3]   A. Yariv, *Optical Electronics*, 3rd edition (Holt, Rinehart and Winston, New York) (1985).

[4]   M. Sargent III, M. O. Scully, and W. E. Lamb, Jr., *Laser Physics* (Addison-Wesley, Reading, MA) (1974).

[5]   A. E. Siegman, *Lasers* (University Science Books) (1986).

[6]   Y. R. Shen, *The Principles of Nonlinear Optics* (John Wiley, New York) (1984).

[7]   A. C. Baldwin, *An Introduction to Nonlinear Optics* (Plenum, New York) (1974).

[8]   N. Bloembergen, *Nonlinear Optics* (Benjamin, New York) (1965).

[9]   A. C. Newell, *Solitons in Mathematics and Physics* (SIAM, Philadelphia) (1985).

[10]   G. L. Lamb, *Elements of Soliton Theory* (John Wiley, New York) (1980).

[11]   P. G. Drazin and R. S. Johnson, *Solitons: An Introduction* (Cambridge University Press, New York) (1989).

[12]   M. J. Ablowitz and H. Segur, *Solitons and the Inverse Scattering Transform* (SIAM, Philadelphia) (1981).

[13]   J. Kervorkian and J. D. Cole, *Perturbation Methods in Applied Mathematics* (Springer-Verlag, New York) (1981).

[14] A. H. Nayfeh, _Perturbation Methods_ (Wiley-Interscience, New York) (1973).

[15] N. Guckenheimer and P. Holmes, _Nonlinear Oscillations, Dynamics, and Bifurcations of Vector Fields_ (Springer-Verlag, New York) (1983).

[16] M. Tabor, _Chaos and Integrability in Nonlinear Dynamics: An Introduction_ (Wiley-Interscience, New York) (1990).

[17] D. F. Nelson, _Electric, Optic and Acoustic Interactions in Dielectrics_ (John Wiley, New York) (1979).

[18] G. P. Agrawal, _Long Wavelength Semiconductor Lasers_ (Van Nostrand Reinhold, New York) (1980).

[19] H. Haug and S. W. Koch, _Quantum Theory of the Optical and Electronic Properties of Semiconductors_ (World Scientific, Singapore) (1990).

[20] R. R. Alfano, editor, _The Supercontinuum Laser Source_ (Springer-Verlag, New York) (1989).

[21] P. N. Butcher and D. Cotter, _The Elements of Nonlinear Optics_ (Cambridge University Press, New York) (1990).

[22] T. B. Benjamin and J. F. Feir, "The disintegration of wave trains on deep water," _J. Fluid Mech._ **27**: 417–430 (1967).

[23] D. J. Benney and A. C. Newell, "The propagation of nonlinear wave envelopes," _J. Math Phys._ **46**: 133–139 (1967).

[24] G. B. Whitham, _Linear and Nonlinear Waves_ (Wiley-Interscience, New York) (1974).

[25] V. E. Zakharov, "Collapse of Langmuir waves," _Sov. Phys. JETP_ **35B**: 908–914 (1972).

[26] V. E. Zakharov and A. B. Shabat, "Exact theory of two-dimensional self-focusing and one-dimensional nonlinear media," _Sov. Phys. JETP_ **34**: 62–69 (1972).

[27] B. Crosignani and P. Di Porto, "Self-phase modulation and modal noise in optical fibers," _J. Opt. Soc. Am._ **72**: 1554 (1982).

[28] O. M. Phillips, _The Dynamics of the Upper Ocean_ (Cambridge University Press, New York) (1966).

[29] D. J. Kaup, A. Rieman, and A. Bers, "Space-time evolution of nonlinear three wave interaction I, Interaction in a homogeneous medium," *Rev. Mod. Phys.* 275–309 (1979).

[30] A. Hasegawa, *Optical Solitons in Fibers*, Springer Tracts in Modern Physics Vol. 116 (Springer-Verlag, Berlin) (1989).

[31] A. Hasegawa and Y. Kodama, "Signal transmission by optical solitons in monomode fibers," *Proc. IEEE* **69**: 1145–1150 (1981).

[32] F. P. Kapion and D. B. Keck, "Pube transmission though a dielectric optical waveguide." *Applied optics* **10**(7), 1519–23 (1971).

[33] T. C. Cannon, D. L. Pope and D. D. Sell "Installation and performance of the Chicago lightwave transmission system." *IEEE Transactions on Communications COM–*, **26**(7), 1056–60 (1978).

[34] R. J. Mears, L. Reekie, I. M. Jauncey and D. N. Payne, "Low noise Erbium doped fibre amplifier operating at 1.54 μm." *Electronic Letters* **23**(19), 1026–28 (1987).

[35] A. Hasegawa and F. Tappert, "Transmission of stationary nonlinear optical pulses in dispersive dielectric fibers, I; Anomalous dispersion," *Appl. Phys. Lett.* **23**: 142–144 (1973).

[36] L. F. Mollenauer, M. J. Neubelt, S. G. Evangelides, J. P. Gordon, J. R. Simpson, and L. G. Cohen, "Experimental study of soliton transmission over more than 10,000 km in dispersion-shifted fiber," *Optics Lett.* **15**: 1203–1205 (1990).

[37] L. F. Mollenauer, J. P. Gordon, and M. N. Islam, "Soliton propagation in long fibers with periodically compensated loss," *IEEE J. Quantum Electron*. **22**: 157–173 (1986).

[38] K. Smith and L. F. Mollenauer, "Experimental observation of soliton interaction over long fiber paths: discovery of a long-range interaction," *Optics Lett.* **14**: 1284–1286 (1989).

[39] L. F. Mollenauer and K. Smith, "Demonstration of soliton transmission over more than 4,000 km in fiber with loss periodically compensated by Raman gain," *Optics Lett.* **13**: 675–677 (1989).

[40] J. N. Elgin. "Inverse Scattering theory with random initial potentials." *Phys. Lett. A 19* **110**, 441–443 (1983).

[41]   J. P. Gordon and H. A. Haus, "Random walk of coherently amplified solitons in optical fiber transmission," *Optics Lett.* **11**: 665–667 (1986).

[42]   L. F. Mollenauer, J. P. Gordon and S. G. Evangelides, "The sliding frequency guiding filter; an improved form of soliton jitler control." *Optics Lett.* **17**(22), 1575–1577 (1992).

[43]   P. V. Mamyshev and L. F. Mollenauer, "Stability of soliton propagation with sliding frequency guiding filters." *Optics Lett.* **19**(24) 2083–2085 (1994).

[44]   L. F. Mollenauer, P. V. Mamyshev and M. J. Neubelt, "Demonstration of soliton WDM transmission at 70 Gbits, error free over transoceanic distances." *Electronic Letters* **32**(5) 471–473 (1996).

[45]   D. Le Guen, S. Del-Burgo, L. Moulinard, D. Giot, M. Henry, F. Faure and T. Georges, "Narrow band 1.02 T bits soliton DWDM transmission over 1000 kms. of standard fiber with 100 km. amplifier spans. Communication conf. on Integrated optics and optical Fiber Communications." Cat. No. 99CH36322 (1999). IEEE, Piscataway, N.J. USA.

[46]   G. P. Agrawal, *Nonlinear Fiber Optics* (Academic Press, San Diego) (1989).

[47]   Special Issue on: "Nonlinear Waveguides," *J. Opt. Soc. Am. B* **5** (February 1988).

[48]   W. J. Tomlinson, J. P. Gordon, P. W. Smith, and A. E. Kaplan, "Reflection of a Gaussian beam at a nonlinear interface," *Appl. Optics* **21**: 2041–2051 (1982).

[49]   N. N. Akhmediev, V. I. Korneev, and Y. V. Kuz'menko, "Excitation of nonlinear surface waves by Gaussian light beams," *Sov. Phys. JETP* **61**: 62–67 (1985).

[50]   A. Aceves, J. V. Moloney, and A. C. Newell, "Theory of light-beam propagation at a nonlinear interface," *Phys. Rev. A* **39**: 1809–1823, 1824–1840 (1989).

[51]   A. Aceves, P. Varatharajah, A. C. Newell, E. M. Wright, G. I. Stegeman, D. R. Heately, J. V. Moloney, and H. Adachihara, "Particle aspects of collimated light channel propagation at nonlinear interfaces and in waveguides," *J. Opt. Soc. Am. B* **7**: 963–974 (1990).

[52] P. Varatharajah, A. C. Newell, J. V. Moloney, and A. Aceves, "Transmission, reflection, and trapping of collimated light beams in diffusion Kerr-like nonlinear media," *Phys. Rev. A* **42**: 1767–1774 (1990).

[53] P. Varatharajah, *Propagation of Light Beams at the Interface Separating Nonlinear Diffusive Dielectrics* (Ph.D. Thesis, University of Arizona) (1991).

[54] N. N. Akhmediev, "Novel class of nonlinear surface waves: Asymmetric modes in a symmetric layered structure," *Sov. Phys. JETP* **56**: 299 (1982).

[55] D. Mihalache, G. I. Stegeman, C. T. Seaton, E. M. Wright, R. Zanoni, A. D. Boardman, and T. Twardowski, "Exact dispersion relations for transverse magnetic polarized guided waves at a nonlinear interface," *Optics Lett.* **12**: 1187–1189 (1987).

[56] A. C. Newell, *"Envelope equations,"* in *Nonlinear Wave Motion*, pp 157–163 (American Mathematical Society, Providence) (1974).

[57] W. Kaiser and M. Maier, "Stimulated Rayleigh, Brillouin, and Raman spectroscopy," in *Laser Handbook*, edited by F. T. Arecchi and E. O. Schulz-Dubois, Vol. 2: p. 1077 (North-Holland, Amsterdam-New York) (1972).

[58] M. Born and J. R. Oppenheimer, "Zür Quantentheorie der Moleküln," *Ann. Phys.* **84**: 457 (1927).

[59] A. S. Davydov, *Quantum Mechanics* (Pergamon Press, New York) (1965).

[60] K. Shimoda, *An Introduction to Laser Physics* (Springer-Verlag, Berlin) (1986).

[61] C. O. Weiss and J. Brock, "Evidence for Lorenz-type chaos in a laser," *Phys. Rev. Lett.* **57**: 2804 (1986).

[62] A. C. Fowler, J. D. Gibbon, and M. J. McGuinness, "The complex Lorenz equations," *Physica* **4D**: 139–163 (1982).

[63] C. Sparrow, *The Lorenz Equations: Bifurcations, Chaos and Strange Attractors* (Springer-Verlag, New York) (1982).

[64] H. M. Gibbs, *Optical Bistability: Controlling Light with Light* (Academic Press, San Diego) (1985).

[65] K. Ikeda, "Multiple-valued stationary state and its instability of the transmitted light by a ring cavity system," *Opt. Commun.* **30**: 257 (1979).

[66] S. M. Hammel, C. K. R. T. Jones, and J. V. Moloney, "Global dynamical behavior of the optical field in a ring cavity," *J. Opt. Soc. Am. B* **2**: 552 (1985).

[67] M. W. Derstine, H. M. Gibbs, F. A. Hopf, and L. D. Sanders, "Distinguishing chaos from noise in an optically bistable system," *IEEE J. Quantum Electron.* **21**: 1419 (1985).

[68] D. W. McLaughlin, J. V. Moloney, and A. C. Newell, "Solitary waves as fixed points of infinite dimensional maps in an optical bistable ring cavity," *Phys. Rev. Lett.* **51**: 75 (1983).

[69] Y. Silberberg and I. Bar-Joseph, "Optical instabilities in a nonlinear Kerr medium," *J. Opt. Soc. Am.* **1**: 662 (1984).

[70] W. J. Firth and C. Páre, "Transverse modulational instabilities for counterpropagating beams in Kerr media," *Optics Lett.* **13**: 1096 (1988).

[71] H. G. Winful and G. D. Cooperman, "Self-pulsing and chaos in distributed feedback bistable optical devices," *Appl. Phys. Lett.* **40**: 298 (1982).

[72] S. L. McCall and E. L. Hahn, "Self-induced transparency by pulsed coherent light," *Phys. Rev. Lett.* **18**: 408–411 (1967), and "Self-induced transparency," *Phys. Rev.* **183**: 457–485 (1969).

[73] M. J. Ablowitz, D. J. Kaup, A. C. Newell, and H. Segur, "Method for solving the sine–Gordon equation," *Phys. Rev. Lett.* **30**: 1262–1264 (1973).

[74] M. J. Ablowitz, D. J. Kaup, A. C. Newell, and H. Segur, "The inverse scattering transform—Fourier analysis for nonlinear problems," *Stud. Appl. Math.* **53**: 249–315 (1974).

[75] G. L. Lamb, Jr., "Analytical description of ultra-short optical pulse propagation in a resonant medium," *Rev. Mod. Phys.* **43**: 99–129 (1971).

[76] M. J. Ablowitz, D. J. Kaup, and A. C. Newell, "Coherent pulse propagation, a dispersive irreversible phenomenon," *J. Math. Phys.* **15**: 1852 (1974).

[77] N. L. Komarova and A. C. Newell, "The competition between nonlinearity, dispersion and randomness in signal propagation." *IMA J. Appl. Math.* **63**, 267–286 (1999).

[78] N. Skribanwitz, I. P. Herman, J. C. MacGillivary, and M. S. Field, "Observation of Dicke superradiance in optically pumped HF gas," *Phys. Rev. Lett.* **30**: 309 (1973).

[79] A. C. Newell, "The inverse scattering transform," in *Topics in Current Physics*, **17,** edited by R. Bullough and P. Caudrey, pp. 177–242 (Springer-Verlag, New York) (1980).

[80] D. J. Kaup and L. R. Sciacca, "Generation of OTT pulses from a zero area pulse in coherent pulse," *J. Opt. Sci. Am.* **70**: 224–229 (1980).

[81] A. Penzkofer, A. Lauberau, and W. Kaiser, "High intensity Raman interactions," *Prog. Quantum Electron.* **6**: 55–140 (1979).

[82] C. M. Bender and S. A. Orszag, *Mathematical Methods for Scientists and Engineers* (McGraw-Hill, New York) (1978).

[83] J. A. Simmonds and J. E. Mann, *A First Look at Perturbation Theory* (Krieger, Florida) (1986).

[84] N. Bleistein and R. A. Handlesman, *Asymptotic Expansions of Integrals* (Holt, Rinehart and Winston, New York) (1976).

[85] F. W. Olver, *Asymptotic and Special Functions* (Academic Press, New York) (1974).

[86] E. T. Copson, *Asymptotic Expansions* (Cambridge University Press, New York) (1965).

[87] A. C. Newell, "The dynamics and analysis of patterns," in *Lectures in the Sciences of Complexity*, Vol. I, pp. 107–173 (Addison-Wesley, Reading, MA) (1989).

[88] A. C. Newell, T. Passot and J. Lega, "Order Parameter Equations for Patterns." *Ann. Rev. Fluid Mech.* **25**, 399–453 (1993).

[89] P. Coullet, L. Gil, and F. Rocca, "Optical Vortices," *Optics Commun.* **13**: 403 (1989).

[90] J. T. Stuart and R. DiPrima, "The Eckhaus and Benjamin–Feir instability resonance mechanisms," *Proc. Roy. Soc. London A* **362**: 97 (1978).

[91] C. G. Lange and A. C. Newell, "A stability criterion for envelope equations," _SIAM J. Appl. Math._ **27**: 441 (1974).

[92] P. S. Hagan, "Spiral waves in reaction–diffusion equations," _SIAM J. Appl. Math._ **42**: 762 (1982).

[93] J. Lega, J. V. Moloney and A. C. Newell, _Universal description of linear dynamics near threshold._ Physica D 83 478–498 (1995).

[94] C. S. Gardner, J. M. Greene, M. D. Kruskal, and R. M. Miura, "Method for solving the Korteweg–de Vries equation," _Phys. Rev. Lett._ **19**: 1095–1097 (1967).

[95] C. S. Gardner, J. M. Greene, M. D. Kruskal, and R. M. Miura, "The Korteweg–de Vries equation," _Comm. Pure Appl. Math._ **27**: 97–183 (1974).

[96] D. J. Kaup and A. C. Newell, "Solitons as particles and oscillators and in slowly varying media; A singular perturbation theory," _Proc. Roy. Soc. London A_ **361**: 413–446 (1978).

[97] S. V. Lewis, "Semiclassical solutions of the Zakharov–Shabat scattering problem," _Phys. Lett. A_ **112**: 99 (1985).

[98] J. Satsuma and N. Yajima, "Initial value problem of one-dimensional self modulation of nonlinear waves in dispersive media," _Supp. Prog. Theor. Phys. (Japan)_ **55**: 284 (1974).

[99] Y. Kodama, "Optical solitons in a monomode fibre," _J. Statis. Phys._ **39**: 597 (1985).

[100] H. G. Schuster, _Deterministic Chaos: An Introduction_ (Physik–Verlag) (1984).

[101] W. H. Press, B. P. Flannery, S. A. Teukolsky, and W. T. Vetterling, _Numerical Recipes: The Art of Scientific Computing_ (Cambridge University Press, London) (1986).